# ON THE SCREEN

FILM AND CULTURE SERIES

**FILM AND CULTURE**

Edited by John Belton

A series of Columbia University Press

For a complete list of titles, see page 301.

# ON THE SCREEN

*Displaying the Moving Image, 1926–1942*

---

ARIEL ROGERS

Columbia University Press
*New York*

Columbia University Press
*Publishers Since 1893*
New York Chichester, West Sussex
cup.columbia.edu



Library of Congress Cataloging-in-Publication Data
Names: Rogers, Ariel, author.
Title: On the screen : displaying the moving image, 1926–1942 / Ariel Rogers.
Description: New York : Columbia University Press, [2019] | Series: Film and culture | Includes bibliographical references and index.
Identifiers: LCCN 2018058579 (print) | LCCN 2019004423 (ebook) | ISBN 9780231548038 (e-book) | ISBN 9780231188845 | ISBN 9780231188845 (cloth : alk. paper) | ISBN 9780231188852 (paperback : paper) | ISBN 9780231548038 (e-book)
Subjects: LCSH: Motion pictures—United States—History—20th century. | Motion picture industry—Technological innovations.
Classification: LCC PN1993.5.U6 (ebook) | LCC PN1993.5.U6 R6 2019 (print) | DDC 791.430973—dc23
LC record available at https://lccn.loc.gov/2018058579

Columbia University Press books are printed on permanent and durable acid-free paper.
Printed in the United States of America

Cover design: Guerrilla Design
Cover image: George H. Clark Radioana Collection, Archives Center, National Museum of American History, Smithsonian Institution

# CONTENTS

# ACKNOWLEDGMENTS

In the process of writing this book, I have been buoyed by the inspiration, advice, encouragement, and various forms of support, both large and small, that I have received from many sides. John Belton has been providing me guidance, offering feedback on my writing, and sharing his own archival materials with regularity and incredible generosity since I was a graduate student. Francesco Casetti and two anonymous readers for Columbia University Press offered invaluable feedback on the manuscript, which helped to strengthen it considerably. Sarah Keller and Katharina Loew also read the entire manuscript, some parts more than once, with unflagging intelligence, care, and good humor. Doron Galili, Carter Moulton, Theresa Scandiffio, Lynn Spigel, Charles Tepperman, and Julie Turnock provided crucial feedback on sections.

My work was assisted by an ACLS Fellowship from the American Council of Learned Societies, which provided a valuable year of research leave. I am also grateful to Dean Barbara O'Keefe of the Northwestern University School of Communication and David Tolchinsky in the Department of Radio, Television, and Film at Northwestern for offering the time and resources that made the writing of the book possible. In addition, I thank Wendy Wall and Tom Burke of the Alice Kaplan Institute for the Humanities at Northwestern for supplying subvention funds. I began this project while at the University of Southern Maine, where a Faculty Senate Research Grant permitted me to begin my research.

My colleagues in the Screen Cultures program at Northwestern—Scott Curtis, Hamid Naficy, Miriam Petty, Jeffrey Sconce, Jacob Smith, Lynn Spigel, Neil Verma, and Mimi White—have contributed to this project in many ways, not least through their own scholarship. I am also grateful for the conversation and camaraderie of many other colleagues, including Thomas Bradshaw, Aymar Jean Christian, Nick Davis, Zayd Dohrn, Dilip Gaonkar, Rebecca Gilman, Kyle Henry, James Hodge, Anna Parkinson, Spencer Parsons, Eric Patrick, Dassia Posner, Ozge Samanci, Freda Love Smith, Shayna Silverstein, Liz Son, Debra Tolchinsky, Domietta Torlasco, and Harvey Young. Beyond Northwestern, I have benefitted from discussions, exchanges, and collaborations with Dudley Andrew, Robert Bird, James Chandler, Michael DeAngelis, Tiago de Luca, Zoë Druick, Allyson Field, Oliver Gaycken, Bernard Geoghegan, Tom Gunning, Erkki Huhtamo, James Lastra, Laura Lee, Diane Wei Lewis, Joshua Malitsky, Caitlin McGrath, Ross Melnick, Daniel Morgan, Jennifer Peterson, Dana Polan, Jennifer Porst, Michelle Puetz, D. N. Rodowick, Zoran Samardzija, Salomé Skvirsky, Jacqueline Stewart, Haidee Wasson, Kristen Whissel, Jennifer Wild, Pamela Wojcik, and Steve Wurtzler.

I have had the opportunity to present portions of my research at Northwestern University, Yale University, and Princeton University as well as at the Society for Cinema and Media Studies Conference, *Screen* Studies Conference, Modernist Studies Association Conference, Society for the History of Technology Conference, Console-ing Passions Conference, Visible Evidence Conference, and Ends of Cinema Conference at the Center for 21st Century Studies, University of Wisconsin–Milwaukee. The discussion and feedback provided by copanelists and members of the audience in these contexts contributed greatly to what has become this book.

For help with research, I thank Ellen Keith and Lesley Martin at the Chicago History Museum Research Center; Jenny Romero, Bob Dickson, Louise Hilton, and Kristine Krueger at the Margaret Herrick Library of the Academy of Motion Picture Arts and Sciences; Brett Service at the Warner Bros. Archives; Mark Quigley at the University of California at Los Angeles Film and Television Archive; Cassie Blake at the Academy Film Archive; Ashley Swinnerton at the Museum of Modern Art Film Study Center; Rebecca Poulson at the Northwestern University Device Lab; Patrick Seymour at the Theatre Historical Society of America Archives; Escott Norton at the Los Angeles Historic Theatre Foundation; Edward Baney at the Los Angeles Theatre; and the staffs at the National Museum of American History Archives Center, the Paley Center for Media, the University of California at Los Angeles Libraries Special Collections, the Columbia University Rare Book and Manuscript Library, the Columbia University Avery Architectural and Fine Arts Library, the Billy Rose Theatre Division of the New York Public Library for the Performing Arts, and the Library of Congress Manuscript Reading Room. For images and permissions, I thank Roni Lubliner of NBCUniversal; Betty Uyeda at the Natural History Museum of Los Angeles County; Steve Wilson at the Harry Ransom Center, University of Texas at Austin; Buddy Weiss at Photofest; Jennifer Willis at Crytek GmbH; and Joan Adria D'Amico, executor of the Mary and Alfred Crimi estate.

It has been a pleasure working with Philip Leventhal, senior editor at Columbia University Press, who has been guiding the development of this project with both patience and persistence for several years. I also thank Michael Haskell and Monique Briones at Columbia for their work on the project and Annie Barva for her meticulous editing of the manuscript. The book has benefited in addition from the excellent research assistance of Crystal Camargo, Ian Hartman, Lauren Herold, and Carter Moulton.

Versions of parts of the book will appear or previously appeared as "Screen Practices and Hollywood Cinema in the 1930s," *Screen* 60, no. 2 (Summer 2019); "'Taking the Plunge': The New Immersive Screens," in *The Excessive Screen: Optical Media, Environmental Genealogies*, edited by Rüdiger Campe, Francesco Casetti, and Craig Buckley (Amsterdam: Amsterdam University Press, 2019); "Die Konstruktion eines 'synchronen Feldes': Benjamin Schlangers Experimente mit der Gestaltung von Leinwänden und Kinosälen in den 1930er Jahren," translated by Guido Kirsten, *Montage AV* 25, no. 2 (2016): 167–80; and "Classical Hollywood, 1928–1946: Special/Visual Effects," in *Editing and Special/Visual Effects*, edited by Kristen Whissel and Charlie Keil (New Brunswick, NJ: Rutgers University Press, 2016), 68–77. I am grateful for feedback from the reviewers and editors on these publications, which has also strengthened my work here.

I am especially grateful for the friends and family who have supported me in and beyond the pursuit of this project. Carly Detterman, Tricia Har, and Joseph Maida have been helping to keep me grounded for a very long time. My parents have been unfailingly enthusiastic, compassionate, and willing to help whenever I have needed it. Michael continues to be one of my favorite interlocutors. Anthony has made this book possible in more ways than I could hope to enumerate, and Senya has filled its writing with music. They have both brightened every day I have spent with them.

# ON THE SCREEN

# INTRODUCTION

In an article appearing in both the *Atlantic Monthly* and *Life* magazine in 1945, Vannevar Bush outlined his vision for a "future device" consisting of a desk topped with translucent screens. Modeling this device on the mental process of selection by association and dubbing it the "memex," Bush envisioned it as a tool for "associative indexing" (figure 0.1).[1] Bush's description of the memex has been considered a forerunner to and catalyst for developments in computing in the second half of the twentieth century, especially hypertext.[2] Indeed, Bush's ideas were grounded in his own work with analog computing at MIT since the 1920s (and in his role directing the U.S. Office of Scientific Research and Development, which coordinated academic, industrial, and military efforts during World War II, Bush played a significant part in forging the connection between computing and warfare at midcentury).[3] First conceived in the early 1930s, the memex also has clear roots in microfilm, a long-standing technology that would take on new utility during the war.[4] The memex's use of multiple display screens—small, light emitting, and situated atop a desk—would, however, also seem prescient of the contemporary media landscape, which features a proliferation of small, light-emitting screens, many destined to sit on desks and many more with graphical user interfaces designed to simulate desktops.[5]

It might seem anomalous that the idea for the memex would emerge at the height of the era of so-called "classical" theatrical cinema since the contemporary small screens and associated viewing arrangements that it evokes are often considered a radical departure from the screens and viewing arrangements of

**0.1** Vannevar Bush envisioned the memex as a desk topped with translucent screens.

*Source*: Illustration by Alfred D. Crimi. *Life*, September 10, 1945. By permission of Joan Adria D'Amico.

the classical era. Contemporary viewing arrangements featuring a profusion of small screens seem to diverge in particular from a configuration of theatrical cinema that coalesced early in the twentieth century and centered around a single, large, reflective screen in a darkened auditorium. Often considered the dominant mode of presenting moving images at least until the popularization of television in the 1950s, this conception of theatrical exhibition, together with the purported representational unity of classical cinema, fueled 1970s apparatus theory's claims about the centering and fixation of the spectator.[6] Such a view of classical-era theatrical cinema continues to exert considerable influence on scholarship despite a general rejection of the conclusions that apparatus theory drew about it.

This book challenges that conception of twentieth-century theatrical cinema—and, concomitantly, its relation to contemporary screens—by examining screen practices in the United States during what is often considered Hollywood's golden age.[7] Although I diverge from apparatus theory's methods and polemic, I draw on its fundamental claim that the ensemble of technologies composing cinema is bound up with particular material, social, and psychic configurations, or *dispositifs*. In this regard, the book takes part in the ongoing effort to explore how revisiting the notion of the apparatus as *dispositif* can open up new approaches to the historiography of cinema and related media. That project can be aligned with particular approaches to media archaeology, as

Thomas Elsaesser has argued, especially if we attend to the synchronic and diachronic diversity of cinema's *dispositifs* and their entwinement with other media and cultural formations.[8] Inspired especially by the call to employ cinema's digital turn as what Elsaesser describes as a "reflexive turn in thinking about cinema,"[9] I propose harnessing the sense of upheaval associated with contemporary screens by employing these objects as a matrix through which to reconsider American cinema and its *dispositifs* within the historical context in which they adopted their most hegemonic form. Doing so makes it clear that the small-screen device envisaged by Vannevar Bush not only pointed toward the "new" digital media of the late twentieth and early twenty-first centuries but also emerged within a media landscape that was, both within and beyond the movie theater, already dominated by pervasive and heterogeneous screens.

In its focus on screens, this book participates in a burgeoning area of scholarship, which has been fueled by the ubiquity of these objects in contemporary life. Within cinema studies, the screen had historically been treated predominantly as a passive component of the cinematic arrangement, functioning as a metonym for cinema as a whole and, as such, the object upon which to pin metaphors for the medium (as window, frame, mirror, etc.). The effort to elucidate the significance of particular screens is rooted in the critiques of apparatus theory that arose in the 1980s and 1990s, which—from historical and theoretical perspectives alike—emphasized the socially, culturally, and materially contingent conditions of the exhibition and experience of cinema.[10] The proliferation of small screens since around the same time has only increased the urgency of that effort since cinema's migration to personal computers and mobile devices (not to mention its more long-standing manifestation on television sets) has challenged long-held beliefs about what cinema is, where and when it takes place, and how it addresses viewers. In doing so, the new screens have helped to destabilize the very notion of a singular or coherent cinematic apparatus by making it clear that cinema takes multiple forms, intersects with other media, and readily exists outside the theatrical context often considered central to it.

Within the rash of scholarship on "new" digital media that appeared in the 1990s and early 2000s, the perceived pervasiveness of screens contributed to the notion that people were becoming immersed in the dematerialized, fabricated realm of "cyberspace."[11] Since that time, as screens have continued to multiply, scholars have combated the sense of rupture previously attributed to the notion of virtuality in particular by emphasizing the role the user's body plays in the experience of digital media and by highlighting the materiality of screens in their own right.[12] In connection with that effort, a growing body of historical work has also made it clear that the concepts and attributes often associated with the new screens—from virtuality and immersion to mobility, multiplicity, and a small scale—have long-standing precedents.[13] Such attention to the historical diversity of screen practices has revealed that the cinematic apparatus, as both technological and psychosocial arrangement, has long been, in Haidee Wasson's words, "multiply articulated" both diachronically and synchronically.[14] Indeed, work in film and art history has elucidated the heterogeneity of cinema's *dispositifs* by showing how screen practices in areas such as early cinema, useful cinema, and the historical avant-garde diverged from or repurposed elements of the theatrical exhibition of Hollywood cinema in the classical era.[15]

This book builds on that work by examining screen practices within as well as beyond Hollywood during what I am calling the "long 1930s," a period spanning from the mid-1920s to the early 1940s. I focus on the ways in which a range of screen practices contributed to the *dispositifs* of moving-image media at that time, especially insofar as such practices promoted particular configurations of represented and built space. In my attention to these spatial configurations, I am indebted to Anne Friedberg's observation that screens bring together the virtual spaces of moving images and the actual environments in which those images are displayed, forming what she calls an "architecture of spectatorship" that incorporates both realms.[16] As I hope to show, a focus on the spatial formations associated with diverse screen technologies in the long 1930s reveals relationships not only among areas of film practice that are often addressed in isolation (such as production and exhibition) but also across what are often considered divergent modes of cinema (including theatrical and extratheatrical as well as mainstream and avant-garde modes) and beyond the borders of cinema itself (extending to the nascent technology of television).[17] In exploring the ways in which screen technologies bridged these areas, the book thus charts how a crucial component of the apparatus participated in diverse but overlapping sets of discourses and practices. In doing so, it illuminates the fluidity and porousness of that apparatus even in the context of Hollywood cinema at the height of the classical era. And it thereby presents the heterogeneity and openness of cinema's *dispositifs* as a pervasive and persistent part of film history rather than a product of Hollywood's others (such as avant-garde or extratheatrical modes) or an outcome of digital technologies.

Although the period under consideration here is usually remembered as one in which screen practice, especially within mainstream theatrical cinema, was singular and fairly static, it actually witnessed a surprising range of experiments with screen technologies in the areas of production, theatrical exhibition, and extratheatrical exhibition alike, as I detail in this introduction and throughout the book. In addressing this range of experiments, I adhere to the conviction that what renders a particular object or formation a "screen" is not a specific material quality but rather, as Francesco Casetti has argued, its role within a certain type of assemblage.[18] Broadly speaking, the various objects and formations deemed screens—including audiovisual screens as well as windscreens, window screens, folding screens, fire screens, and smoke screens—supply means of sheltering, concealing, filtering, partitioning, or revealing. In this regard, as Friedberg, Casetti, and others have suggested, screens work to organize spaces.[19] What distinguishes audiovisual screens is that the spaces they organize are composed of both actual and virtual realms. It should be emphasized, however, that even audiovisual screens do not simply reveal virtual views but also and often simultaneously work to shelter, conceal, filter, and partition actual and virtual spaces. Indeed, these functions underlie film theory's long-standing conceptualization of these objects not only as windows but also as thresholds, barriers, masks, frames, and mirrors. The material qualities of screens, in conjunction with the films they display, contribute to these functions. The screen's borders, for instance, work to enclose, obscure, reveal, or demarcate the spaces within and surrounding its edges, enabling the screen to function as a frame, mask, aperture, or connector. The screen's surface serves as the foundation for its function as a threshold, barrier, reflector, membrane, or vehicle for light and sound, thus joining, separating, or reconfiguring the spaces in front of and behind it. And the screen's mobility, associated with the capability both to move and to display movement, renders these spatial mediations dynamic.[20]

As the book elucidates, audiovisual screens in the long 1930s took a variety of forms. Even theatrical cinema screens were made from a range of materials (from reflective metals to translucent silk), installed in diverse arrangements (including for both front and rear projection), framed by different kinds of borders (from static black masking to moving, reflective, and even luminescent image-surrounds), and available in various sizes. Moreover, these objects collaborated with onscreen imagery to organize spaces in diverse ways, working variously to shelter, conceal, filter, partition, and reveal actual and virtual realms. An exploration of film and television screens in this period thus illuminates connections among heterogeneous objects, configurations, and practices that are nevertheless united in their application to the mediation of actual and virtual domains. Such connections, it should be noted, extend beyond what are commonly recognized as film and television screens to include certain uses of objects such as windows and mirrors. For the sake of focus, my discussion revolves primarily around audiovisual screens accommodating the display of moving imagery through the projection and scanning operations associated with film and television. But I do not suggest that there is a clear-cut qualitative distinction between the forms of spatial mediation achieved by these screens and those enacted, for instance, by mirrors. Indeed, as I discuss, mirrors were included in certain film-projection arrangements and television displays. I do not propose a clean break between screens used for still images and those employed for moving imagery, either, as my invocation of the memex has already revealed. In fact, as we will see, several experimental screens of this period were used for both purposes.

An examination of audiovisual screen practices in the long 1930s most basically highlights the pervasiveness of these surfaces in that period and the variety of spaces structured by them. Such spaces ranged from studio soundstages (where screens were increasingly being employed on the set) to theaters (which were experimenting with multiple projections and auxiliary screens), homes (with home movies and the anticipation of television), spaces of transit (such as stores, hotels, and train stations), and many other places of work, commerce, and education. Exploring the use of screens in these diverse realms, however, also reveals certain spatial logics traversing and uniting them. In particular, the proliferation of screens across the areas of production, theatrical exhibition, and extratheatrical exhibition formed spaces that were profoundly synthetic. In this regard, screen technologies collaborated with concurrent practices in the production, reproduction, and transmission of sound, ranging from film sound to radio.

In the domain of film production, for instance, the early 1930s saw Hollywood embrace the special-effects technique of rear projection. This technique employed screens to combine previously filmed background imagery with action shot in the studio. While editing is usually considered the most significant means by which the classical Hollywood cinema marshaled filmed fragments into new spaces, compositing techniques such as rear projection performed a similar function within the shot itself, joining ontologically distinct elements, filmed at different times and places, within the frame.[21] This form of spatial amalgamation functioned analogously to and in conjunction with concurrent approaches to film sound. As James Lastra has shown, this period witnessed a transformation in the way Hollywood engineers conceived the spatiality of diegetic sound as they began to employ dubbing and rerecording to construct hierarchically layered soundtracks.[22] In both special-effects and sound practice, the resulting composites could read as homogeneous entities; however, even the most unified visual and sonic spaces thus constructed were nevertheless deeply synthetic.

Exhibition screens extended this logic of spatial synthesis to the viewer's material environment and in doing so reduplicated it. In the realm of theatrical exhibition, the 1930s saw a widespread effort to integrate onscreen images more fully with their physical surroundings. Throughout the decade, architects and engineers outlined several methods for achieving such integration, marshaling numerous components of screen and theater design, including the size, material, positioning, masking, and brightness of the screen as well as the lighting, decoration, layout, and material construction of the auditorium. Experiments included the use of light surrounding the screen, "daylight" projection, and multiple projections, both within the auditorium and throughout the building by means of auxiliary screens. In different ways, such practices formed theatrical spaces that were, in the architect Benjamin Schlanger's words, "tuned to" or even "synchronized with" the screen presentation.[23] They thus also yielded new spatial syntheses. In this regard as well, screen technologies functioned similarly to and in tandem with sound technologies. As Emily Thompson has demonstrated, the movie theaters wired or built for synchronized sound in the late 1920s and early 1930s favored a model of synthetic sound space similar to and functioning in collaboration with that increasingly found on the soundtrack.[24] With screen technologies, the synthetic spatialities arising through exhibition also worked in conjunction with those achieved in production. The synthetic form of space structured by the situation of screens in the theater reflected and extended the form of composite space that was, by virtue of special-effects techniques such as rear projection, also increasingly evident onscreen.

Recognizing the role that spatial synthesis played in the theatrical exhibition of classical Hollywood cinema challenges the suggestion, put forth most prominently by Lev Manovich, that the current pervasiveness of compositing (together with animation) distinguishes contemporary images from the mainstream cinema of the twentieth century and links the new images instead with the products of the historical avant-garde.[25] Contrary to Manovich's assertion that compositing was marginalized in the classical era, a focus on historical screen practices reveals its pervasiveness in both representation and exhibition.[26] When people went to the movie theater to view Hollywood films in the long 1930s, they entered a cinematic space structured through the synthesis of multiple spatiotemporal fragments. Insofar as screens were harnessed to marshal such fragments into new composites—both on the screen, thanks to rear projection, and in the theater, thanks to the screen's integration within the auditorium—the use of these objects endowed cinematic space, even as shaped by Hollywood cinema, with the forms of multiperspectivality and spatiotemporal reorganization associated with concurrent practices in modern art and architecture, which in turn reflected the social and sensorial changes arising with urban-industrial modernity, including the rise of cinema.[27] Screen practices of the long 1930s, in short, contributed to what Miriam Hansen termed the "vernacular modernism" of the so-called classical Hollywood cinema, providing its mass audience an aesthetic experience that both reflected and contributed to the perceptual transformations attendant to the new configurations of space and time (if not always the forms of shock) associated with modernity.[28]

This use of screens, moreover, links the production and theatrical exhibition of Hollywood cinema with contemporaneous extratheatrical and even extracinematic moving-image practices, which likewise contributed to the spatiotemporal reconfigurations marking modernity. In the domain of extratheatrical exhibition, the long 1930s witnessed a proliferation of film screens made to accompany 16mm and, by 1932, 8mm projectors. This period also saw the development and (limited) dissemination of mechanical and, by the mid- to late 1930s, fully electronic television systems. Engineers and practitioners in both arenas worked to integrate the screens associated with these systems into a range of private and public spaces, including the domestic space of the home, commercial spaces such as stores, and the mobile spaces formed by vehicles such as ships, trains, and airplanes. Extratheatrical cinema and television screens were often (though not always) smaller than theatrical cinema screens; however, their integration into such spaces was achieved through several of the same strategies. As a result, extratheatrical cinema and television screen practices extended the forms of spatial synthesis found onscreen and in the theater into and across several other realms of everyday experience. In this regard, these practices also functioned similarly to and in connection with radio, which was likewise permeating and forging new spatial amalgamations within and across everyday domains.

Throughout this book, I map the specific contours of the spatial reorganizations produced by screen technologies in this period by exploring discourses and practices pertaining to the construction, installation, and employment of these objects, whether for use on the set, in the theater, in the home, or in transit. In pursuing this project, I examine discussions of screen technologies in professional journals devoted to filmmaking, exhibition, broadcasting, engineering, and architecture as well as in the popular press. I also examine films that these discourses flag as making exemplary use of particular screen technologies and production records associated with those films. Because very little television programming from this period remains in existence, I have had to rely primarily on descriptions in that area. Technical discourses supply notoriously teleological accounts of media history and reductive instantiations of media theory. As historical objects, however, these discourses can help to contextualize particular modes of representation as well as particular conditions of spectatorship, illuminating historically specific approaches to cinematic and televisual technologies—and, indeed, to cinema and television themselves. When considered in conjunction with the films and (where possible) television programs they address, these discourses thus offer insight into the conceptual and material framing of mediated experience within particular social and cultural contexts.[29]

I have chosen to focus primarily on the context of the United States in order to provide an anchor for what is in other regards this project's centrifugal thrust. Like the Hollywood industry itself in the long 1930s, however, the screen practices of this period participated in what can be identified, drawing on Arjun Appadurai's description of contemporary global cultural flows, as a transnational circulation of people, media, technologies, capital, and ideas.[30] Not only did émigrés (many from places at the center of the emerging geopolitical crisis, such as Germany) contribute significantly to the technological and aesthetic experiments taking place in the United States, but American engineering discourses also maintained an eye on international practices, especially, coincidentally or not, in countries with which the United States would soon be at war, specifically Germany and to a lesser extent Japan.[31] In the book's chapters, I mention these connections as they become pertinent to my discussion in the hope that they might illuminate scholarly intersections and possible paths for further exploration.

The period that I examine and dub the "long 1930s" extends roughly from 1926 to 1942. By the time this period commenced, cinema screens had undergone several permutations, and television screens had long been envisioned.[32] Movies had been shown as both peep show and projection, and they had moved through phonograph parlors, penny arcades, amusement parks, opera houses, vaudeville theaters, and storefront theaters, installing themselves in the picture palaces that reached their pinnacle in the mid-1920s. In the course of these movements, cinema had aligned itself with a range of other technologies and practices, from phonographs and roller coasters to illustrated lectures, vaudeville, the magic theater, and other theatrical traditions such as melodrama.[33] Film projectors had long been marketed for home use as well, thus also aligning movies with the views and sounds offered by other domestic devices such as stereoscopes, magic lanterns, and (again) phonographs.[34] The screen and its use transformed with these permutations and in conjunction with cinema's protean sites and cultural contexts.[35] In storefront theaters around the turn of the twentieth century, the screen, often a muslin sheet, was affixed to the back wall—an arrangement continued in the nickelodeons that had appeared by 1905.[36] With the reform efforts taking hold during the nickelodeon era, there were experiments with "daylight" projection, achieved through practices ranging from the use of downlighting in the theater to the deployment of highly reflective—or even luminescent—screens.[37] By 1912, as picture palaces were emerging, screen materials ran the gamut from muslin to the plaster wall itself, matte aluminum, and even plate-glass mirror.[38] By 1913, screens were being situated within theatrical sets and placed toward the rear of a stage.[39] Rear-projection screens made of glass were also used experimentally in the production of films as early as 1913, joining other early compositing practices such as glass painting and mirror shots.[40]

Although each of the practices I address has a longer history, the confluence of high-profile experiments with screens in 1926 and 1927 is especially notable. As this book discusses, those years saw not only Hollywood's embrace of synchronized sound (and with it the introduction of permeable sound screens) but also the debut of the Magnascope system for large-screen theatrical exhibition, AT&T's widely publicized demonstration of television, and the high-profile appearance in American cinemas of rear projection as a special effect thanks to the German import *Metropolis* (Fritz Lang, 1927), which employed a rear-projection screen to simulate a television screen. Roughly concurrent with the opening of such spectacular picture palaces as the Paramount and Roxy Theatres in New York in 1926 and 1927, moreover, was the founding of the Amateur Cinema League, which capitalized on the capacity for extratheatrical exhibition (as well as amateur filmmaking) facilitated by the recently introduced 16mm film gauge and the projectors and screens associated with it. With these diverse experiments, a variety of novel (if not precisely new) screens—from the very small to the very large—thus entered a range of spaces nearly simultaneously, endowing cinematic diegeses, movie theaters, and homes alike with certain shared spatial formations.

Both rear-projection and television technologies and the practices associated with them were developed throughout the 1930s and standardized by 1942, when attention and resources were redirected to the war effort. Likewise, extratheatrical cinema practices expanded and burgeoned in the 1930s. After a short-lived widescreen-cinema boom following in the wake of Magnascope, theatrical exhibition moved away from both large screens and lavish revival décor in favor of the effort to integrate smaller screens more fully into functionalist modern spaces. The latter effort, however, also became standard over the course of the 1930s. The period that I have designated the long 1930s thus witnessed the ascent of the classical Hollywood cinema's most familiar configuration of film style and technology, with the historical consolidation of the modes of representation and exhibition that would be critiqued by apparatus theory—and against which previous, concurrent, and subsequent divergences are often measured. But at the same time an exploration of the screen practices undertaken during this period illuminates intersections between that configuration and others that are usually identified as divergences, such as extratheatrical cinema and television.

Such an exploration reveals a particular set of strategies according to which screens were devised and employed to synthesize spaces across these domains in the long 1930s. Specifically, the various technicians, filmmakers, and architects developing and using screens in that period experimented with a fluid collection of qualities shaping how these objects worked to join, divide, and demarcate actual and virtual realms. This included the surfaces' qualities of transparency, scale, scope, mobility, multiplicity, and flexibility. These qualities manifested through variable conjunctions among the imagery displayed on screens, the material construction of the screens, and the physical and social sites of the screens' installation. In exploring the exploitation of these qualities in the classical era, this book challenges the tendency to distinguish between historical and contemporary screens by virtue of material distinctions such as large/small, static/mobile, single/multiple, or light reflecting/light emitting, since it shows that the qualities often considered new (such as small scale, mobility, multiplicity, and light emission) have long been components of even mainstream screen practice. In place of such distinctions, I locate the historical specificity of screen practices in multifaceted synchronic relationships among diverse screens as they are harnessed in particular contexts.

Attending to the exploitation of screens' qualities of transparency, scale, scope, mobility, multiplicity, and flexibility highlights the various forms of spatial synthesis achieved in the classical era and the diverse purposes to which such synthesis was put. However, it also illuminates certain consistent approaches to spatial synthesis moving across otherwise divergent sites and institutions. As I discuss, for instance, practitioners in the long 1930s often harnessed the screens' transparency to explore the capacity to assert proximity and distance, employed their scale to experiment with the formation of mediated environments, used mobility to direct or position bodies, and exploited multiplicity to display feats of coordination and expansion. Such uses were by no means universal or rigid, but they offer a means by which not only to parse the heterogeneity of Hollywood screen practice but also to map the multiple vectors linking that practice to media and cultural formations usually considered opposed to it. Indeed, such deployments of screens moved across and aligned otherwise distinct realms (production and exhibition, the mainstream and the avant-garde, theatrical and extra-theatrical, cinema and television) and even materially dissimilar screens (from the largest to the smallest), contributing to the formation of overlapping configurations traversing those realms.

I argue, in particular, that a historically contingent approach to the construction of time contributed to the effort at spatial synthesis evident in a range of screen practices in the long 1930s. As I elaborate over the course of the book, various screen practices—involving special-effects screens, large auditorium screens, multiple auditorium screens, small auxiliary screens, and even television screens—shared an emphasis on synchronizing the screen image with other components of the apparatus. Through such practices, screen images were synchronized not only with recorded sounds (as with synchronized sound) but also with cameras (as with rear projection), screen borders (as with Magnascope and similar systems), other projections (as with various forms of multiple projection), and transmission apparatuses (as with television). The pervasive interest in synchronization, to be sure, was rooted in discourses on electricity and capitalized in particular on the technology and popularity of synchronized sound. Beyond this, however, the persistent emphasis on such synchronicity, I contend, joined a variety of screen technologies with sound technologies in employing a particular and particularly modern form of temporality to structure new mediated spaces.

Mapping these spatiotemporal formations, moreover, makes it possible to chart how screen practices have also promoted particular social formations, which could traverse disparate realms as well. Insofar as screens are harnessed to divide, connect, and demarcate spaces, both virtual and actual, their employment contributes to the social organization of both diegetic and built environments. As I discuss, the use of screens, together with cognate surfaces such as windows and mirrors, bolstered and even anchored ideological constructions of space in the long 1930s, especially via constructions of gender and race. Hollywood, rightly or wrongly, considered (white) women its most important audience at this time, and assumptions about that audience drove production decisions (e.g., the woman's film), efforts at regulation (e.g., concerns over representations of the "fallen woman"), and approaches to publicity (e.g., via links to fashion).[41] Scholars such as Anne Friedberg and Mary Ann Doane have compared cinema to both a shop window and a mirror since it addresses female viewers as virtual flâneuses and consumers, serving as what Friedberg identifies as a "site of identity construction" that rests on what Doane describes as the viewers' own objectification.[42] Movie theaters, especially picture palaces, contributed to such constructions in the designation of spaces for women, such as powder rooms, nurseries, and cry rooms, which could be structured around actual windows, mirrors, and—in at least one set of plans—movie screens. Extratheatrical exhibition arrangements associated with both cinema and television also anticipated an address to women not only in conjunction with the technologies' entry into domestic spaces but also through their installation in department stores and concomitant links to fashion, which literalized female viewers' role as flâneuses and consumers.

Hollywood cinema of the long 1930s also maintained a vexed relationship with performers and audiences of color, especially African Americans. Sound film proved a boon to blackface performance; the roles available to African American actors were deeply limited and stereotyped; and both de jure and de facto segregation continued to constrain and shape viewers' experience of movie theaters. As scholars such as Arthur Knight and Miriam Petty have shown, however, Black performers and audiences in this period engaged with Hollywood cinema in complex ways, negotiating and challenging its racism.[43] As we will see, several experiments with screens were tethered to the depiction of African Americans. *Gone with the Wind* (Victor Fleming, 1939), in particular, emerges as a recurring example in the book since the spectacular nature of that production was staked not only on the use of Technicolor but also on the exploitation (both actual and anticipated) of screen-related practices such as rear projection and multiple-projector exhibition. Insofar as that film, like *The Birth of a Nation* (D. W. Griffith, 1915) before it, harnessed aesthetic and technological experimentation to construct a view of American national identity resting on the assertion of white supremacy, even as it offered a (profoundly problematic) platform for African American actors such as Hattie McDaniel and Butterfly McQueen, it reveals the role that screens could play in shaping Hollywood cinema's politics of representation.[44]

Exploring the ways in which screen practices have contributed to such formations and dynamics sheds new light not only on the history of audiovisual media but also on the ways in which their *dispositifs* can be conceptualized. Before elaborating this component of my argument, it will be useful to provide an account of the concept of *dispositif* and its value for contemporary work on media technologies, especially given the debates that have surrounded its use. Jean-Louis Baudry introduced this concept to cinema studies in the 1970s, around the same time that Michel Foucault was also articulating his own notion of it. While Baudry's investment in psychoanalysis distinguished his notion of *dispositif* from Foucault's, in both cases it referred to a heterogeneous collection of elements producing subjectivity.[45] For Baudry and other film theorists drawing on his ideas, the concept facilitated the mapping of connections among cinema's technological, psychic, and social arrangements and among its material configurations, textual operations, and spectatorial address. Although the Althusserian–Lacanian notion that the cinematic machine worked, in Baudry's words, to "[constitute] the 'subject'" fueled charges of apparatus theory's technological determinism, theorists such as Jean-Louis Comolli, Christian Metz, and even Baudry himself also emphasized the social construction of film technology and articulated a more multifaceted and reciprocal relationship between the cinematic institution and its spectators than the charge of determinism suggests.[46]

Taking issue, however, with the monolithic and passive conception of spectatorship attributed to apparatus theory and attuned specifically to the dimensions of social and cultural difference and spectatorial agency that theorists such as Baudry did elide, subsequent scholarship would, as Will Straw has put it, slowly detach "questions of spectatorship from those of the apparatus," introducing "an account of viewer-subjects as produced in long processes of identity-formation which occur at least partly outside of the cinematic institution."[47] A second, related critique, in this case arising in connection with historical work on early cinema, had to do with apparatus theory's equally monolithic account of cinema itself. Insofar as historical research revealed early cinema to have operated according to a different logic than classical cinema, with a presentational mode of address that diverged sharply from the representational address taken for granted by apparatus theory, scholars working in this area emphasized the historical variability and cultural grounding of film style, exhibition practice, and the modes of spectatorship they conspired to invite.[48]

Despite apparatus theory's conception of a monolithic and static cinematic machine and spectator, the concept of *dispositif* can serve as a methodological tool for parsing their variability. Such a project has become especially pressing insofar as the new technologies and media popularized since the 1980s have prompted changes in filmmaking, exhibition, and intermedial dynamics reminiscent of the upheavals that took place around the turn of the twentieth century—if not resuscitating earlier modes of address, then at least recalling the multidimensional transformations associated with another period of seismic media change.[49] Noting the way in which English translations of key works of apparatus theory, including Baudry's essays, used the word *apparatus* to encompass and thereby conflate the French terms *dispositif* and *appareil*, scholars have worked to disambiguate the notion of *appareil*, which refers to the more literal significance of apparatus as a kind of machine, from that of *dispositif*, which suggests the more abstract concept of an arrangement.[50] In conjunction with that effort and informed by the historical work on early cinema as well as reconsiderations of Foucault's concept by philosophers such as Gilles Deleuze and Giorgio Agamben, scholars such as Frank Kessler, Thomas Elsaesser, and Francesco Casetti have, from both historical and theoretical perspectives, emphasized the multiplicity and dynamism of cinematic (and, more broadly, mediatic) *dispositifs* as both technological and spectatorial arrangements.[51]

The concept of *dispositif*, in short, provides a means to reassess media and mediated experiences as simultaneously protean and historically situated. In particular, it supplies a model for parsing flexible connections among the heterogeneous components of cinema, offering a way of mapping shifting relationships between production and exhibition and across representation, technology, and culture. It thus also offers a framework for linking cinema with other mediatic and cultural configurations. The project of exploring screens' role in such dynamic assemblages resonates in many ways with the overlapping concerns of actor-network theory, new materialism, and recent work on the social and political functioning of space.[52] Although the notion of *dispositif* does not emphasize the forms of social and cultural difference underlying—and rendering heterogeneous—individuals' encounters with particular formations, it can accommodate the recognition of such heterogeneity if we follow Agamben in understanding living beings not simply as subject to a particular apparatus (and, for a consideration of film viewers, not only to the cinematic apparatus) but rather as engaging in a "relentless fight" with a plurality of apparatuses (indeed, Agamben argues that subjectivity emerges from that relation).[53] Such a formulation thus also reveals how the notion of *dispositif* can account for spectatorial agency. As Casetti puts it, "When we enter within the field of a dispositive, we are assigned a precise profile—an identity, a self—which inevitably categorizes us, but which also opens to us a field of action," making it possible for spectators, in turn, to fuel transformations in the *dispositifs* within which they are simultaneously situated.[54]

Whereas the *dispositifs* of audiovisual media are usually conceived via material configurations anchored by particular screen formations, close attention to a range of screen practices paradoxically reveals how these *dispositifs* can be conceptualized (and concomitantly historicized) in terms of the spatiotemporal and social logics traversing heterogeneous material configurations. Elsaesser has proposed mapping variable conjunctions of space/place, time/duration, and agency/subject as a means of reassessing cinema's *dispositifs* in light of their protean formations.[55] As this book shows, examining screens as components of otherwise disparate *dispositifs* reveals how these conjunctions move across diverse technological and institutional arrangements. Indeed, beyond (and contributing to) their role within larger assemblages, screens, as I have been suggesting, also operate as dynamic apparatuses in their own right, giving rise to particular spatiotemporal and social configurations.[56] They can even be considered mediums, understood, as Giuliana Bruno advocates, via the etymological root of *medium* as "a condition of 'betweenness' and a quality of 'becoming' as a connective, pervasive, or enveloping substance."[57] Insofar as we view screens not as inert objects but rather as enacting a particular process of mediation—a process that can be conceptualized as "screening" with that term's multiple valences—we can consider them to have long operated as interfaces, which, as Alexander Galloway contends, are "not simply objects or boundary points" but rather "zones of activity" or "processes that effect a result of whatever kind."[58] The spatiotemporal and social configurations that screens thereby form thus come into variable relation with other components of audiovisual media and their contexts.

In its claim that mainstream theatrical cinema effected the centering and fixation of the spectator, apparatus theory highlighted the technological arrangement of exhibition—notably, a single large film screen—as well as the unity of the films projected on that screen. A close examination of the screen practices of the long 1930s shows the historical consolidation of material strategies for ensuring such centering, fixation, and unity. But, as the foregoing discussion has already begun to suggest, such an examination also reveals that those forms of positioning and representation were not exclusive or hermetic. Indeed, they were fluidly related to configurations that facilitated decentering, distraction, and fragmentation—qualities that are usually attributed to contemporary media and the historical avant-garde. The pervasive emphasis on spatial synthesis and temporal synchronization in particular transcended distinctions such as centering/decentering, attention/distraction, and unity/fragmentation. In doing so, these spatiotemporal formations, together with the social formations they undergirded, united diverse screen practices in endowing various domains of spectatorial experience with broadly shared but historically contingent contours.

The chapters in this book chart this terrain by exploring the development and use of screens for the production and exhibition of moving images in the long 1930s. Chapter 1 examines the employment of screens in film production, particularly for the special-effects technique of rear projection. I argue here that the increasing use of screens on the set worked in conjunction with concurrent practices in film sound to render diegetic space pervasively synthetic in this period. Much as the practices of dubbing and rerecording contributed to the construction of layered soundtracks, rear projection promoted the layering and arraying of spatiotemporal fragments within the shot. The chapter maps several ways in which technicians and filmmakers devised and employed rear-projection screens to synthesize such fragments. Work on the material construction of screens addressed their surfaces' quality of transparency and concomitantly their situation within deep space. Work on screen scale explored relations within and beyond the frame of the background screen(s). And work addressing the stability of the rear-projected image explored how the quality of mobility could be transferred among diverse components of the production apparatus. As I show, these material foci contributed to particular forms of representation and address in a wide variety of Hollywood films employing rear projection in the long 1930s, from Westerns structured on the separation and transgression of image planes to films that harnessed rear projection to immerse characters in fiery, watery, and aerial environments.

Chapters 2 and 3 show how such forms of spatial synthesis extended to the material space of exhibition through the employment of theatrical screens. In the context of theatrical exhibition, as mentioned earlier, the historical period under consideration spans from the height of the picture-palace era through the move to a functionalist approach to theater design, which had become the norm by the beginning of the 1940s. Chapter 2 addresses theatrical screen practices from 1926 through 1931, and chapter 3 explores theatrical screen practices from 1931 through 1940. The earlier period, I argue, set the stage for the later one.

As chapter 2 elaborates, the period 1926–1931 encompassed not only the adoption of synchronized sound and, with it, permeable sound screens but also widespread experimentation with large-screen and widescreen systems such as Magnascope and Fox's Grandeur system. As I show, technicians and filmmakers in this period explored how the increased scale and scope of the new large and wide screens could transform the relationship between diegetic and theatrical space, especially by filling the auditorium with spectacles such as large landscapes and masses of moving bodies. Deployments of these systems also experimented with the flexibility of the screen frame made possible by the inclusion of screen modifiers, which altered the dimensions of the image either during the film, as with Magnascope, or over the course of an evening's program, as with Grandeur. Although these large-screen and widescreen systems had been abandoned by mid-1931, the experimentation with the scale, scope, and flexibility of their screens would, I contend, guide approaches to the integration of screen and theater space in the subsequent years.

Chapter 3 examines how work on screen and theater design through the remainder of the decade pursued alternative ways of endowing the cinematic spectacle with such qualities. Here I explore engineering work that addressed the positioning of the screen in the auditorium as well as the brightness of the screen, the borders surrounding it, and its relationship to the illumination of the theater. The chapter also includes several case studies that address more unusual experiments with screens, including the use of rear projection and multiple screens within the auditorium as well as the installation of auxiliary screens in disparate parts of the building. In the wake of the abandonment of widescreen, I argue, architects and engineers attempted to endow the cinematic spectacle with the qualities of scale, scope, and flexibility less through manipulation of the screen itself than through its integration into theater space. Harnessing diverse components of screen and theater design, from the masking of screens to the material construction of the auditorium, practitioners calibrated screens with their theatrical surroundings to form dynamic audiovisual syntheses. In this regard, theatrical screens functioned analogously to and in conjunction with the special-effects screens being employed in production. Scholarly work has tended to portray the integration of screen and theater space that occurred in the 1930s as offering viewers an experience of immersion. Attention to the diverse ways in which theatrical screens were employed and envisioned in that decade, however, reveals that screens' integration into theatrical space took a variety of forms and performed a range of functions. As I show, screens were incorporated into theatrical space not only by virtue of their scale but also through the qualities of flexibility, luminescence, transparency, and multiplicity. Although this incorporation in many cases worked to immerse and transfix viewers, it could also serve other purposes, such as to encourage the viewers' mobility throughout the theater.

Chapter 4 continues my focus on exhibition in the long 1930s, shifting attention to extratheatrical and extracinematic screens. This includes the screens marketed for use with the small-gauge film formats (especially 16mm) that were helping to foster cinema's presence in a range of spaces beyond the movie theater—from homes, schools, and churches to stores, hotels, and office buildings to trains, ships, and airplanes. It also includes a range of television screens being designed for installation in some of the same locations. Focusing primarily on film and television screens designed for use in the home and in spaces of transit, the chapter shows how extratheatrical and extracinematic screen practices exploited and reframed the qualities of transparency, multiplicity, scale, and mobility mapped throughout the book. Doing so reveals connections between theatrical and extratheatrical sites as well as between cinema and television in this period. In particular, it indicates how extratheatrical cinema and television screen practices collaborated with theatrical screen practices in harnessing a multiplicity of screens to promote spatial syntheses, in this case within and among everyday spaces. The multiple film screens populating domestic spaces were available primarily for serial rather than simultaneous use. Television, however, promoted a variant of simultaneous multiscreen exhibition, offering a form of temporality that dovetailed with that of concurrent cinematic practices.

The contemporary proliferation of screens has been associated with the rise of electronic and digital media, expanded cinema, and, with them, the interactions among media encapsulated by terms such as *multimedia*, *transmedia*, and *convergence*. This book, however, makes the case that at the height of the classical era the cinematic apparatus was also characterized by a proliferation of screens functioning within an intermedial context. This characterization suggests an expanded historical genealogy for multiscreen media, moving beyond the context of the avant-garde with which it has been more often associated. But rather than attempting to trace a particular *dispositif* diachronically and far from imposing a concept such as virtuality on the earlier context, I harness my investigation of the screen technologies of the classical era to propose means of assessing the historical specificity of the *dispositifs* opened up within particular contexts. Whereas multiscreen practices were employed to achieve spatial synthesis and temporal synchronization in the long 1930s, contemporary multiscreen practices, as I argue in the coda, manifest a different spatiotemporal logic.

CHAPTER 1

# PRODUCTION SCREENS IN THE LONG 1930s

*Rear Projection and Special Effects*

One of the most iconic shots in *King Kong* (Merian C. Cooper and Ernest B. Schoedsack, 1933)—perhaps the most well-known instance, together with *The Wizard of Oz* (Victor Fleming, 1939), of Hollywood's use of special effects in the sound era—has the large creature plucking clothing from a terrified Ann Darrow (Fay Wray). The heterogeneity in scale that makes an 18-inch model of Kong seem to tower over Wray was achieved through miniature rear projection, wherein a live-action shot of the actress collapsed inside a giant prosthetic hand was projected at a small scale behind the model of Kong and advanced forward frame by frame to synchronize with the stop-motion photography (figure 1.1). Many other shots in the film, such as those in which Wray cowers in a tree while Kong and other beasts loom in the background, also feature full-scale rear projection, in such cases positioning live actors in front of a large screen upon which moving images of the animated creatures were projected (figure 1.2).[1] These moments can be viewed through the numerous frameworks that have, as befitting the film's infamy, been erected around it, from divergent analyses of the visible seams produced by the special effects to a range of allegorical readings (framing Kong, most notably, as a colonialist figure for a racialized other or as an emblem of economic adversity during the Depression).[2] The film's employment of special effects, moreover, contributes to its politics of representation—in particular the construction and containment of Kong's threat, figured as the threat of miscegenation—since the compositing of ontologically distinct figures

**1.1** Through the use of miniature rear projection in *King Kong* (RKO, 1933), an 18-inch model of the creature seems to tower over Fay Wray.

*Source*: Screen grab by author.

**1.2** Instances of full-scale rear projection in *King Kong* (RKO, 1933) also made it possible for animated beasts to threaten live actors.

*Source*: Screen grab by author.

(animated and live) renders the prospect that Kong and Wray may touch simultaneously thrilling and impossible.[3]

A focus on the practice of rear projection illuminates the construction of cinematic space underlying this dynamic. The use of rear projection structured the production of *King Kong*, like that of many other Hollywood films throughout the 1930s and beyond, around what Julie Turnock terms "the screen on the set."[4] As discussed in the introduction, screens work to unite, divide, and demarcate spaces in a variety of ways, functioning simultaneously as surfaces, apertures, frames, and barriers.[5] As Anne Friedberg argues, screens operate at the interface of virtual and actual space, joining represented and built spaces to form an "architecture of spectatorship" that spans both domains.[6] Although screens are usually conceived as components of media such as film and television, these surfaces are thus sites of mediation in their own right. Most theorizations of screens emphasize their role in exhibition, but the screens used in production work in similar ways to mediate virtual and actual spaces. Indeed, the rear-projection screens employed in the classical era encouraged production spaces to become exhibition spaces simultaneously. Displaying prefilmed imagery as an extension or even replacement of a physical set, these screens connected prefilmed and live-action spaces on the soundstage to form diegeses constituted from ontologically distinct domains. Such mediation occurred in countless films—from fantastic spectacles such as *King Kong* to the most quotidian Hollywood fare—making it seem as though the actors, without actually straying from the studio, occupied more exotic or dangerous (or even simply motion-filled) locales.

The example of *King Kong* shows how close attention to the use of screens in production can cast American cinema of the long 1930s—and even the most canonical of its products—in a new light. The 1930s are often considered the decade in which Hollywood's industrial organization and mode of representation reached their apotheosis, culminating in "Motion Pictures' Greatest Year," an appellation that was coined as part of a promotional campaign in 1938 but that many believe serves as an apt designation for 1939.[7] The decade was marked by the continued rationalization of labor practices and consolidation of the studios' monopolistic structure, especially in the wake of the National Recovery Administration's Code of Fair Competition for the Motion Picture Industry of 1933, which accommodated vertical integration and sanctioned practices such as block booking. This period was also characterized by the standardization of sound-film techniques such as multitrack recording and mixing as well as the development, with the implementation of the Production Code, of a system of conventions according to which films could be seen to uphold dominant cultural values (while also making possible a multiplicity of readings).[8] At the same time, the 1930s saw cinema permeating areas of daily life beyond the movie theater. Not only did Hollywood cinema intersect with other components of mass culture including radio, but—both drawing on and contributing to recognition of the movies' social force and cultural prominence—cinema itself also burgeoned within a variety of institutions beyond Hollywood, with schools, museums, political organizations, government agencies, and corporate entities exploring its function as a means of education, form of art, and mode of persuasion.[9]

In its employment of rear projection, *King Kong* not only exemplifies how special effects could contribute to Hollywood cinema's mode of representation in this period but also reveals how this mode of representation could collaborate with the arrangement of theatrical exhibition to form cinema's address. With the pervasive presence of the sheets of glass, cellulose, and stretched rubber employed to composite live actors with animated monsters (via glass painting, full-scale rear projection, and miniature rear projection, respectively), *King Kong*'s diegetic space is constituted from a series of stacked planes. Echoing the construction of the film's soundtrack, which employs Max Steiner's score as background music, this spatial formation exemplifies and furthers what Dominique Païni describes as the "layered effect" of rear projection itself.[10] In *King Kong*, the stacking of image planes not only unites monsters and humans within the frame but also consigns them to what Noël Carroll identifies as different "zones," located at different depths, within the represented world.[11] In doing so, it supports a form of representation in which meaning is conveyed through the separation and transgression of spaces. While the surfaces of the rear-projection screens in particular act as ontological barriers between the different spatial zones, the film conveys the monsters' threat to the humans, as in the scene where Kong undresses Darrow, by suggesting that these barriers can be breached.

The address of films that feature such layering, including *King Kong* and many others of the period, is illuminated by considering the fact that the exhibition screen constitutes one more layer in the series of stacked planes receding into the depth of cinematic space. In this regard, cinematic space is taken, following Friedberg, to encompass the represented and built realms mediated by the theatrical screen, contributing to what Thomas Elsaesser identifies as an "expanded concept of diegesis."[12] If *King Kong* revolved around the threat that dangerous figures could transgress the rear-projection screen and harm the characters on the other side, it simultaneously introduced the prospect that these figures could cross the ontological divide also marked by the theatrical screen, posing a felt, if not actual, peril to the film's audience. Insofar as *King Kong* activated racist and colonialist tropes through the figure of the ape and depiction of stereotypically primitive inhabitants of the jungle, it aligned that threat with contemporaneous constructions of social difference. In doing so, the film collapsed different practices and social groups, evoking both the slave trade and ethnographic expeditions (recalling Merian C. Cooper and Ernest B.

Schoedsack's own films of the 1920s, especially via Cooper's reflection in the character of Carl Denham) and casting African American actors (including prominent "race film" producer and star Noble M. Johnson) as natives of a fictional island in or near Indonesia.[13]

If rear-projection screens worked to mold diegetic space to the dimensions of ideology by creating separations and unions that formulated racial difference as a threat, their mirroring in the theatrical screen extended that ideological organization of space to viewers' actual environment. James Snead has argued that *King Kong* implicates its own viewers in the "optical colonialism" perpetrated by the characters, a form of voyeurism that links seeing to capturing and killing.[14] Since the film's rear-projection screens bridged the spaces they demarcated through the prospect of physical (sexual and/or violent) contact across a permeable surface, attention to this special-effects practice suggests that the film's ideological project—conveyed through the dual address to spectators within and of the film—extended beyond the realm of vision and into the domain of touch. In having Kong threaten to transgress the rear-projection screen and come into tactile contact with the various audiences assembled to ogle him within the fictional world (a transgression that, the film suggests, justifies the countervailing breaches represented by Kong's capture and execution), the film also extended the creature's ideologically rooted threat to its own viewers by proffering the sensation that he could reach across the barrier of the theatrical screen.[15] The film's use of rear projection thus supports Paul Young's claim that Kong served as a figure for the transgression of boundaries in the radio age.[16] *King Kong*, in short, harnessed rear-projection screens to structure its address to viewers as well as its diegesis around cultural fears of spatial and social border crossing, endowing the prospect of bodily contact from across an ontological—and, allegorically, racial—divide with a sense of titillation and terror.

*King Kong* thus exemplifies how the employment of screens in production could contribute to a film's mode of representation and address by encouraging particular formations of diegetic space. In classical Hollywood films, the use of editing facilitated the construction of fictive spaces from the assembly of shots.[17] However, special-effects techniques such as rear projection, which were also pervasive in the sound era, made it possible to marshal filmed fragments into new spaces within the shot as well.[18] Indeed, rear projection functioned analogously to editing, with the connective tissue of the screen's surface and frame enacting spatially a form of intermediation similar to that which the cut produced temporally.[19] Such intermediation could, of course, entail rupture as well as linkage. The exhibition screen carried out a similar operation on the viewer's material environment as its own surface and frame marked what Friedberg identifies as "a separation—an 'ontological cut'—between the material surface of the wall and the view contained within the frame's aperture."[20]

With the rise of rear projection, the virtual spaces depicted on the theatrical screen were, as in *King Kong*, also constructed with screens. Insofar as these screens often constituted the background within a shot, they contributed to the layering of image planes within deep space. At least one technician went so far as to advocate nesting rear-projected images within rear-projected images, raising the prospect of a *mise-en-abyme* of screened images receding into the depth of the diegesis.[21] As my discussion of *King Kong* has indicated, moreover, an expanded view of diegetic space as encompassing the conjunction of built and represented realms reveals that the theatrical screen contributed to such layering as well. Viewers within the theater, in other words, confronted a series of screens, including the exhibition screen in the actual foreground and, within and beyond it, the rear-projection screen(s) in the projected background. In many cases, rear-projection practice also entailed the arraying of multiple planes laterally so that they enacted a kind of spatial montage.[22] *King Kong*, for instance, includes shots juxtaposing multiple miniature rear-projection screens. As discussed later in this chapter, other films achieved a similar effect through the use of dual-screen backgrounds or simply the inclusion of a small screen (often picturing the view out a window) within a larger set. As I elaborate in subsequent chapters, this practice, too, found an echo in exhibition with the arraying of multiple theatrical screens within and beyond the auditorium.

In this chapter, I explore the range of ways in which rear-projection screens were harnessed in the long 1930s to structure the spaces of production and, with them, representation. As I show, the diegetic spaces of numerous Hollywood films in this period were structured around the proliferation of image planes. Attending to this proliferation, I contend, compels us to recognize the profoundly synthetic nature of classical Hollywood diegeses, which brought together fragmentary profilmic material not only through editing but also through compositing within the shot. *King Kong* exemplifies how the resultant spatial formations could contribute concomitantly to the shaping of particular social formations. In the remainder of the book, I discuss the screens upon which those synthetic spaces were projected, showing how the concurrent proliferation of screens within and beyond the movie theater endowed places of exhibition with similar spatial—and temporal—logics.

## Approaching Rear Projection in Hollywood of the Long 1930s

Despite the reputed novelty of digital visual effects' capacity for simulation and spectacle, and contrary to widespread assumptions about the relative authenticity of predigital mise-en-scène, cinematic images in the classical era were—as *King Kong* and many other films evidence—subject to heavy manipulation and even construction. Indeed, a growing body of scholarship has begun to challenge the perception of contemporary digital effects' radical novelty, revealing continuities in effects practice and, in examining the special effects of the classical era, drawing attention to what had been a largely neglected area in film history.[23] An exploration of how screens were used for rear projection in the long 1930s contributes to that discussion by situating this special-effects technique within a broader field of screen practice and thus elucidating the role of special effects in a more widespread restructuring of cinematic space.

Addressing the use of special effects in the classical Hollywood cinema raises a nexus of issues related to realism, reflexivity, and notions of classicism and modernism.[24] In designating Hollywood output from around 1917 to 1960 as "classical," David Bordwell emphasizes the way in which that concept evokes "notions of decorum, proportion, formal harmony, respect for tradition, mimesis, self-effacing craftsmanship, and cool control of the perceiver's response," as well as the aesthetic qualities of "elegance, unity, [and] rule-governed craftsmanship."[25] Aligning special effects with this notion of classicism thus entails situating them within an aesthetic of transparency that would efface the materiality of the image in favor of a seemingly unmediated view into a unified diegesis. Indeed, Bordwell and Kristin Thompson argue that special-effects techniques such as rear projection and optical printing—together with other technological developments embraced in the 1930s, such as synchronized sound, Technicolor, and deep-focus cinematography—were assimilated into the classical Hollywood cinema's effort to create a unified sense of diegetic space and to present narrative information in a clear, redundant, and unambiguous way.[26] In this regard, the use of rear projection and optical printing to integrate component parts into unified images can be understood to fall in line with the classical Hollywood style's aim of encouraging viewers, in Bordwell's words, to "construct a spatial whole out of bits."[27] Although Bordwell and Thompson's approach to the classical Hollywood cinema marks a significant departure from the apparatus theory of the 1970s, their emphasis on the unity and transparency of Hollywood cinema is consistent with the qualities that theorists such as Jean-Louis Baudry and Stephen Heath claimed underlay its ideological function.[28]

By contrast, several scholars have recently argued that classical-era rear-projection practice introduced forms of fragmentation and reflexivity that worked against the aesthetic of unity and transparency valued by Hollywood. Centering primarily on the close analysis of films (and overwhelmingly on the oeuvre of Alfred Hitchcock), this work has shown in particular how rear projection renders the film frame spatially, temporally, and ontologically heterogeneous by bringing together, for instance, foregrounds shot on studio soundstages by the primary crew and backgrounds filmed earlier on location by a second unit. Rear projection is here taken to be obtrusive or even distancing rather than embodying the form of "self-effacing craftsmanship" that

Bordwell associates with classical cinema. In an essay on Hitchcock's use of rear projection, for example, Dominique Païni explores the ways in which Hitchcock's films deploy the technique to create disjunctures within the image, resulting in what Païni identifies as "a certain dreaminess, an element of oneiric reserve" that he attributes to rear projection in general.[29] Drawing on Païni, Laura Mulvey argues that these disjunctures produce what she calls a "clumsy visibility" and put rear-projection shots "at odds with the principles of transparency and associated realism to which Hollywood cinema generally aspired" by foregrounding the materiality of the image and baring the device in a way that "seems to smuggle something of modernism into the mass medium of modernity."[30]

Mulvey's argument exemplifies the way in which the classical Hollywood cinema, particularly via its emphasis on transparency, has been construed in opposition to the reflexivity associated with modernism. The perceptibility and reflexivity of special effects, however, need not be considered at odds with Hollywood practice, as Christian Metz already suggested in 1972.[31] Here I also contend that rear projection—which can, in Païni's words, "be seen as a kind of duplication within the film itself, a symmetry of pictured pictures"[32]—functioned reflexively, embedding a form of cinema within cinema. But I do not want to suggest that this produced the form of estrangement associated with baring the device in the Russian Formalist tradition cited by Mulvey.[33] Although classical-era rear projection can be distancing for contemporary viewers, there is little evidence that filmmakers or moviegoers treated it as such in any pervasive way in the 1930s.[34] Considering rear-projection practice as a form of vernacular modernism addresses the way in which the nesting of virtual production screens within the actual theatrical screen offered viewers a sensory experience of the spatiotemporal reorganizations associated with the larger mediascape, while also acknowledging that this sensory encounter was not necessarily obtrusive and was indeed often fully commensurate with the forms of narrative and visual realism to which practitioners in Hollywood aspired.[35]

Drawing on the observation that rear projection contributed to the construction of diegeses from spatially, temporally, and ontologically heterogeneous "bits," I thus move away from the concern with evaluating the realism of the resultant composites, whether in terms of unity/fragmentation or transparency/opacity. I focus instead on how screens worked to synthesize these "bits," providing viewers—of conspicuous and inconspicuous composites alike—a sensory encounter with particular configurations of technologically mediated space. Parsing the specific ways in which rear projection contributed to such formations of cinematic space in the long 1930s can be seen as one answer to Julie Turnock's call for attention to the historically contingent and protean nature of Hollywood effects practice.[36] In an essay on rear projection in *Detour* (Edgar G. Ulmer, 1945), Vivian Sobchack has shown the analytic value of a focus on the spatiality of rear projection, arguing that this technique concretizes film noir's emphasis on spatial entrapment.[37] My analysis of rear projection in the long 1930s expands on that observation by exploring the various and multifaceted ways in which the increasing use of screens in production contributed to the construction of cinematic space across a broad range of films.

In its contribution to this construction of cinematic space, particularly through the emphasis on spatial layering, rear projection worked in concert with several other techniques and technologies of filmmaking. Indeed, Bordwell identifies the stacking of planes as a key strategy in the classical Hollywood cinema's effort to efface the two-dimensional surface of the screen by representing three-dimensional space. In this case, lighting, focus, costuming, and set design are taken to work together to distinguish figures in the foreground from separate background planes within the mise-en-scène.[38] In the 1930s, such spatial layering also played a role in animation practice (culminating with the development of the multiplane camera), stereoscopic 3D (especially in the wake of MGM's release of *Audioscopiks* [Jacob Leventhal and John Norling] in 1936), and the reemergence of deep-focus cinematography.[39]

The employment of spatial layering within film sound, however, offers an especially apt analog to rear projection.[40] As scholars such as Rick Altman and James Lastra have shown, the 1930s witnessed a transformation in the ways technicians in Hollywood conceived the spatiality of diegetic sound. Specifically, they moved away from modeling film sound on the perspective of an imaginary auditor, a perspective that had emphasized fidelity to a profilmic event—matching, for instance, the scale of images with the volume of sound. They instead began constructing soundtracks that were hierarchically layered to privilege narrative clarity, emphasizing, in particular, the intelligibility of dialogue.[41] Lastra suggests that the hierarchical organization of what he terms "planes" or "layers" of sound emulated the separation of classical Hollywood images into foreground and background.[42] The form of image construction on which sound was modeled, to be sure, often relied on straight photography in conjunction with the approaches to lighting, focus, costuming, and set design mentioned earlier. But as was pointed out at the time (and as both Lastra and Altman have indicated), special-effects techniques for creating composite images provide a particularly close parallel to the practice of constructing a layered soundtrack from sources recorded at various times and on different strips of film, which had emerged by 1932.[43] In other words, like special-effects technicians and almost simultaneously, film-sound technicians also came to embrace the amalgamation of fragmentary material into synchronic spatial composites. At the same time that special-effects technicians were employing rear-projection screens on the set to create a synthetic, layered space within the shot, sound engineers were compiling heterogeneous sonic fragments into foreground and background layers on the soundtrack.

Screen and sound technologies not only functioned in similar ways but also worked in concert to render cinematic space increasingly synthetic in the long 1930s. Scholarship on film sound can therefore provide a guide in assessing screen practices in this period, particularly since it has worked through some of the complexities regarding the issues of realism, reflexivity, classicism, and modernism that I mentioned earlier. An article from 1930 written by RCA sound technician John L. Cass raises this nexus of issues and figures as a point of reference for much scholarly work on sound space. In it, Cass contended that the "blend of sound" achieved through mixing among multiple microphones "may not be said to represent any given point of audition, but is the sound which would be heard by a man with five or six very long ears, said ears extending in various directions."[44] In that Cass took issue with what he identified as a lack of coordination between this indefinite (and monstrous) sonic experience and the "very definite" visual perspective provided by the camera, the article exemplifies an attitude toward sound space that would be supplanted over the course of the 1930s.[45] In the context of apparatus theory, the eventual embrace of mixing has been viewed as another means by which the classical Hollywood cinema hid its work and functioned ideologically, and Cass's depiction of bodily fragmentation and confusion has been taken as an indication of a potential early disruption to that process.[46] As Emily Thompson has argued, however, the blending of sound from heterogeneous sources also contributed to film sound's modernity insofar as it transformed traditional spatial relationships. In this regard, Cass's many-eared man—whose sonic experience no longer involved a definite perspective but instead entailed a form of spatial amalgamation that Thompson likens to cubist painting—serves as a figure for the perceptual changes accompanying this transformation.[47]

Like Cass's "blend of sound" simulating the auditory experience of a person with several long ears extending in different directions, shots featuring compositing techniques such as rear projection blended multiple images together and could similarly be said to mimic the visual experience of a person with eyes directed toward different spaces (and times). Unlike editing, which presents varying visual perspectives sequentially, compositing techniques such as rear projection display multiple views simultaneously and thus could be open to similar charges of spatial ambiguity and perceptual monstrosity. This spatiotemporal hybridity and the potential for defamiliarization underlie Mulvey's suggestion that rear projection smuggles an element of modernism into Hollywood cinema.[48] In line with Thompson's argument, however, we can also read rear projection and other compositing techniques as proffering a sensory encounter with the spatiotemporal reconfigurations associated with modernity, reflecting an urban experience also identified with the collapse of actual and virtual spaces, without necessarily distancing the viewer.[49]

In what is easily the most well-known discussion of cinema's relationship to modernity, "The Work of Art in the Age of Its Technological Reproducibility," first composed in 1935–1936, Walter Benjamin focused on the processes of cinematography, sound recording, and—especially—editing, together with slow and fast motion, as the means by which films reflected and contributed to the technological reconfiguration of experience, enabling cinema to perform what he considered to be its social function: "to establish equilibrium between human beings and the apparatus" of urban-industrial modernity.[50] As I have been suggesting, however, compositing techniques such as rear projection also participated significantly in that process. Crystallizing within the shot the capacity for technological reproducibility that Benjamin attributed to the proliferation of photographic images more broadly, these techniques epitomized the way in which, as Benjamin argued, film's "manifold parts" were "assembled according to a new law," presenting an "equipment-free aspect of reality" precisely "on the basis of the most intensive interpenetration of reality with equipment."[51] When projected onscreen, shots featuring rear projection thus rendered viewers' sensory environment profoundly synthetic, inviting a haptic encounter with a technologically reorganized (and ideologically structured) *physis*.

In the remainder of this chapter, I map that reorganization, first situating rear projection within the larger landscape of special-effects practice and subsequently undertaking a close examination of the development and use of rear-projection screens. This examination reveals several specific ways in which the use of screens on the set contributed to a reshaping of diegetic space, centering on the screens' qualities of transparency, scale, and mobility. It is my larger contention that this technological reconfiguration reflected and contributed to the wider context of spatial reconfiguration in modernity, wherein exhibition screens also played an important role.

## Rear Projection and Hollywood Effects Practice

Rear projection worked in conjunction with several other special-effects techniques to produce heavily mediated cinematic spaces in the long 1930s. Beyond their obvious contribution to fantastic films such as *King Kong* and *The Invisible Man* (James Whale, 1933), special effects were an integral component of Hollywood cinema in the sound era, pervading even run-of-the-mill productions. In such cases, special effects often served the practical purpose of replacing location shooting with the creation of composite images that combined principal action shot on studio grounds with background scenery, whether photographed or painted.[52] The ability to film dialogue in the controlled environment of the studio was particularly useful in the wake of the coming of synchronized sound; however, special-effects practices accorded generally with Hollywood's rationalized business model, offering the most economical, efficient, and safe means of producing imagery that nevertheless displayed a form of production value associated with perilous situations and exotic land-, sea-, and cityscapes. As Paramount technician Farciot Edouart put it with regard to rear projection, special effects conformed "to the industry's ideal of getting the best possible picture under the most completely controllable conditions, and with a minimum of time, expense, and danger."[53]

Although spectacular trick effects continued to appear throughout the 1930s and beyond, by the late 1920s technicians were moving away from an emphasis on "tricks" and toward a focus on effects that, as G. A. Chambers put it in 1932, "unrecognized as 'trick' shots, are inserted in a picture to lend production value."[54] As Edouart explained in advocating such unrecognizable effects work, "Such shots are resorted to and attained by means of special methods because economic limitations, production difficulties, or mechanical reasons make it impossible or impractical to secure the required results by straight ordinary photography."[55] In this regard, as Turnock suggests, effects were deemed "special" in this period not by virtue of their spectacle but rather because of the specialized labor needed to achieve them.[56] A corollary to the emphasis on unrecognizability is the lack of attention devoted to effects, despite their pervasiveness, in studio publicity, reportedly "lest the 'secrets' broadcast to the world might take the glamour from the screen."[57]

Designated in the 1930s by terms such as *process projection*, *background projection*, and *transparency process*, rear projection entailed the projection of previously photographed background imagery (either static or moving) through a translucent screen on the set. Actors performed in front of the screen so that the composite camera could capture their action and the projected background simultaneously (figure 1.3).[58] The concept has precedents predating cinema—rear projection was already employed, for instance, in the phantasmagoria and to insert backgrounds in portrait photography—and it was, as mentioned in the introduction, used in the movies as early as 1913.[59] Hollywood, however, embraced rear projection between 1930 and 1932, when the coming of synchronized sound had made location shooting impractical, the deepening Depression favored the technique's economy, and several technological developments made it newly feasible. The latter included adequate means for synchronizing the background projector with the composite camera in order to avoid flicker (achieved with electrical hook-ups that had been developed for sync sound), faster and more sensitive negative emulsions to photograph the rear-projected image (achieved with the introduction of supersensitive panchromatic film stock), and more powerful light sources, paired with improved optics, to make the rear-projected image brighter (developed to accommodate larger theaters and widescreen projection).[60]

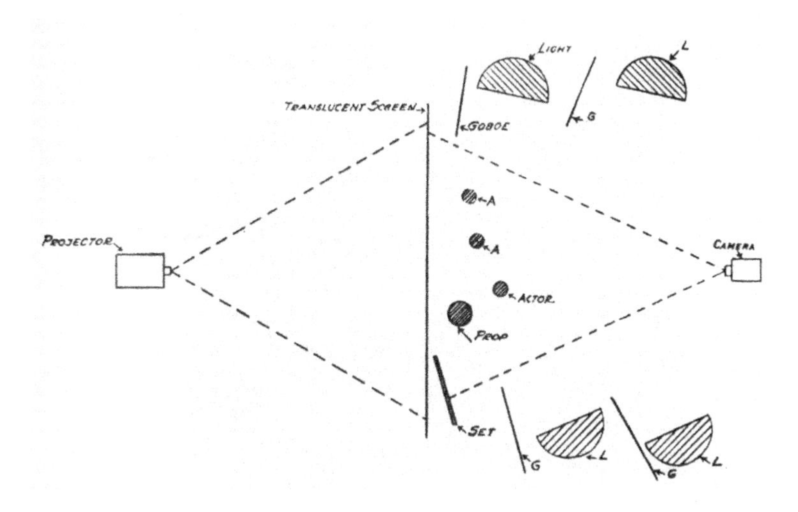


1.3 Rear projection entailed the projection of previously photographed background imagery through a translucent screen, with actors performing in front of it.

*Source*: G. G. Popovici, "Background Projection for Process Cinematography," *Journal of the Society of Motion Picture Engineers* 24, no. 2 (February 1935): 103.

Probably the most familiar special-effects technique of this era, rear projection became ubiquitous, providing the moving backgrounds for countless images of actors, who from the safety and comfort of Hollywood soundstages would pretend to be driving cars, riding trains, sailing ships, flying airplanes, confronting fires, or touring exotic locations (figure 1.4). The embrace of rear projection was reputedly initiated at Fox with the productions of *Liliom* (Frank Borzage, 1930) and, a few weeks later, *Just Imagine* (David Butler, 1930).[61] As early as 1932, Edouart, who came to be one of the most prominent technicians to develop and employ the process, contended that the technique had "come into general use in practically all of the major studios" and that "scarcely a picture is released by any such major producer which does not include at least two or three such scenes—and some productions have used this process to make 75 or 80% of their total footage."[62] Fox's film *State Fair* (Henry King, 1933), which pictures the principal actors against background images of the fair, for instance, reportedly included rear projection in about 65 percent of its scenes, employing a new fine-grain background negative material allegedly developed as a result of the technique's rising popularity.[63]

Rear projection emerged in dialogue with traveling-matte techniques such as the Williams and Dunning processes. Also innovated in the silent era, these techniques, like rear projection, made it possible for a moving figure to be composited with a separate background—in this case, by creating mattes that shifted from frame to frame. The Williams process took footage of the primary

1.4 *Gone with the Wind* (Selznick International Pictures, 1939) used rear projection to composite live action with previously filmed fire footage.

*Source*: Screen grab by author.

action, shot against a black or white background (usually black cloth), to create a matte that made it possible to composite that action with separately photographed background imagery during printing.[64] The Dunning process and related techniques such as that developed by Max Handschiegl had actors perform in front of a colored (usually blue) background and then used color-separating filters or film either to produce mattes, as in Handschiegl's method, or to "self-matte" the action footage onto prefilmed background footage at the time of shooting, as in the Dunning process.[65]

The terms used to describe the traveling-matte and rear-projection processes overlapped significantly. Although the term *process* eventually came to signify effects achieved at the time of production (as opposed to those created in post-production), and although it has often been used to designate rear projection specifically, it was used in the early 1930s to describe optical effects in general.[66] In this context, *process* was employed not only as a noun but also as a verb, and effects-heavy films were designated "processed" for much the same reason that industrially prepared food is: the end product is the result of a process of de- and recomposition.[67] Caroll Dunning, for instance, explained that *The Subway Express* (Fred Newmeyer, 1931) had imagery of the speeding train "processed into" it through the use of color separation, and he described the film as "a processed picture from its main title to its end."[68]

Although the term *transparency process* would also come to designate rear projection, it also initially referred to the color-separation process. Edouart, whose title at Paramount as of 1932 was "head of transparency process photography" and who worked with both processes at that time, used the terms *transparency process* and *transparency projection process* for the Dunning and rear-projection processes, respectively. The shared term *transparency*, he contended, resulted from that fact that with both processes the background images were rephotographed by transmitted light rather than reflected light. In the color-separation process, the blue screen placed behind the live actors acted as a printing light, causing the colored film in the camera, which contained the prefilmed background image, to imprint that image onto the negative only in the areas surrounding the actors. With rear projection, because the projector was directed toward the composite camera, the background image also transmitted rather than simply reflecting light.[69] As I discuss later, the concepts of "transparency" and "process" prove useful in analyzing the functions of screens more broadly.

Rear projection was simpler and more economical than the Williams and Dunning processes, and it had largely supplanted them by the mid-1930s, by which point the prevalence of effects shots was also increasing dramatically. In a report on the process department "at one of the major studios" (likely Edouart's department at Paramount), *American Cinematographer* compared that studio's creation of composite shots in 1930, when such shots were achieved with the color-separation process; in 1931, the year the studio adopted rear projection; in 1932; and in 1933.[70] While the number of such shots rose from 146 to 340 to 503 to 658, the average cost per shot dropped from $314.95 to $184.61 to $140.59 to $116—a reduction attributed to both volume and the reuse of background film from the library amassed by the studio in those years.[71] By 1942, Paramount was reportedly making between 1,600 and 2,000 rear-projection shots per year.[72]

Nearly simultaneous with the rise in the use of rear projection, an increased reliance on optical printing facilitated the creation of composite shots in post-production. Optical printing involves the rephotography of carefully controlled portions of footage projected within the optical printer.[73] Like rear projection, optical printing was not a new technique, but it became newly plausible with the popularization after 1926 and refinement in the early 1930s of duplicating negative and positive film stocks, which mitigated the loss of quality in rephotographed images.[74] As Carl Louis Gregory put it in 1928, optical printers were "an ideal instrument for selecting from films already made the components desired for the construction of composite or 'trick' pictures, and for the recombining of these components into new pictures."[75] Optical printers took over effects previously achieved in the camera—ranging from transitions such as fades and dissolves to matte shots and split-screen effects—and contributed to the vogue for wipes in the 1930s. In 1932, Warner Bros. technician Fred Jackman contended that all fades and dissolves were being made on optical printers, and by 1936 RKO optical-printing pioneer Linwood Dunn could proclaim (though perhaps exaggeratedly) that "during the past four or five years hardly a single production has been released that did not utilize the services of the optical printer to some extent."[76]

Although other techniques (notably the Schüfftan process) also permitted the incorporation of miniatures into composite shots, both rear projection and optical printing facilitated that practice.[77] For the flood scene in *Noah's Ark* (Michael Curtiz, 1928), achieved before the rise of the latter techniques, Fred Jackman's effects department at Warner Bros. integrated a life-size foreground featuring three hundred live actors with a background city, built in miniature, that would be destroyed by the water. The scene was filmed in one take, which was recorded simultaneously by eleven high-speed cameras and four normal-speed cameras. Pulling off the effects required the coordinated work of 105 technicians on the set. The miniature was reported to have taken four months and $40,000 to build, and the level of stress provoked by the need to record its destruction in a single take allegedly caused the "supervising technician" (i.e., Jackman) to suffer a nervous breakdown.[78]

With the embrace of compositing techniques such as rear projection and optical printing in the 1930s, miniatures could be filmed separately from the live action and even in component parts. When miniatures were employed to depict earthquakes toppling large buildings in *The Last Days of Pompeii* (Ernest B. Schoedsack, 1935) and *The Rains Came* (Clarence Brown, 1939), for instance, rear projection, traveling mattes, and optical printers allowed the technicians to join several separately filmed elements within the frame so that, as Fred M. Sersen put it in relation to the latter film, all of the pieces "fit together like a jig-saw puzzle."[79] For *The Rains Came*, which won the first Academy Award for Best Special Effects (beating fellow nominees *The Wizard of Oz* and *Gone with the Wind* [Victor Fleming, 1939]), Sersen's department at Fox employed optical printing to combine live-action shots of fleeing people, filmed on a street set in the back lot, with images of crumbling buildings filmed in miniature. Sersen's team, moreover, used rotoscoping to create special mattes for pieces of the crumbling buildings so that it looks as though parts of the miniature set are actually falling onto the live actors.[80]

As we have already seen with *King Kong*, miniatures could grace rear-projection backgrounds. Paramount's film *Cleopatra* (Cecil B. DeMille, 1934), for example, used miniatures to stage a nautical battle, which was then combined with the live-action foreground through both rear projection and optical printing.[81] Rear-projection screens were, moreover, incorporated into miniatures, as *King Kong* also exemplifies.[82] That film featured miniature jungle sets arranged in several planes, separated from one another by partially painted sheets of glass and including small stretched-rubber screens for miniature rear projection.[83] *Gone with the Wind* also used miniature rear projection, particularly for the "burning of Atlanta" sequence, where it provided some of the background images of the fire. In a shot of the slow march away from the burning city, for example, the fire in the background is a miniature projection, and the cliffs, trees, and part of the building at frame right were painted on glass. Paint was scraped off the glass painting in the spaces where the fire appears, and these elements were exposed twice, first to capture the glass painting and then to capture the projection behind it.[84]

Together with techniques such as traveling mattes and optical printing, rear projection thus contributed to the increasingly "processed" nature of cinematic space in the long 1930s. As the remainder of this chapter shows, an examination of the discourses and practices associated with rear projection in that period reveals three broad ways in which rear-projection screens were harnessed to mediate actual and represented realms on the set. Each of these is connected with particular material qualities of screens as they were developed and employed by technicians and filmmakers. The first approach relates to the screens' transparency—their capacity, as surfaces, to act as membranes and barriers, both joining and separating the spaces in front of and behind them. The second relates to the screens' scale—and their concomitant ability as frames, masks, and apertures to enclose, obscure, and reveal the spaces within and surrounding their edges. The third relates to the screens' mobility—their simultaneous capability to move and to display movement. Experimentation along these vectors shaped Hollywood films in this period by encouraging particular deployments of the image planes that proliferated on the set.

These planes included screens as well as related surfaces such as sheets of glass, painted backings, and mirrors. Although film theory has long identified windows, frames, and mirrors as metaphors for the screen, an exploration of effects practice and, as discussed in subsequent chapters, of exhibition practice reveals how these objects and the qualities of transparency, delineation, and reflectivity associated with them have actually functioned in tandem with one another.[85] An archaeology of rear-projection screens and related surfaces thus shows how these proliferating expanses of glass, cloth, and cellulose, with their differing qualities of transparency, scale, and movement, marshaled heterogeneous profilmic fragments into particular formations of diegetic space.

## An Archaeology of Rear-Projection Screens

Technicians worked throughout the 1930s to address the purported problems with the rear-projection process, especially reputed limitations on the quality and scale of the rear-projected image.[86] The degradation of image quality derived primarily from the fact that rear-projected images were always at least second generation. Because rear-projected footage was photographed twice (once to create the background and then again to composite it with a live-action foreground), it tended to display less definition than straight photography. In addition, the quality of rear-projected images in this period suffered from their unsteadiness and the unevenness of their illumination.[87] As a result, rear-projected portions of the frame did not blend seamlessly with straight footage—a significant problem for effects technicians, who, as Turnock argues, viewed the achievement of seamlessness as their goal.[88] Though technicians did not ever fully ameliorate the loss of definition entailed in rephotographing rear-projected images, the twin goals of achieving the steady, even, bright rear-projected images considered necessary for sufficient quality and meeting the constantly increasing demands for larger background sizes drove experiments with different components of the rear-projection apparatus throughout the 1930s, especially through screen construction and projector design. In this section, I explore the ways in which these goals contributed to experimentation with the screens' qualities of transparency, scale, and mobility. As I show, this experimentation shaped the ways in which screens united the actual and virtual spaces on the set into composite diegeses.

## Transparency

My discussion so far has treated transparency as a concept associated with the screen's function as what Friedberg terms a "virtual window."[89] The actual permeability of rear-projection screens, however, was crucial to their construction and use, and it served as a foundation for their representational functions. In the design of rear-projection screens, a major concern was the surface's capacity simultaneously to transmit and to diffuse light. Both adequate transmission and sufficient diffusion were considered necessary for image quality. A high level of transmission rendered rear-projected images bright enough to rephotograph well at a large scale, and diffusion guarded against the uneven illumination ("hot spots" at the center of the screen and "fall off" at the edges) that disrupted these images' uniformity.[90] In this subsection, I detail how the technical work on rear projection that took place over the course of the 1930s addressed these qualities through experimentation with screen-construction materials. The effort to construct screens that would transmit light (rather than reflecting it, as did the majority of exhibition screens) aligned rear-projection screens with other transparent, partially transparent, and translucent surfaces ranging from windows to cellophane wrapping. The use of transparent surfaces such as rear-projection screens on the set contributed to the layering within three-dimensional diegetic space that I have already identified. Considering the fluid relationship between rear-projection screens and other contemporaneous experiments with transparency, however, illuminates the shifting valences of this spatial organization.

Rear-projection screens were initially made of sand-blasted glass, which had a high level of transmission but was susceptible to hot spots. Attempts to eliminate hot spots by increasing the screens' diffusion, however, sacrificed their capacity for transmission.[91] In addition, glass screens were breakable and expensive, making them both dangerous to employees and impossible to insure.[92] An alternative idea proposed in 1932 was to replace the glass screens with mirrors, which were more affordable. The mirror would be positioned at an angle to the composite camera, reflecting a standard screen (as used in theatrical exhibition) situated to the side of the camera.[93] This plan was a variant of the Schüfftan process, which used mirrors to composite profilmic and offstage objects, including projected images, and it was put forth as a means of replacing the troublesome transmissive screens with conventional reflective surfaces.[94] By eliminating the need for a translucent screen, this proposed arrangement also anticipated the more economical front-projection process developed after World War II, which used a beam splitter (often a two-way mirror) positioned at an angle in front of the composite camera (and which was also informed by the Schüfftan process).[95] With front projection, the camera shot through the transparent side of the beam splitter, while a projector situated to the side of the camera was aimed at the reflective side, which bounced the background image onto a reflective screen behind the actors.[96]

Beginning in 1932, however, glass rear-projection screens started being replaced by translucent surfaces made from what *American Cinematographer* described as "a cellulose composition somewhat akin to the familiar 'Cellophane' used in the wrapping of many commercial products."[97] The screen developed for RKO by Sidney Saunders was of this type. As Vernon Walker, head of RKO's special-effects department, put it, the cellulose screen "resembles a large sheet of waterproofed canvas; in use, it is stretched in a frame much as a sheet of canvas would be."[98] Such screens were inexpensive, nonbreakable, rumored to reduce hot spots, and capable of being made in larger sizes than glass screens. Cellulose screens had as their base either cellulose nitrate or cellulose acetate, which as a liquid was sprayed onto a polished surface. That coating was then sprayed with a polarizing material such as glass powder to make it appropriately diffusive before being peeled off.[99] Screens made of cellulose nitrate were highly flammable, but screens constructed from cellulose acetate were not. Although the trade press had touted "non-inflammable [*sic*]" cellulose screens since 1932, both sand-blasted glass screens and cellulose nitrate screens remained in use several years later.[100]

In addition to the screens made of the cellulose materials that came to be favored, screens made of gelatin-infused silk were also used for rear projection in the 1930s. The screens made by the aptly named Trans-Lux Corporation were of this type.[101] These screens were created on a special table that imprinted "prismatic ridges" on the surface of the screen in order to facilitate diffusion.[102]

In 1935, the Trans-Lux screen was reported to have "very good diffusion but low transmission, thereby handicapping scenes requiring big picture sizes."[103] We might note that the interest in using larger and larger rear-projection screens for the creation of special effects, which posed the primary obstacle to employing Trans-Lux screens on the set, existed simultaneously with, in theatrical exhibition, a move away from massive picture palaces and toward the smaller scale of neighborhood houses.[104] Trans-Lux rear-projection screens were also employed in that exhibition context, wherein the cinemas operated by the Trans-Lux Corporation used rear-projection screens as part of a strategy to make theater space more compact, as I discuss in greater detail in chapter 3.

The quality of transparency shared by glass, cellulose, and silk rear-projection screens emphasized the surfaces' capacity to join or separate the spaces they bisected by conducting or obstructing light—their capacity, in other words, to act as permeable membranes that mediated the spaces in front of and behind them. Rear-projection screens enacted such mediation in a dual fashion. Within the actual space of the set, these screens mediated the realm of the performance, which took place in front of the screen surface, and the domain of the projection apparatus positioned behind that surface. In this sense, the screens took on a filtering function, allowing light to pass through but obscuring the camera's view of the projector. At the same time, as we have seen, technicians developed rear-projection screens to provide an aperture to represented space, rendering them *virtual* windows. The practice of using these screens to supply views out of windows within the set served to underscore and naturalize that function.

This employment of rear-projection screens echoes other practices in this period, both within and beyond the domain of special effects. Within the realm of special effects, there are significant sites of overlap between rear projection and the practice of glass painting. A long-standing special-effects technique, glass painting was popular into the early 1930s, although it fell out of favor by the middle of the decade.[105] This technique entailed the use of a partially painted pane of glass positioned in the foreground in front of the action to be filmed. The camera would capture both the image painted on the glass and the action unfolding behind the transparent area.[106] Like rear projection, this technique thus facilitated the creation of composite imagery at the time of shooting by positioning a sheet of glass within the set. *King Kong*, for instance, used glass painting to create the background in the long shot of the crew's arrival at Skull Island, with the faraway images of the great wall and Skull Mountain painted on a sheet of glass that was actually positioned a few feet away from the camera.[107]

Whereas rear-projection practice positioned the glass (or, later, cellulose) surface in the background, glass paintings were positioned in the foreground on the set while often functioning as background within the diegesis, as in the

Skull Mountain example. In both cases, though, the combination of live action with imagery gracing the glass surface, whether painted or projected, endowed the composite image with a layered spatiality. In both cases, moreover, the glass sheet delineated foreground and background by serving simultaneously as an image plane and as the opening to a view beyond the surface, whether revealing the actual set (as in glass painting) or a virtual environment (as in rear projection).[108] *King Kong* went so far as to multiply these surfaces and views by employing glass painting in conjunction with rear projection, as in the shot of Carl Denham (Robert Armstrong) and his group approaching Kong's unconscious body on the beach, where glass painting was used for the background in the live-action portion of the scene and the resultant shot was then projected in miniature behind the 18-inch model of Kong to achieve the final composite (figure 1.5).[109]

The experimentation with transparency undertaken vis-à-vis both rear projection and glass painting can be seen as part of a larger focus on that concept within a range of modernist practices, from modern architecture to cubist painting, where transparency serves, among other roles, as a literal and figurative vehicle for spatiotemporal reorganization and for the intersection of virtual and actual presence.[110] The special form of transparency associated with these special-effects practices also finds echoes in consumer culture with the introduction of cellophane—to which, as we have seen, cellulose rear-projection screens were compared—which was employed as a wrapping for merchandise and even as a material for women's clothing. As Judith Brown has argued in

**1.5** *King Kong* (RKO, 1933) employed glass painting in conjunction with miniature rear projection.

*Source*: Screen grab by author.

relation to the use of cellophane wrapping, "Transparency relies on the logic of consumer desire: the clean, see-through material that appeared to consumers in the early thirties offered both an unimpeded view of the product and an additional sheen that improved its appearance; its smooth, glossy surface created a relationship of desire, acting as tantalizingly flimsy barrier, between consumer and good." While Brown contends that the use of cellophane in art deco sets contributed to the engineered gloss of Hollywood films such as *Swing Time* (George Stevens, 1936) in the 1930s, the investigation undertaken here indicates more pervasive connections.[111] Like the cellophane wrapping encasing goods such as cigarettes, cellulose rear-projection screens—also functioning as "tantalizingly flimsy" barriers—at once proffered unimpeded views to a space beyond the surface and endowed that space with an extra sheen, in this case by rendering it through the glossy medium of film itself.

The widespread practice of using rear projection to lend films "production value" by proffering exotic or spectacular backgrounds reflects the modernist notion of transparency as a figure for spatiotemporal reorganization and virtual presence as well as the more popular but equally modern employment of transparency as a means of provoking consumer desire. That practice is also imbricated with the colonialist project that I have attributed to *King Kong.* Indeed, the use of rear projection to display exotic backgrounds also echoes other modern means of reshaping social space, especially modes of transportation such as trains and ships, which functioned, in Ella Shohat and Robert Stam's words, as "engines of empire."[112] We might note, for instance, that in *King Kong* it is the use of rear projection and other special-effects techniques that allows the live actors to inhabit purportedly exotic spaces without enduring the kind of actual journey undertaken by Denham and his crew in the story (or by Cooper and Schoedsack in making their earlier films). In this regard, *King Kong*'s employment of rear projection and other special effects mirrors the ways in which cinema had long contributed to a colonialist worldview by (in part) proffering virtual global travel for audiences in Europe and North America.[113]

The class-A Westerns that reemerged toward the end of the 1930s often used rear projection in a similar manner, harnessing background screens both to display and to distinguish Native American characters as well as the desert landscapes with which the films associate them.[114] Films such as *The Plainsman* (Cecil B. DeMille, 1936), *Geronimo* (Paul Sloane, 1939), *Stagecoach* (John Ford, 1939), and *Union Pacific* (Cecil B. DeMille, 1939) employed rear projection to separate the European American protagonists in the foreground from groups of attacking Native American warriors occupying background landscapes. Describing the use of rear projection to reenact "a battle between Custer's soldiers and thousands of Indians" in *The Plainsman*, for instance, director Cecil B. DeMille explained that in "filming scenes of this type, the Director has two dramatic elements to coordinate. . . . In the foreground, he has his principals and from twenty to fifty extras to consider. In the [projected] background, he may have five or six thousand Indians and a regiment of cavalry, none of whom are at all picture-wise."[115] For *Union Pacific*, DeMille again employed rear projection to combine close shots of the principal actors made in the studio with backgrounds depicting Native Americans shot on location by the second unit (figure 1.6). As the second-unit director Arthur Rosson explained, for the "Plum Creek Massacre" scene in this film, his unit could set up their camera "on the inside door of the [train] car, shooting through it out to the prairie; see the Indians doing their stuff. . . . Then Mr. deMille [*sic*] would do the faces of the people in the studio, only having to build a door on the transparency stage."[116] DeMille ultimately ordered that the backgrounds with the attacking Native American warriors, originally shot in Utah, be retaken "due to insufficient number of Indians and unsatisfactory action" and requested that the number of "mounted Indians" be raised from fifty to seventy-five.[117]

At a time when European American actors were increasingly occupying Native American roles in Hollywood films, especially B Westerns, such second-unit location shoots may have offered more opportunities for Native American performers (even if at the expense and against the objections of those working in Hollywood).[118] But the use of rear projection literally relegated many Native American performers to the background, contributing to the segregation of both production space (through the division of first- and second-unit shoots) and diegetic space (through the ontological break between foreground

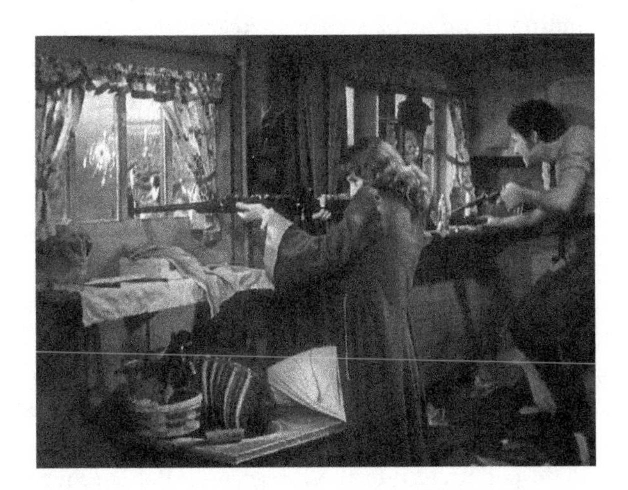

**1.6** *Union Pacific* (Paramount, 1939) employed rear projection to combine close shots of the principal actors, shot in the studio, with backgrounds depicting Native Americans, shot on location.

*Source*: Screen grab by author.

and background). Insofar as rear projection thus separated Native American performers spatially and socially while simultaneously putting them (as well as a mythologized and purportedly timeless Native American culture) on display, it contributed to what Michelle H. Raheja identifies as Hollywood cinema's function as a "virtual reservation." Noting that, from "their creation, reservations have been often-perverse tourism sites where non-Indians would travel to experience a glimpse into a purportedly 'vanished' culture," Raheja contends that "film became a virtual reservation for a viewing public eager for Indigenous images, but lacking the spare time and money to visit a geographical one."[119] In this way as well, rear projection perpetuated the kind of colonialist worldview forged through travel, simulating the transcontinental journeys undertaken by the films' colonizing characters (and by tourists and second-unit crews seeking Indigenous images) while alleviating European American viewers and actors of the need to endure actual travel. At the same time, promotion for *Union Pacific* and *Dodge City* (Michael Curtiz, 1939) enacted such journeys by launching the films' stars and executives on high-profile train tours.[120]

As Raheja points out, however, for Native American actors such as the activist Luther Standing Bear, who passed away on the set of *Union Pacific*, "acting in films permitted a socially acceptable way of talking back to colonialism." Indeed, Raheja's argument that virtual reservations not only "sequester Native Americans (as visual and imagined representations) but also provide Indigenous people with an organizing principle that allows communities and individuals to coalesce around commonly held as well as contested ideas and objectives" suggests the way in which Hollywood's representational and industrial practices, including those associated with rear projection, have prompted action.[121] As Raheja and others have recounted, Native American actors—led by figures such as Standing Bear and Victor Daniels (known as Chief Thunder Cloud), who performed in *The Plainsman*, *Union Pacific*, and *Geronimo*, among many other films—actively organized in this period. In doing so, these actors not only contested the studios' portrayals of Native Americans but also advocated for better labor conditions, including challenging the practice of bypassing the Native American actors working in Hollywood in favor of European American actors or—in a trend associated with second-unit shoots—of Native American performers on reservations.[122]

At the same time that the screen surface separated foreground and background spaces, scenes featuring rear projection often attempted to bridge these spaces by suggesting, as in *King Kong*, that elements from the projected background could penetrate into the foreground. Battle sequences, in particular, often conveyed the threat that rear-projected antagonists could reach through the background screen to harm the protagonists inhabiting the foreground. This is the case, for example, with the naval battle in *Reap the Wild Wind* (Cecil B. DeMille, 1942), in which a rear-projected ship is seen firing toward the foreground and, after a delay, a cannonball splashes in the live-action set, apparently having penetrated the plane demarcated by the rear-projection screen. Such transgressions also feature in Westerns displaying rear-projected Native American warriors, sometimes as seen through the window of the white protagonists' train or stagecoach, firing their weapons into the foreground. Here, too, as Adrian Danks observes with regard to *The Plainsman*, the firing of a weapon "in one plane has an appropriate effect in the other."[123] In all of these cases, the prospect that cannonballs, bullets, or arrows will cross the diegetic threshold marked by the rail of a ship or the window of a vehicle suggests that these projectiles can also penetrate the ontological divide marked by the rear-projection screen. Like *King Kong*, the Westerns in particular thus harness the spatial organization associated with permeable surfaces (an actual permeability in the case of diegetic windows, mirroring a virtual permeability in the case of rear-projection screens) to construct the prospect of boundary crossing, both literally and as a figure for interracial contact, as a danger.

## Scale

Screen size represented another persistent concern in the engineering discourses on rear projection in the 1930s. At the outset of the decade, the process was limited to relatively small screens of around 6 or 8 feet wide, which, in turn, posed limitations on the camera distances possible for rear-projection shots. Rear projection was often consigned to a small portion of the image, such as the view out a window, but if the projected image was to constitute the background fully, actors had to be photographed at fairly close range.[124] As Edouart explained, when the process was first introduced, technicians were happy when they could "perhaps make scenes showing our foreground actors from head to ankles."[125] Here I explore the work that was done over the course of the decade to address that limitation. Attempts to increase the scale of rear-projected images entailed various forms of multiple projection, employing multiple screens, multiple projected images, or both. These experiments thus achieved a form of proliferation different from the layering I have discussed so far. In this case, images also proliferated through lateral juxtaposition. The backgrounds gracing the larger rear-projection screens, moreover, tended to favor particular kinds of imagery, specifically dangerous environments such as water, fire, and sky.

There was some variation in the screen sizes employed by different studios throughout the decade. By 1932, Fox, MGM, and Paramount were reportedly using 18- to 20-foot-wide ground-glass screens for rear projection—a size that, it should be noted, closely matched contemporaneous averages for exhibition screens.[126] The cellulose screen adopted by RKO in 1932 also measured 20 feet wide.[127] That screen was used in *King Kong*, which had begun with an 18-foot-wide ground-glass screen that broke after a week.[128] By October 1932, however, Vernon Walker at RKO claimed that an even more recent cellulose screen installation measured 23 feet wide.[129] By 1936, Fox reportedly had a 26-foot-wide screen.[130] At Paramount, by contrast, technicians did not find it possible to use 20- to 24-foot-wide screens until around 1938.[131] Despite this variation, the drive toward larger screens fueled work at several studios, including Paramount and Warner Bros. in addition to RKO. In justifying this work, Edouart explained that "the dramatic and economic usefulness of the [rear-projection] process is dependent upon the physical scope of the process being sufficient to allow the director freedom closely comparable to what he would enjoy if his company was working upon the actual location. It is of very little use to have a process that can put Gary Cooper in Paris, or Barbara Stanwyck in Wyoming, if such scenes must be restricted to close shots of one or two players, or if the movements of the actors must be restricted."[132]

Three factors contributed to limitations on the scale of rear-projected images: (1) the glass screens used at first were difficult and expensive to make in large sizes; (2) larger screens necessitated more light than projectors could initially muster; and (3) the projection throws necessary to fill large screens were constricted by the size of the soundstages in which they were employed. Replacing glass with cellulose surfaces marked a first step in addressing these issues, after which the most significant developments entailed strengthening the illuminating power of rear projectors.[133] The latter was achieved by multiplying the projectors, whether to juxtapose their beams (filling multiple adjacent screens), to superimpose their beams (creating a brighter background on a larger screen), or to do both (as was happening at Paramount by the early 1940s). By 1936, Paramount was employing a dual-projector, dual-screen apparatus to expand the scale of its background images, a system that garnered an Academy Award in 1937.[134] By 1938, both Paramount and Warner Bros. had developed triple-head rear projectors, which superimposed three identical versions of the background image and thus supplied more light, making it possible to increase the screen size significantly.[135] At Paramount, where the design of the triple-head projector grew out of the dual-screen setup, the production of *Spawn of the North* (Henry Hathaway, 1938) began with a 24-foot-wide rear-projection screen but ended up, with the new triple-head projector, using a 36-foot-wide screen.[136] By 1939, the Academy of Motion Picture Arts and Sciences' Process Projection Equipment Committee could recommend standard rear-projection screen sizes that ranged from a minimum of 7 feet wide to a maximum of 36 feet wide.[137]

The revamped Technicolor process popularized mid-decade necessitated more illumination and thus posed greater difficulties for large rear-projection screens.[138] Prior to the introduction of triple-head projectors in 1938, the maximum size for a rear-projected color image was 10 or 12 feet wide; the new projectors made it possible to employ 15- to 18-foot-wide screens for that purpose.[139] At Warner Bros., it was reportedly in the interest of using larger rear-projection screens for the Technicolor film *Gold Is Where You Find It* (Michael Curtiz, 1938) that the studio developed its triple-head projector, for which it received an Academy Award in 1938.[140] Both Warner Bros. and Paramount used their triple-head projectors for subsequent films in Technicolor, such as *The Adventures of Robin Hood* (Michael Curtiz and William Keighley, 1938) and *Dodge City* at Warner Bros. and *Dr. Cyclops* (Ernest B. Schoedsack, 1940) at Paramount.[141]

In 1942, new rear-projection equipment, built to industry-wide standards that had been adopted in 1939, was put into service. This new equipment had single-head projectors with equivalent power to the earlier triple-head models. Further, three of the new units could be combined to create an even more powerful new triple-head projector.[142] Winton Hoch of Technicolor claimed in 1942 that "we very frequently photograph screens in color more than 20 feet wide, and have photographed, in color, process screens approximately 28 feet wide."[143] Paramount used such larger screens for *Reap the Wild Wind*, a Technicolor film featuring hurricane scenes that, according to production executives, necessitated especially large backgrounds.[144] Paramount went so far as to use two of the new triple-head projectors with dual screens to achieve a 48-foot-wide Technicolor background image of a forest fire for *The Forest Rangers* (George Marshall, 1942) (figure 1.7).[145] Even in 1942, however, Fred Sersen of

1.7 For *The Forest Rangers* (Paramount, 1942), Paramount used two triple-head projectors with dual screens to achieve a 48-foot-wide Technicolor background image of a forest fire.

*Source*: Courtesy of the Film and Television Archive, University of California at Los Angeles, and Universal Studios Licensing LLC.

Fox still complained that rear projection lacked "the scope to produce satisfactorily all the special photographic effects demanded in modern picture production."[146]

The spaces in which the studios conducted their process work also constrained the size of the backgrounds they were capable of using. In 1932, the practice at Fox was to cart glass rear-projection screens, mounted on dollies, to whichever set required them. Ralph Hammeras at Fox explained, "We can take either our 8 by 10 foot or our largest glass anywhere on the lot and put either one behind a window or porch or wherever they may want it."[147] By that time, Paramount already had a dedicated "transparency stage" where it was doing about 85 percent of its process shots. This stage was set up to prepare as many as four rear-projection shots at once. According to Edouart, "We may have set up closest to the projector a small shot through a little window on a small glass, such as an automobile shot. We may then pull that out and have already set up next in line an 8 by 10 foot glass, pulling that out afterwards, and so on."[148]

These stage spaces constrained the size of the rear-projected image by limiting projection throws, which needed to be somewhat long since short-focal-length projection lenses also posed the problem of hot spots. As of 1932, Fox was using a 60-foot projection throw and 4- to 8-inch projection lenses for screens up to 20 feet wide. MGM was using a 104-foot throw and 5.5-inch projection lenses for 18-foot screens—the longer throws and lenses representing an attempt to eliminate the hot spot.[149] Fred Pelton at MGM explained that he and his colleagues had "decided that we shouldn't attempt this work on a stage under 200 feet in length, where we can allow 125 feet for projection and 75 feet for the cameras. As a matter of fact, we would like 50 feet more in length."[150] The building that housed Paramount's transparency stage, Edouart claimed, was not "more than 135 feet long, which naturally limits us as to our throw"; he contended that he would prefer to "have a stage 250 or 300 feet long."[151] One means of addressing this issue, adopted by Columbia Pictures in 1934, was to use a mirror to increase the throw. The 8-foot mirror was placed 50 feet from the projector and threw the image back to the screen. With this arrangement, it was possible to fill an 18-foot screen using a 50-foot throw and a 6-inch projection lens.[152]

Effects technicians continued to advocate for the construction of larger process stages in the ensuing years, citing the need for long projection throws and the capacity to have multiple rear-projection shots set up at once.[153] In 1939, for the production of *Geronimo*, Paramount was using a 195-foot-long stage, with the composite camera and dual background projectors placed at opposite ends. With this arrangement, the dual screens could be as far as 70 feet from the camera, and the space in front was large enough to accommodate forty to fifty horses running through.[154] By 1942, Paramount was creating its large-screen rear-projection shots in an outdoor space that was about 300 feet long, where, in shooting *The Forest Rangers*, the new triple-head projectors were positioned from 100 to 150 feet behind the adjoining 24-foot screens.[155]

The arrangement of the transparency stage to accomplish such large-scale rear projections exemplifies the way in which rear-projection practice reconfigured production space, structuring it around screens whose scale required positioning in deep space. Developments in rear-projection screen size also contributed to the construction of diegetic space. Together with the use of transmissive materials, the move to larger screens and longer shots (as well as to the brighter projectors and larger stages that made them possible) contributed to an emphasis on deep space within the cinematic image. Like the optical printing that famously facilitated the deep-focus aesthetic of *Citizen Kane* (Orson Welles) in 1941, rear projection was heralded as making possible sharper focus in background as well as foreground planes since it could bring together elements shot with different focusing distances (as long as the actor and screen were positioned sufficiently close together that the composite camera could keep both in focus).[156] At the same time, developments associated with large-scale rear-projection screens, such as the triple-head projector, both facilitated and demanded an increased depth of focus for the composite camera—facilitated because the increased image brightness supplied by the projector made it possible to stop down the lens of the camera; demanded because the capacity for large-scale rear projection, as we have seen, meant a concomitant ability to increase the distance between actor and screen.[157]

The movement toward larger and larger rear-projection screens was accompanied by the production of films that capitalized on that scale in other ways as well. *Dr. Cyclops*, for instance, is thematically focused on the issue of scale, and its production employed Paramount's new large-screen setup, including triple-head projectors and dual screens, as well as miniature rear projection in order to explore that issue. In this regard, director Ernest B. Schoedsack reprised and expanded on the work with large screens and miniature rear projection done at RKO for *King Kong*. In *Dr. Cyclops*, Albert Dekker plays the evil Dr. Alexander Thorkel, who uses an experimental device to shrink a group of visitors to the size of small rodents. The production created the resultant distinction in scale between the life-size Thorkel and the shrunken visitors in three ways. Straight shots of the actors playing the visitors were filmed in overscale sets, which had been built on a leased stage on the Warner Bros. lot. Shots of the visitors were also filmed on the transparency stage at Paramount, where triple-head projectors and dual screens allowed massive rear-projected images of Thorkel to loom over the live actors (figure 1.8). Finally, miniature rear projection was used to insert previously shot footage of the visitors via miniature screens into life-size spaces.[158]

More pervasively, large-scale rear projection, which enabled projected backgrounds to fill the frame even in long shots, emphasized the creation of

**1.8** In *Dr. Cyclops* (Paramount, 1940), rear projection allows Albert Dekker to tower over the other actors.

*Source*: Screen grab by author.

immersive virtual space. This dynamic is also exemplified by *The Plainsman*, which made early use of Paramount's dual-screen background setup for the culminating battle sequence, which DeMille described at the time as "the longest and biggest and most dramatically important sequence that has ever been done entirely by the Transparency process." The 50-ton foreground set for this sequence was mounted on a wheeled support so that it could be turned. DeMille explained: "Set up on the stage, we had not one screen, but two, behind this set, with the space between carefully masked by a dead tree in the set. Two projectors threw their images on these screens. The background-plates were made by two cameras, side by side, shooting at predetermined angles." For reverse-angle shots, the crew "simply turned our set around (though the 50-ton weight made it no small task), re-aligned the screens, and carried on with different backgrounds."[159] This use of multiple screens in production—which pairs the live-action foreground with multiple projected views of a prefilmed background landscape, shot from opposing angles—results in a hermetically sealed diegetic space (figure 1.9). Such enclosure illustrates the form of entrapment or "claustration" that Vivian Sobchack attributes to the use of rear projection in film noir, wherein the use of prefilmed background imagery "works to temporally forestall and spatially foreclose any sense of the characters' existential 'freedom' and to make this construction sensually, as well as cognitively, intelligible to viewers."[160] Ironically, *The Plainsman* and other Westerns of the late 1930s harnessed this means of spatial entrapment to envisage the open space of the frontier.[161]

**1.9** *The Plainsman* (Paramount, 1936) made early use of a dual-screen background setup, with a dead tree in the set masking the seam between the rear-projected images.

*Source*: Screen grab by author.

Many other films deployed rear projection to position actors in front of engulfing background images of water, fire, and sky. MGM, for instance, used rear projection in *Mutiny on the Bounty* (Frank Lloyd, 1935) and *Captains Courageous* (Victor Fleming, 1937) to make the principals appear to be at sea.[162] *Captains Courageous*, in particular, was notable for the sheer amount of rear projection it employed: more than 80 percent of the footage used in the film reportedly included rear projection.[163] Paramount used rear projection for a similar purpose in *Spawn of the North* and *Reap the Wild Wind*, as did Warner Bros. in *Captain Blood* (Michael Curtiz, 1935) (figure 1.10) and *The Sea Hawk* (Michael Curtiz, 1940).[164] Warner Bros.' film *Gold Is Where You Find It* deployed the technique in the culminating sequence in which a dam breaks and floods the mine, threatening to engulf the protagonists in water.[165] Selznick International Pictures (in *Gone with the Wind* and *Rebecca* [Alfred Hitchcock, 1940]) and Paramount (in *Aloma of the South Seas* [Alfred Santell, 1941] and *The Forest Rangers*) also favored using rear projection to immerse characters in fiery landscapes.[166] The Paramount film *Typhoon* (Louis King, 1940)—which production executives considered to entail "a tremendous amount of Transparency [*sic*] work"—employed rear projection to composite actors with both a rushing tidal wave and a forest fire.[167] MGM used rear projection in several films portraying the aerial experience of airplane pilots, including *Test Pilot* (Victor Fleming, 1938), which was promoted as "the *Captains Courageous* of the air";

**1.10** In *Captain Blood* (Warner Bros., 1935), rear projection and miniatures make it seem as though Errol Flynn is at sea.

*Source*: Screen grab by author.

*A Guy Named Joe* (Victor Fleming, 1943); and *Thirty Seconds Over Tokyo* (Mervyn LeRoy, 1944).[168]

In all of these cases, rear projection made it possible seemingly to immerse actors in what would otherwise be a dangerous environment, whether "hostile" frontier, stormy ocean, flaming inferno, or wild blue yonder. It is hardly coincidental that many of the pioneering experiments with large-scale rear-projection screens—including *The Plainsman*, *Spawn of the North*, *Gold Is Where You Find It*, *Geronimo*, *Reap the Wild Wind*, and *The Forest Rangers*—were undertaken in the service of such spectacles. John Durham Peters has argued that water, fire, and sky should in their own right be understood as media, insofar as media are conceptualized as "vessels and environments, containers of possibility that anchor our existence and make what we are doing possible." Peters considers the ships and airplanes that people use to survive at sea or in the air as media also since these vessels mediate humans and hostile environments, serving as the condition of human existence in such spaces.[169] We can view the large-scale rear-projection screens used to transport such environments to Hollywood soundstages as proffering similar forms of mediation. These surfaces not only made it possible to present water, fire, and sky as diegetic environments but also functioned similarly to a ship's bow or an airplane's windscreen (or, relatedly, a fire screen) in forming an interface between humans and these environments. Indeed, the practice of placing rear-projection screens beyond the railings of on-set ships or the windscreens of on-set planes aligned the screens with these interfaces—an alignment already suggested in the wind*screen*'s function as a screen. As I discuss in subsequent chapters, a range of exhibition screens have also been employed to similar ends in conjunction with spectacles of water, fire, and sky.

## Movement

Obtaining and maintaining steady background images presented another persistent challenge for technicians. The stability of these images could be compromised by inaccurate perforations in the negative as well as by shrinkage of the developed negative. It could also be undermined by unintended movements of the background camera or projector. Thus, engineers worked throughout the 1930s to develop accurately perforated film, to design projectors aimed at ensuring the stability of the film, and to engineer mounts that would stabilize the background camera and projector.[170] As with the issues of screen construction and screen size, these technical challenges and the practices associated with them contributed to particular formations of production space and diegetic space. In particular, the instability of the rear-projected image contributed to the emphasis on movement in rear-projection practice in the 1930s. Whereas the introduction of synchronized sound initially limited the mobility of the actors and, to an extent, the camera, the moving backgrounds particularly amenable to rear projection endowed even the most static staging—the prototypical example being two actors seated in a cut-away car frame on a soundstage—with a sense of motion.[171] This sense of motion could be achieved by coordinating the moving images that graced the rear-projection screen with the movement of other components of the production apparatus, including the composite camera and the background projector itself.

Concerns about the stability of rear-projected images guided choices about what and how to film for backgrounds, contributing to a preference for moving backgrounds. As G. G. Popovici of Eastern Service Studios in Astoria, New York, reported to the Society of Motion Picture Engineers (SMPE) in 1934, "When buildings or other subjects that must appear steady on the screen are photographed, the camera must stand very steadily upon its tripod, which may have to be tied down. When photographing from bridges, long focal length lenses must be avoided because of the vibration of the bridge. Wind also causes unsteady negatives, and it is suggested that plates not be photographed on windy days."[172] Although the studios amassed libraries of footage for rear projection, concerns about stability also guided which kinds of library images could be used. Popovici explained that he did not use library footage for shots that needed to be steady due to the shrinkage that afflicts negatives over time; however, he admitted that he did use library shots for "scenes of the kind required for taxicab shots or running automobiles or trains, where steadiness is not required."[173]

Because unintended image movement could result from film shrinkage or from the vibration of the background camera or projector, in other words, a moving background mise-en-scène posed less difficulties than did a static one, proving amenable even for the use of library footage.

Like backgrounds featuring dangerous environments, moving backgrounds also facilitated the creation of shots that would otherwise be difficult to achieve, and they thus boasted spectacular appeal. One visitor to Fred Jackman's effects department at Warner Bros. in 1934, for instance, was impressed to witness the creation of a rear-projection shot for *Six Day Bike Rider* (Lloyd Bacon, 1934) in which "a conversation took place between two riders supposedly tearing around the track at some 20 miles per hour." Citing this shot as evidence that rear projection "is employed for other purposes than to save the railroad fares otherwise necessary to place actors against more or less distant backgrounds," the visitor enthused, "The difficulty of recording such a conversation on an actual race-track, with both camera and microphones sweeping around the track ahead of the riders, may be easily appreciated."[174] Rear projection's reputed suitability to movement, owing both to the instability of background images and to the sense of spectacle that moving backgrounds provided, rendered the car, train, boat, and airplane scenarios that dominate so much rear-projection footage of this period particularly apt for the technique. In order to match the foreground action to such moving backgrounds, various devices for simulating movement were employed on the set, such as treadmills, car and airplane mounts, vehicle rollers, mechanical horses, wind machines, and water tanks with wave machines.[175] For *Six Day Bike Rider*, stationary bicycles on the set were supported by swiveled mounts that swayed back and forth to match the movements around the track shown in the background.[176]

In addition to encouraging shots featuring the movement of the background camera (i.e., the camera used to shoot the background footage on location), rear projection was also paired with movement of the composite camera (i.e., the camera shooting the actors against the rear-projection screen on the set). The capacity of the composite camera to move, albeit in a limited fashion, was indeed cited as one of the benefits of the rear-projection process over the color-separation traveling-matte process.[177] In *Captains Courageous*, for example, the stationary ship in the foreground set was made to appear as if it is rocking by pairing rear-projected backgrounds with the motion of a boom-mounted composite camera (figure 1.11).[178]

The capacity to expand the range of camera movement on the set represented a significant benefit to the use of larger rear-projection screens. One early suggestion was to use one of the widescreen processes that had recently been developed, such as Fox's 70mm Grandeur system, for rear-projection backgrounds in order to facilitate movement of the composite camera. As one writer in *American Cinematographer* proposed in early 1932, employing the Grandeur system

1.11 The stationary ship in *Captains Courageous* (MGM, 1937) was made to look as if it is rocking by pairing rear-projected backgrounds with the motion of a boom-mounted composite camera.

*Source*: Screen grab by author.

would make "it possible to use a background sufficiently large so that the camera man can pan and follow his action just as he does now in ordinary production."[179] When Fox began experimenting with rear projection, in fact, it had considered using the studio's wide-film cameras for the backgrounds; however, it embraced rear projection just as it was abandoning wide film, and by the time the idea was floated, the studio had reportedly already retired its developing machines and printers for the larger gauge.[180] Discussing his use of Paramount's dual-screen setup to shoot the battle sequence in *The Plainsman*, DeMille similarly emphasized that the larger background area allowed "plenty of room" not only for long shots but also "for panoramic and dolly-shots, without exceeding the scope of our background-screens."[181]

Although technicians sought to stabilize the rear-projected image in the 1930s, they also worked to increase the mobility of the projector and screen. Portable projector installations were considered desirable since they liberated process shots from their confinement to suitably equipped stages. By 1933, it could be difficult to schedule process shots due to the congestion afflicting the stages that could accommodate them, and technicians worked to develop portable projectors that balanced flexibility with rigidity and precision. At that time, George Teague, one of the designers of the rear-projection equipment used at Fox since 1930, introduced a mobile process projector that, at 30 inches wide, could go "anywhere a blimped camera can be used."[182] In 1934, Teague introduced improvements to that device. In addition to promising greater rigidity, it had its projection head and lamphouse mounted on a tilting base, and the projection head could also be rotated a full 360 degrees. One benefit of this flexibility was that it was possible to adjust the projector to match a slightly off-kilter foreground or background. However, it also presented possibilities for motion effects. As *American Cinematographer* proposed, scenes "apparently laid on a boat can be given the proper roll by slowly rocking the projector; this can be done either by hand, or by adding a simple semi-automatic device which is being designed."[183] This idea represents a correlate and precursor to the strategy of simulating the rocking of the boat in *Captains Courageous*. The combination of moving background images with either a moving projector or a moving composite camera promised to marshal multiple elements of the rear-projection apparatus to endow a static foreground set with a sense of motion.

Technicians also experimented with various means of moving the screen and related surfaces. Screens were tilted to accommodate high- and low-angle shots, in which case slanting screen frames were employed in conjunction with elevator-mounted process projectors so that the rear projector and screen could remain at a right angle to one another. For extreme camera angles, however, technicians sometimes tilted the foreground rather than the screen, as MGM did for a 45-degree high-angle airplane shot in *Test Pilot*.[184] *Captain Blood* featured a foreground set with ships built on the regular stage floor, and it used both rear-projected backgrounds and painted backings. Rear projection was used for battle scenes, with the backgrounds filmed in miniature; however, for less action-filled shots, painted backings representing the sea and sky were manually rocked and pulled up and down to make it seem as though the ships were rocking.[185] Such sea backings were a persistent focus for Warner Bros. production head Hal Wallis, who during production of *Captain Blood* urged director Michael Curtiz to ensure that these backings were properly lit and visible, "so we can see . . . the movement."[186] *Captain Blood*, made two years before *Captains Courageous*, thus also achieved the illusion of motion by moving an element of the effects apparatus surrounding a stationary foreground.

Together, these examples show how motion within the background image could collaborate with movement of the composite camera, the rear projectors, or the background surface. Such collaboration among elements of the rear-projection apparatus worked to make motion contagious, endowing even static components with a sense of dynamism. Scholars such as Anne Friedberg and Nanna Verhoeff have shown how exhibition screens mediate motion and stasis, especially insofar as moving images (as in early panoramic films) have provided immobile spectators with a virtual mobility.[187] My exploration of rear projection shows how on-set screens enacted a similar form of mediation, creating a virtual mobility within the diegesis. As we have seen, the stacking of image surfaces created relationships between live-action and projected components through spatial depth, employing the rear-projection screen as a permeable membrane. And the arraying of image planes produced graphical relationships mediated by the screen's frame. With the motion effects described here, the rear-projection screen operated as a kind of infuser, allowing the quality of movement to seep out into an otherwise stationary environment.

I argue later in this book that attention to the range of screen practices of the long 1930s compels us to recognize the ways in which moving images have not only addressed immobile viewers but also appealed to viewers in motion. The use of screens in production, however, while facilitating a mobile diegesis, simultaneously contributed to the stasis of actors, both on the set and in the frame. The many actors stationed in replica cars, trains, and boats—and on various other perches—backed by rear-projected land-, sea-, and cityscapes, not only remained ensconced in the studio but also often stayed centered in the frame, immobile except for the jostling that could simulate the activity of travel. As shots of Fay Wray in *King Kong*—clinging to a tree in front of large rear-projected beasts—and of any number of car-bound female stars attest, such arrangements thereby also contributed to the way in which Hollywood cinema of this period spectacularized white women, not only limiting their range of action within the frame but also expanding the application of soft glamour lighting by making it possible to shoot a wide variety of scenarios in the controlled environment of the soundstage.[188]

## Proliferating Screens

This exploration of rear-projection practice in the long 1930s shows how screens came to pervade—and in many cases to dominate—the spaces of production, both through their increasing size and through their sheer proliferation. The growing scale of screens was, indeed, related to the proliferation of screened images since the larger backgrounds were constituted from multiple juxtaposed screens (the dual-screen arrangement at Paramount), multiple superimposed projections (the triple-head background projectors), or both (as became the practice at Paramount). The proliferation of screens within production left its mark on the films' diegetic spaces, which were constituted from multiple screened images filmed in different places, at different times, and often by different camera crews. As we have seen, this practice entailed the layering of image planes receding into the depth of the diegesis, achieved through the stacking of screens and other transmissive surfaces, as in the conjunction of glass painting and miniature rear projection in *King Kong* and *Gone with the Wind*. Many times, as in Paramount's dual-screen backgrounds, the proliferating screened images were also arrayed laterally within the shot—an arrangement that could also be achieved through smaller full-scale rear projections (e.g., juxtaposing the rear-projected view out of a window with an interior set), miniature rear projection (e.g., inserting a small figure into a larger space, as in *Dr. Cyclops*), and optical printing (e.g., bringing together live-action and miniature footage, as in *The Rains Came*).

As I suggested in relation to *King Kong*, the layering of image planes within a film was redoubled by the exhibition screen, which acted as yet another image plane stacked in deep space. A particular shot in *Notorious* (Alfred Hitchcock, 1946) provides a representation of this kind of mediated space constituted by stacked planes (figure 1.12). That film heavily features rear-projected images of Rio de Janeiro, but in this shot the painting of a seascape creates the background. The window frame and the curtains flanking it make this painted surface a figure for the various screens used in the film's production and exhibition, particularly since the aperture left by the curtains approximates the Academy ratio, while the shape of the window as a whole evokes a widescreen aspect ratio. If Cary Grant occupies the position of a film viewer, the movie puts the actual viewer into the familiar position of apprehending a series of stacked planar images, starting with the theatrical screen itself.

The arraying of screen planes within films also echoed the ways in which screens functioned in exhibition. For instance, the rear-projection arrangement used at Paramount for *The Plainsman* and many other films—employing, as it did, two cameras shooting at angles from one another to capture the background footage as well as dual rear projectors and side-by-side screens on the set—echoed experiments with multiple-camera, multiple-projector widescreen

**1.12** *Notorious* (RKO, 1946) represents the construction of cinematic space from stacked planes.

*Source*: Screen grab by author.

cinema that, as I discuss in chapter 3, were being undertaken around the same time. Indeed, the use of a revolving set with changeable projected backgrounds rendered *The Plainsman*'s production space, like the spaces envisioned in contemporaneous approaches to multiple-projector exhibition, a realm enclosed by multiple screens.

Insofar as rear-projection screens mediated disjointed ontological realms within Hollywood films of the long 1930s—combining into a single image, for example, a studio-bound shot of Barbara Stanwyck and location-shot footage of a desert landscape—these films reflected and contributed to a larger mediascape in which screens of all sizes were pervading daily life and endowing it with a similar ontological heterogeneity. These screens appeared everywhere from cinemas to homes, from ocean liners to drive-ins, and from brokerage houses to schools, as the following chapters elaborate.

# CHAPTER 2

# THEATRICAL SCREENS, 1926–1931

*Transforming the Screen*

The "burning of Atlanta" sequence in *Gone with the Wind* (Victor Fleming, 1939), as we have seen, featured a heavy use of special effects. Indeed, that sequence employed full-size and miniature rear projection as well as optical printing to combine footage of the principal actors, shot on a soundstage in 1939, with background footage of the fire, filmed on the back lot of the former RKO-Pathé studio—using the great gate from *King Kong* (Merian C. Cooper and Ernest B. Schoedsack, 1933) and other old sets as kindling—on December 10, 1938, before the role of Scarlett O'Hara had even been cast.[1] Culminating a film that was in its own right a culmination, of sorts, of the sound era in Hollywood (as that era's biggest commercial success), the "burning of Atlanta" sequence thus also exemplifies the form of spatial synthesis that had ascended in the 1930s through the use of these effects techniques, as discussed in chapter 1.[2] The sequence drew together spatially, temporally, and ontologically distinct fragments (glass painting, the physical effect of the fire, and principal photography) into the kind of composite space, constituted by proliferating planes, that accompanied the rising use of screens on the set. The prospect that screens could structure cinematic space, however, guided plans for the theatrical exhibition of this sequence as well.

Producer David O. Selznick—who had overseen the development of *King Kong* while working at RKO in the early 1930s—acquired the rights to Margaret Mitchell's novel in the summer of 1936, just after its publication. By this time, Selznick had established his own production company, Selznick International Pictures, devoted to prestige pictures, and *Gone with the Wind* became its centerpiece project. Selznick mounted a huge publicity campaign while the film was still in development, launching a nationwide talent search for an actress to play Scarlett. The announcement that Vivien Leigh had been selected finally came on January 13, 1939, shortly before principal photography began.[3] In conjunction with this publicity stunt, Selznick also pursued means of rendering his *Gone with the Wind* spectacular, including the use of Technicolor as well as elaborate special effects.[4] That effort also entailed experimentation with a new screen setup for exhibiting the film. In late 1937, Selznick learned about the experiments that special-effects artist turned inventor Fred Waller was doing with multiple-projector exhibition for the 1939 World's Fair (Waller would eventually streamline this system into Cinerama, one of the cinematic sensations of 1952),[5] and he seriously explored the possibility of using a similar multiple-projector arrangement for the exhibition of *Gone with the Wind*, a pursuit that continued through the length of the film's production.[6]

The idea for multiple-projector exhibition was initially devised in connection with the "burning of Atlanta" sequence, although Selznick eventually urged his staff to explore the possibility of using it more generally for the film's "spectacle sequences," including montage sequences, Scarlett's "oath" just before intermission, and her return to Tara at the end of the film.[7] Effects artist Jack Cosgrove made a multiple-camera test shot of the fire, which was deemed particularly appropriate for multiple-projector exhibition not only because of its spectacularity but also because its movement would hide the seams between the images (figure 2.1).[8] (Although Selznick and production manager Raymond A. Klune discussed plans for a "triple-screen" setup, the apparatus employed for the test footage featured two cameras and two projectors.[9]) As Selznick later recounted, he wanted to have the "screen expand to the entire width of the stage, and perhaps up the side walls, accompanied by multiple sound effects, to give the audience the feeling that it was almost inside the Atlanta fire."[10]

The "burning of Atlanta" sequence, in short, was imbricated through and through with the actual and anticipated use of cutting-edge screens in both production and exhibition. The planned use of multiple projected images in the theater echoed concurrent innovations in rear projection on the set, including the use of dual projectors, dual screens, and triple-head projectors to expand the scope of background images. In fact, the *Gone with the Wind* effects team used a pair of background projectors to screen the multiple-camera test of the fire.[11] The innovative forms of projection devised for production and exhibition, moreover, also produced dovetailing modes of representation and address. Rear projection helped convey the threat that the massive fire in the diegesis could engulf the characters; Selznick's plan for multiple-projector exhibition extended the prospect of engulfment in virtual flames to viewers in the theater. The film, in other words, promised to provide an experience of engulfment by

**2.1** On *Gone with the Wind* (Selznick International Pictures, 1939), the special-effects team used a multiple-camera setup (*pictured*) to shoot test footage of the fire.

*Source*: From the David O. Selznick Collection, Harry Ransom Center, University of Texas at Austin.

confronting viewers with a series of nested screens, with the theatrical screen(s) in the auditorium simultaneously framing and echoing the rear-projection screens that structured the diegesis.

In proffering this experience of engulfment, the prospect of employing multiple screens across production and exhibition also promised to further *Gone with the Wind*'s function as what Miriam Petty describes as a "monument" to Lost Cause ideology.[12] During the development of the film, Walter White of the National Association for the Advancement of Colored People (NAACP) had confronted Selznick with the concern that Mitchell's book was "so essentially superficial and false in its emphases" that it would "require almost incredible effort to make a film from the novel which would not be both hurtful and inaccurate picture [*sic*] of the Reconstruction era."[13] Selznick professed sympathy for these concerns, claiming that, "as a member of a race that is suffering very keenly from persecution these days, I am most sensitive to the feelings of minority peoples," and emphasizing that the film would not contain racist epithets or depict the Ku Klux Klan.[14] But he apparently did not follow through on White's suggestion that he hire an African American historian as an adviser.[15]

When the film was released, its reception among African American critics and audiences was mixed, with some lauding its aesthetics, the absence of the most overt kinds of racist violence, and the performances delivered by the African American principals (especially Hattie McDaniel), while others critiqued its more covert racism and distorted view of history.[16] In an article titled "*Gone with the Wind* Is More Dangerous Than *Birth of a Nation*," published in the African American newspaper the *Washington Tribune*, for example, Melvin B. Tolson explained that the film falsified history by failing to depict the brutality of slavery—thus ignoring the reason the Old South was "gone with the wind" and instead evoking sympathy for it.[17] The film's proposed use of screens across production and exhibition promised to contribute to this evocation of sympathy by presenting the razing of Atlanta—and symbolically the demise of the Old South—as a mortal threat (the threat of engulfment in flames). While the employment of rear projection allowed the film to depict this threat to the characters, the use of multiple projection in the theater would have made it especially palpable to viewers as well.

The idea of using multiple projection for the exhibition of *Gone with the Wind* was eventually abandoned for reasons Selznick attributed to the cost and time it would take to equip theaters, and the film's exhibition strategy ultimately emphasized a multipronged approach to ensuring spectatorial immersion in other ways, including directives to exhibitors about the elimination of house lighting and the need for ushers to keep aisle curtains and doors shut.[18] The relationship between rear projection on the set and Selznick's thwarted plan for multiple projection in the theater, however, exemplifies the new connections that are revealed by focusing on the screen as an integral component of both production and exhibition. Selznick's experiment with multiple projection represents one of several ways in which screens morphed and multiplied in theatrical exhibition in the long 1930s. Indeed, this plan for multiple projection in 1939 echoed a range of innovative approaches to theatrical screens that had emerged since the mid-1920s, from widescreen exhibition and variable screen masking to the proliferation of auxiliary screens and related surfaces throughout the theater. As I detail in this chapter and the next, these screens structured theatrical space in much the same way that rear-projection screens structured diegetic space, drawing together heterogeneous realms to form new spatial syntheses.

Both this chapter and the next are devoted to examining the development and employment of theatrical screens, with this chapter focusing on the period 1926–1931 and the next chapter on the years 1931–1940. The first period is most closely associated with the coming of synchronized sound; however, it also witnessed significant experiments with the design, construction, and installation of theatrical screens.[19] These experiments included, most prominently, the effort to design screens that would be conducive to the exhibition of sound films as well as work on large-screen and widescreen exhibition. As scholars such as John Belton and William Paul have shown, the work on sound screens and widescreen that took place in the period 1926–1931 contributed to a transformation in the screen's relationship to both theatrical space and the viewers within it, rendering the screen an integral part of the theatrical architecture and proffering an experience of engulfment.[20] This chapter explores that transformation by mapping the ways in which engineers, filmmakers, and critics conceptualized the new screens' functions and possibilities for mediating virtual and actual spaces.

Although that effort was bound up with the wiring of theaters for sync sound, I argue that it also bore connections to a range of other experiments with screen technologies (as well as sound technologies). As I show, several of the material qualities that fueled experiments with production screens also informed work with theatrical screens. The latter involved the screens' qualities of transparency, scale, scope, and mobility (here figured primarily as flexibility). As with the work on production screens, the work on theatrical screens harnessed these qualities to explore new possibilities for layering and arraying reproduced images and sounds. The work on sound screens harnessed the transparency of screens to produce new forms of spatial layering within the auditorium. Meanwhile, the experiments with large-screen and widescreen cinema addressed means of exploiting the scale, scope, and flexibility of the screen, thus expanding, broadening, and dynamizing the cinematic spectacle. As I discuss in chapter 3, after the abandonment of widescreen cinema around 1931, practitioners would explore how the qualities of scope and flexibility as well as transparency and multiplicity could function to integrate smaller screens more fully into the exhibition space. This integration, I argue, would enact a form of compositing within theatrical space, mirroring and collaborating with the forms of compositing that were also increasingly present onscreen. *Gone with the Wind* exemplifies how the pursuit of such synthesis, both within films and encompassing them, would join the spaces on the screen with the spaces of the screen in a logic of proliferating surfaces.

## Sound Screens

With the introduction of synchronized sound, technicians explored diverse ways of situating the sound-reproduction apparatus in relation to the image-reproduction apparatus, especially the screen. This work entailed not only assessing the positioning of the screen and speakers within the auditorium but also reassessing the screen's material construction. Early experiments with the exhibition of sound films had the speakers positioned beyond the edge of the screen. For contexts in which synchronized sound was used for musical accompaniment, the speakers were placed below the screen in the orchestra pit to simulate the provenance of sound with an actual orchestra. For films with synchronized speech, engineers initially attempted to make the figures onscreen appear to be the source of the sound by positioning additional speakers alongside or above the screen.[21] These practices, however, were quickly abandoned. The most obvious reason for this abandonment was a perceived "loss of illusion" created by the spatial disjuncture between the images onscreen and the speakers from which the sound emanated, but the early arrangements were also considered impractical insofar as they "lacked the mechanical simplicity and flexibility inherent in 'flying' the speakers, or mounting them on towers behind the screen so that they could be removed at will."[22] This suggests that there may have been incentives for abandoning the early arrangements beyond what William Paul identifies as technicians' unquestioned and mistaken assumptions about the need for "absolute unity" between image and sound.[23]

## Transparency

In February 1927, Fox technician Earl Sponable developed a screen for Movietone that was, in his words, "transparent to sound," enabling "the use of loudspeakers directly behind the screen." Voicing the opinion, however misguided, that this new arrangement proved "a great help in improving the illusion," Sponable claimed years later that it "was immediately accepted by the industry."[24] Indeed, a general consensus quickly emerged that in exhibiting films with synchronized sound, speakers should be located behind a permeable screen.[25] Such screens were initially made of loosely woven cotton, so that sound could pass through interstices in the cloth, and then by punching perforations in otherwise-solid materials such as coated fabric.[26] As Paul has shown, the position of screens in the auditorium changed with the adoption of these permeable surfaces. The screen had previously been placed upstage, often within a picture setting (a theatrical set on the auditorium stage). With the coming of synchronized sound—and specifically with the perceived need to position the speakers behind the screen—screens were resituated downstage, just behind the curtain. This new position contributed to the integration of the screen into auditorium space since the screen, no longer separated through encasement within a stage, could, as Paul argues, now be seen as "an architectural object itself."[27]

This arrangement also exploited the screen's transparency in a way that echoed the practice of rear projection, contributing to a similar form of spatial layering. Although rear projection would not come into widespread use as a special-effects technique until the early 1930s, it was being harnessed in extratheatrical exhibition and experimented with in theatrical exhibition by 1927, as discussed later in this chapter and in subsequent chapters. With the employment of both rear-projection screens and permeable sound screens, a key component of the technical apparatus (background projector or speaker) was placed behind the screen, and the screen took on a new role mediating the spaces it bisected—the space of the apparatus behind the screen surface and the space of the set or auditorium in front of it.[28] That function is evident in the names given to such screens, especially "Trans-Lux" and "Transvox."[29] Engineers' overarching goal in designing such surfaces was that they transmit images or sounds from behind without a diminishment of pictorial quality in front. In both cases, this involved significant compromises. As discussed in chapter 1, rear-projection screens' capacity to transmit light interfered with their ability to diffuse it. Sound screens' capacity to transmit sound similarly interfered with their ability to reflect light. Woven screens were deemed unsatisfactory both optically and acoustically.[30] And although perforated screens transmitted sound well, the pores diminished the screens' reflectivity.[31] Various methods were suggested to combat that effect, including the use of very thin screens that would allow sound transmission through a combination of small, widely spaced pores and a vibration of the surface.[32] By the end of 1931, the SMPE Screens Committee contended that the optimal balance between transmissiveness and reflectivity occurred with the use of such small, widely spaced openings that "in the aggregate, offer a comparably small total open area" amounting to 5 percent of the total screen area.[33] Projection expert F. H. Richardson, however, continued throughout the 1930s to bemoan the loss of light caused by perforations as well as the dirt they attracted and to advocate (unsuccessfully) for replacing perforated screens with solid surfaces and moving the speakers back from behind the screen to its periphery.[34]

Claiming in 1929 that perforations produced a "lack of clearness, depth and detail and the presence of blurs and smudge," A. L. Raven of the Raven Screen Corporation contended that when "viewing a picture on such a screen the effect is much the same as when looking at people or objects through a dusty plate glass window."[35] Theatrical sound screens were opaque to light and thus did not actually function as windows, even dusty ones. Their transparency to sound, however, makes the connection to windows more literal than the familiar metaphor employed by Raven (screen as window to a virtual view) suggests. These permeable movie screens actually transmitted sound in much the same way as window screens. Moreover, like glass windows, which transmit light but not weather, permeable sound screens simultaneously united and divided the spaces they bisected by functioning as conduits to certain forms of sensory experience (in this case sound) and barriers to others (in this case vision).[36] These theatrical screens thus functioned similarly to the rear-projection screens employed for special effects in that they exploited the quality of transparency as a means of reconfiguring deep space. In this case, the projected images were reflected from the screen at one plane, while the sounds were transmitted from another, more distant plane. This arrangement positioned viewers (like actors on a soundstage) within the layered space insofar as they found themselves—together with the screen—sandwiched between components of the projection apparatus, the film projector on one side and the "sound projector" (or, on a soundstage, composite camera) on the other.[37] This configuration within the auditorium would, moreover, collaborate with the form of spatiality presented by film sound itself in the 1930s, when engineers also began to arrange soundtracks into hierarchical layers.[38]

## Large Screens and Widescreen

The experimentation with large-screen and widescreen exhibition that took place between 1926 and 1931 also contributed to a reconceptualization of the screen's role within the auditorium space. Increasing the scale of the projected image became particularly desirable with the advent of large picture palaces, and the 1920s witnessed the development of several systems entailing large screens, wide screens, wide-gauge film, and combinations thereof. Important experiments in France in 1926–1927 included Abel Gance's Polyvision system and Henri Chrétien's anamorphic "Hypergonar" lens. Around the same time in the United States, Paramount was putting Lorenzo del Riccio's Magnascope lens to use and George K. Spoor and P. John Berggren's Natural Vision system was employed to shoot short films (figure 2.2). This period of experimentation culminated in 1929–1930, when all of the major Hollywood studios were working with systems that featured wide-gauge film and wide aspect ratios. These systems included Fox's 70mm Grandeur, MGM's 70mm Realife (which employed Grandeur cameras), Warner Bros.' 65mm Vitascope, United Artists' 65mm Magnifilm, RKO's adoption of the 63.5mm Natural Vision, and Paramount's 56mm Magnafilm. Most of these systems provided aspect ratios of 2:1, the exception being Natural Vision, which had an aspect ratio of 1.85:1.[39] The Natural Vision system, in particular, was also heralded as offering a 3D effect (the means by which Natural Vision was said to achieve this effect morphed over the many years of the system's development; however, as of 1930, its purported depth effect was attributed to stereoscopic photography).[40] Whereas the employment of permeable sound screens had involved a change in the way the screen surface was positioned in deep space, the concurrent embrace of large and wide screens entailed a shift in the way the screen frame interacted with contiguous spaces, both the diegetic space within the frame and the auditorium space surrounding it.

**2.2** George K. Spoor (*left*) and P. John Berggren with wide film and their Natural Vision camera.

*Source*: From the Essanay Film Studio Collection of Visual Materials, Chicago History Museum.

The expansion of the frame accomplished with widescreen formats is easily taken for granted as fulfilling the familiar cinematic objective of immersing viewers in virtual space. Indeed, engineering discourses suggested that whereas smaller, narrower projected images (especially the nearly square ones that initially resulted from the adoption of sound-on-film) called viewers' attention to "the black nothingness at each side of the screen," as A. S. Howell and J. A. Dubray of Bell and Howell put it, larger and wider screens made viewers seem to "*live* with the characters of the story and in their ambient."[41] The individuals designing theatrical spaces throughout the 1930s would share the aim, in theatrical architect Benjamin Schlanger's words, of "making the picture appear to fill the field of view of the spectator in the theater, so that the spectator is no longer 'picture conscious.' Rather, he should be made to feel that what is being unfolded before his eyes is very much the same as his natural field of vision in real life."[42] This goal, however, emerged within a historical context marked by a proliferation of screens of various sizes, operating in diverse environments and serving heterogeneous functions. Although many within the film industry considered large theatrical screens the antithesis of (and antidote to) small extratheatrical screens, such as those associated with home movies and television, attention to the range of ways in which all such screens mediated virtual and actual spaces reveals multifaceted relationships among them.

The next sections of this chapter explore how particular large-screen and widescreen systems as well as the modes of filmmaking associated with them harnessed the screen's qualities of scale, scope, and flexibility. These approaches to those qualities align large-screen and widescreen cinema with other screen practices, including practices entailing materially dissimilar screens. Indeed, attending to the diverse ways in which such screens worked to mediate virtual and actual spaces ultimately reveals connections even between practices exploiting the largest and smallest screens.

## Scale

In December 1926—at a time when most theatrical screens, even in the biggest picture palaces, did not exceed 24 feet wide—Magnascope was presented at the Rivoli Theatre in New York, for the exhibition of *Old Ironsides* (James Cruze), with a screen that filled the theater's 40-foot proscenium arch.[43] Rather than simply a large-screen system, Magnascope employed variable screen masking and a specially fitted third projector to alter the size of the screen image over the course of a film. Although Magnascope screens were 36 to 40 feet wide, the screen maskings would cover part of that area during most of the film's running time, leaving a picture aperture that, at about 18 feet wide, approximated the size of a standard screen.[44] For spectacles such as the sailing and battling ship in *Old Ironsides*, the stampeding elephants in *Chang* (Merian C. Cooper and Ernest B. Schoedsack, 1927), and the soaring airplanes in *Wings* (William Wellman, 1927), the exhibitor would switch to the third projector, which was fitted with a special wide-angle lens to fill the expanse of the large screen.[45] The side and top maskings would be withdrawn gradually to expose the full screen (figure 2.3). As showman Harry Rubin explained, "During this operation the portion of the picture for the large screen, which is being projected on the black maskings, does not show up and therefore the illusion is created that the picture is gradually becoming larger."[46] The system thus emphasized the spectacle of inflation as well as the contrast between small and large images. Commenting in the *New York Times* on the Rivoli premiere of *Old Ironsides*, Mordaunt Hall contended that this spectacle "proved so stirring to the spectators that every man and woman in the theatre rose to their feet and applauded."[47] In exhibiting *The Trail of '98* (Clarence Brown) in 1928, MGM employed an adaptation of Magnascope dubbed the "Fantom Screen," which used a roller-mounted screen in conjunction with the wide-angle projection lens so that the screen could be moved forward on the stage while the image was being enlarged. As Paul notes, however, further experiments with such forward movement were

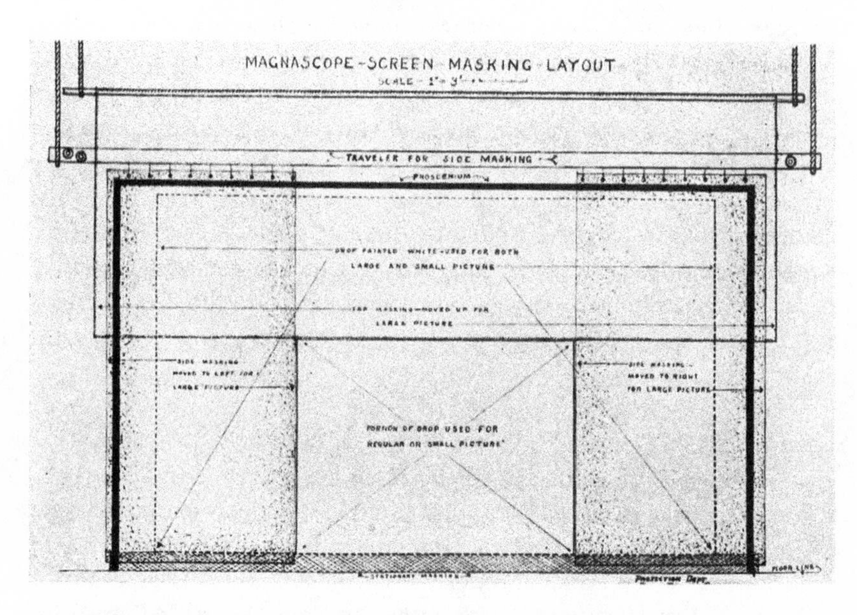


**2.3** With the Magnascope system, side and top maskings were withdrawn gradually to expose the full expanse of the screen.

*Source*: H. Rubin, "The Magnascope," *Transactions of the Society of Motion Picture Engineers* 12, no. 34 (April 1928): 404.

precluded by the new permanent positioning, within a year, of permeable sound screens at the front of the stage.[48]

The scenes used to display the full expanse of Magnascope's large screen created novel forms of spectacle through the display of thrilling and dangerous settings: the rough sea in *Old Ironsides*, the wild jungle in *Chang*, and the open sky in *Wings*. The massive spectacle provided by such settings would find an echo in special-effects practices of the 1930s and early 1940s. As we saw in chapter 1, many films in that period used rear projection to seemingly immerse characters in threatening environments, including stormy oceans, raging fires, and the open sky as well as what was portrayed as the hostile frontier of the American West. This was particularly true of films that made pioneering use of large-scale rear-projection screens, including *The Plainsman* (Cecil B. DeMille, 1936), *Spawn of the North* (Henry Hathaway, 1938), *Gold Is Where You Find It* (Michael Curtiz, 1938), *Geronimo* (Paul Sloane, 1939), *Reap the Wild Wind* (Cecil B. DeMille, 1942), and *The Forest Rangers* (George Marshall, 1942). With Magnascope and rear-projection practice alike, the novelty of enlarged screens was thus applied to the creation of thrilling virtual surroundings, whether for viewers in the theater or for characters within the diegesis.

Deployments of Magnascope not only featured similar forms of spectacle to the rear-projected backgrounds of the subsequent decade but also formed similar kinds of synthetic space. Reviews emphasized the way in which the enlarged screen made represented space seem continuous with auditorium space. One review, for example, portrayed the shift from small to large screen in *Old Ironsides* as an "electrifying effect" in which the ship appears "about to sail over the orchestra pit."[49] Another claimed: "When Old Ironsides lurched nearer and nearer to the orchestra pit on Monday night, to burst with an orchestral roar into what appeared to be life size apparently over the very heads of the audience, a number of naval officers, including the Secretary of the Navy, abandoned their seats for the time being and gave a number of whoops."[50] In a snippet on preparations for *Wings*, the *Boston Daily Globe* reported that Magnascope had been improved "to such a degree, it is said, that audience [*sic*] viewing 'Wings' are to experience the terrifying sensation of actually flying 10,000 feet up among the clouds and engaging enemy planes in a reckless duel for life."[51] In its review of the film several months later, the paper confirmed that the "use of the Magnascope, by which the screen is made twice the size of the ordinary screen, is particularly effective and gives the audience the sensation of actually participating in the aerial battles."[52] The exhibition technology was thus touted as putting audience members into contact with dangerous settings in much the same way as the production technique would composite actors with those same settings.

The films exhibited in Magnascope differed from the later ones insofar as the Magnascope spectacles emphasized location shooting. The fact that *Old Ironsides*, *Chang*, and *Wings* were shot (at least in large part) actually at sea, in the jungle, and in flight—with real danger to the production personnel in the latter two cases—accounted for a great deal of their appeal. One review of *Old Ironsides*, for example, contended that what distinguished this film from "a number of pictures, big and otherwise, is the fact that it was an accomplishment in the obtaining of a fine natural background. . . . Probably seventy-five per cent of the film was taken on location at Catalina Island Isthmus and the immediate vicinity."[53] In relation to *Chang*, described as "at once a magnificent news reel . . . and a gorgeous show," critic Richard Watts Jr. proclaimed, "It is simply incredible, no matter how you try to explain it, how many of those scenes could possibly have been photographed. If ever men were willing to risk death for their art, all the evidence is that Cooper and Schoedsack were the ones."[54] Publicity materials for *Wings* advertised: "All the thrills of the air, made in the air, by airmen. . . . Realism really made by men who risked their lives to bring home to you the 'feel' of flying" (figure 2.4).[55] A film such as *Wings* was thus dramatically different from a film such as *Test Pilot* (Victor Fleming, 1938), wherein the use of rear projection mitigated the dangers associated with the spectacle of flight.[56]

The backgrounds in films featuring rear projection, however, were often shot on location by a second unit. In this regard, a film such as *Old Ironsides* should be aligned not with a film such as *Spawn of the North* as a whole but rather with

**2.4** Marketing materials for *Wings* (Paramount, 1927) emphasized aerial shooting.

*Source*: From the Paramount Pictures Press Sheets, Margaret Herrick Library, Academy of Motion Picture Arts and Sciences, Beverly Hills, CA.

the rear-projected images of the sea that constitute the latter film's background. A location-shot background, as Dominique Païni argues, can be seen as "a miniature documentary in the fictional whole."[57] Indeed, the location-shot aesthetic of *Spawn of the North*'s backgrounds is emphasized by their striking resemblance to a *March of Time* newsreel released the previous year. In fact, the *March of Time* installment in question, "Alaska's Salmon War," played at the Washington, DC, Trans-Lux theater and thus was also exhibited with rear projection.[58] It bears noting that the innovative 36-foot-wide screen employed for the rear-projected backgrounds in *Spawn of the North* closely approximated the scale of the innovative 40-foot-wide screen employed for the Magnascope scenes in *Old Ironsides*.[59] We could say, then, that the diegetic space of *Spawn of the North* emulated not the diegetic space of *Old Ironsides* but rather its exhibition space. Like the actors in *Spawn of the North*—who, thanks to rear projection, found themselves sharing with large-scale images of location-shot oceans a new kind of composite space on the set—the viewers of *Old Ironsides*, thanks to Magnascope, occupied an analogous synthetic oceanic space in the theater. I argue in the next chapter that the Trans-Lux theaters created a similar form of synthetic auditorium space, in that case not through screen scale but rather through an equalization of light levels facilitated by rear projection. Insofar as spectacles of sea and sky were employed to display novel screen technologies in all of these cases, they highlighted the mediating nature of the screens themselves—their function as interfaces between human beings (whether actors or viewers) and otherwise inhospitable environments.

## Scope

Technical discourses tended to portray the widescreen cinema boom of 1929–1930 as an improvement on Magnascope provoked by the adoption of synchronized sound. Engineers considered 22 to 24 feet the maximum screen width for acceptable projection of 35mm film. When film of that gauge was projected on larger screens, as it was with Magnascope, the image became what was deemed intolerably grainy.[60] Sound-on-film systems exacerbated that problem because the inclusion of an optical soundtrack on the filmstrip limited the image area even further and thus threatened to increase the graininess of the projected picture. As a Technical Bureau report from the Academy of Motion Picture Arts and Sciences explained in 1930, "The coming of sound-on-film, by forcing the use of part of the film width for the sound track, cut the width of the picture down almost to a square shape, and thereby forced the problem of the picture size and proportion to the attention of the industry. Theatres made it a frequent practice to restore the 3 × 4 shape by matting, and then to magnify the picture to fill the screen; but this additional magnification tended to give the undesirable effect of graininess in the projected picture—an effect already serious in the largest theatres."[61]

Adding insult to injury was the fact that synchronized sound was reputed to favor films with a particularly large scope just as this was becoming more difficult to achieve. It was suggested that sound lent itself to films modeled on the theatrical stage, including what the Technical Bureau report identified as "musical plays and other productions involving ensembles, choruses, and action by large or small groups," all of which were considered unsuited to small screens.[62] As the *New York Herald Tribune* put it, the "'talkie' bawled its way into vogue, demanding an ever-increasing area for Ziegfeldian choruses and calling for close-ups not of individuals but entire groups of persons in order to transplant the technique of the theater to the screen."[63] In addition, one engineer explained, "as soon as speech was added to the picture it was found that the picture area did not allow enough characters to be included in a scene if the projected images were to appear large enough to be commensurate with a sufficient volume of sound."[64] Engineers within the industry thus held up larger film gauges as the solution to the problem of projecting sound-on-film productions on large screens without "sacrificing pictorial quality."[65] Experiments ran the gamut from 70mm to 50mm gauges, the latter representing the largest gauge compatible with existing projection equipment and thus a more economical compromise.[66]

The new, wider screen dimensions—especially the proposed ratios approximating Euclid's "Golden Ratio," 1.618:1—were touted as more suited to pictorial composition than the old 1.33:1 ratio or the even squarer ratio of about 1.15:1 that had been employed with sound-on-film systems. However, adopting a wider shape was more practically the only way that screens could be enlarged in existing movie theaters, where sightlines put significant limits on screen height.[67] The screens associated with widescreen in this period approximated the scale of Magnascope and were similarly touted as filling "the entire length of the conventional stage proscenium" (whether they actually did so varied by theater).[68] The debut of the Fox Grandeur system at the Gaiety Theatre on September 17, 1929—featuring *Fox Movietone Follies of 1929* (David Butler) as well as the short subject *Niagara Falls* and several *Fox Grandeur Movietone News* items—entailed a 34-foot-wide rubberized screen (figure 2.5). When the second feature exhibited in Grandeur, *Happy Days* (Benjamin Stoloff), opened at the Roxy Theatre on February 13, 1930 (also paired with *Niagara Falls*), the films were projected on a new 42-foot-wide beaded screen.[69] When *Happy Days*

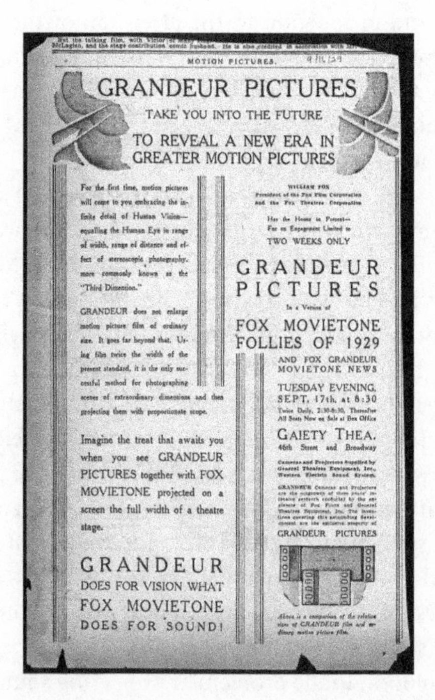


**2.5** The Fox Grandeur system debuted with a screening of *Fox Movietone Follies of 1929* (Fox, 1929) at the Gaiety Theatre.

*Source*: From the Tom B'hend and Preston Kaufmann Collection, Margaret Herrick Library, Academy of Motion Picture Arts and Sciences, Beverly Hills, CA.

opened in Los Angeles at the Fox Carthay Circle Theatre a few weeks later, it again graced a 42-foot-wide screen.[70] Although the 42-foot Grandeur screen only slightly surpassed Magnascope's 40-foot screen, it was celebrated in publicity materials as "a screen twice the width of the old screen!"[71] When the State-Lake Theatre in Chicago was outfitted for Natural Vision that fall, its screen reportedly stretched to 46 feet.[72]

Promotional materials echoed the technical discourses in framing widescreen as an outcome of synchronized sound. The program for the Grandeur screening of *Fox Movietone Follies of 1929* at the Gaiety Theatre, for instance, proclaimed: "The development of the Grandeur system has been an inevitable result of the revolution which Fox Movietone brought to the motion picture industry. The new type of motion picture entertainment which came with sound demanded equal improvement in visual reproduction. Grandeur Pictures do for vision what Fox Movietone does for sound."[73] Contrary to such claims, however, widescreen did not simply follow from synchronized sound. William Fox's pursuit of widescreen, reportedly inspired by Magnascope, was already under way in August 1927, just as his studio was about to exhibit *Seventh Heaven* (Frank Borzage) and *Sunrise* (F. W. Murnau) with Movietone scores.[74] At that time, Fox paid $25,000 for the rights to John D. Elms's Widescope camera and charged sound engineer Earl Sponable (who professed his own skepticism) with developing it into a viable widescreen system for the studio—what would ultimately become Grandeur.[75] As of October 1927, the capacity for Grandeur cameras to accommodate sound was treated as something of a bonus, with Courtland Smith informing Fox that they were "planning these cameras for sound work as well as for silent work" but that if "for the first picture you are only interested in silent work we might be able to shorten the time it would take to turn out two silent Grandeur cameras."[76] By August 1928, however, it had been agreed that no Grandeur films were to be "made without sound."[77]

Rather than causally related, the experiments with synchronized sound, variable screen masking, large screens, and wide-gauge filmmaking in the mid- to late 1920s were mutual participants in a broad field of experimentation with screened images as well as recorded and transmitted sound. As John Belton has pointed out, William Fox attributed his interest in large-screen and wide-gauge cinema to the prospect of competition from television.[78] Indeed, the commencement of work on widescreen at Fox in August 1927 took place in the midst of a swell of enthusiasm for television in the wake of successful demonstrations in 1925 and especially in April 1927.[79] In July 1927, for instance, C. Francis Jenkins had published an article in the *Transactions of the Society of Motion Picture Engineers* proclaiming that "it is time to give the public Radio Movies for home entertainment," contending that "every element necessary to complete an acceptable machine is in hand" and raising (in order, ostensibly, to dispel) the prospect that these "Radio Movies" could "hurt theater patronage."[80] As

Belton explains, "Though television promised to combine the sound of radio with the moving images of the cinema, images on the television screen were both small and lacking sharpness."[81] The enhanced resolution and increased scale and scope touted in conjunction with widescreen cinema were thus made desirable in part vis-à-vis television's perceived deficiencies.

Films made and exhibited in widescreen exploited both the scale and the scope made possible by the format. The Grandeur system was touted as increasing not only the size of the projected image but also the amount of profilmic information that could be portrayed at a large scale. Fox attempted to distinguish the 70mm Grandeur system from Magnascope by virtue of the fact that, in cinematographer Arthur Edeson's words, "any given lens will embrace a considerably wider angle of view on the 70 mm. film than on the smaller standard."[82] Publicity materials similarly proclaimed, "The principle of Grandeur Pictures is not that of magnification. The Grandeur camera is equipped with a new type of lens which records an image of greater scope in width and depth than has ever before been achieved."[83] And they claimed that the system gave "pictures with the detail and definition of a close-up, but covering the wide field of a long shot."[84] The films produced in the Grandeur format exploited these possibilities by filling the frame, often with large land- or waterscapes, masses of bodies, or both. The organization of these spectacles in both width and depth was prescient of the structuring of diegetic and actual space that would come to be associated with screen practices in the 1930s.

One of the criticisms of the Grandeur presentation of *Fox Movietone Follies of 1929*, which was not made expressly for the widescreen format, was that it featured too many close-ups.[85] By contrast, reviews of *Happy Days*, the second Grandeur feature but the first to be conceived specifically for the system, celebrated that the format did "away with the objectionable close-up," opining that "Grandeur film is grand—and especially for the long shots."[86] That film, as well as the subsequent Grandeur release *The Big Trail* (Raoul Walsh, 1930) and the Grandeur shorts and newsreels with which these films were exhibited, put the large scope on display in part by depicting wide expanses of land or water. Most famously, *The Big Trail* used the Grandeur camera to capture the landscapes of the American West (figure 2.6).[87] MGM's Realife Western *Billy the Kid* (King Vidor), which also debuted in October 1930, deployed its widescreen format to similar effect, as did RKO's Natural Vision railroad film *Danger Lights* (George B. Seitz), released in its widescreen (and purportedly 3D) version in November.[88]

The short subjects made with these systems also emphasized such spectacles. The Grandeur short *Niagara Falls*, which was shown with both *Fox Movietone Follies of 1929* and *Happy Days*, for instance, harnessed the wide format to display the scale of the falls, allowing horizontal movements (the flow of the water as well as tracking shots) to correlate the film's subject with the new frame.

**2.6** *The Big Trail* (Fox, 1930) used the Grandeur camera to capture the landscapes of the American West.

*Source*: Screen grab by author.

**2.7** The Natural Vision camera was used to film widescreen footage of Niagara Falls.

*Source*: From the Essanay Film Studio Collection of Visual Materials, Chicago History Museum.

That film recalled the Natural Vision short also titled *Niagara Falls*, which accompanied screenings of *Danger Lights* in 1930 despite having been made (without sound) several years earlier (figure 2.7).[89] The Grandeur short *Hudson River Bridge*, which accompanied screenings of *Let's Go Places* (Frank Strayer) at the Roxy Theatre in early 1930, began with the title "Work on the World's Biggest Span," seemingly referring simultaneously to the bridge (which would subsequently be christened the George Washington Bridge), the wide scope of

**2.8** The Grandeur short *Hudson River Bridge* (Fox, 1930) employed the system's wide scope to emphasize the astonishing scale of the new bridge.

*Source*: Courtesy of the Academy Film Archive, Academy of Motion Picture Arts and Sciences, Beverly Hills, CA.

the Grandeur frame, and the proscenium-spanning screen (figure 2.8).[90] The film employed the system's wide scope to emphasize not only the astonishing length of the new bridge but also the breadth of the land- and seascapes surrounding it. *Fox Grandeur Movietone News* items displaying New York Harbor, the Statue of Liberty, and a New York–Chicago express train featured a similar emphasis on land- and seascapes. In the latter item, the camera was mounted on the train in the tradition of the phantom ride. Such short subjects were well received even when *Fox Movietone Follies of 1929* was not. Fox executive H. F. Jermain reported that at the first showing of Grandeur pictures at the Gaiety Theatre in September 1929, "practically all of the shots of Niagara as well as the numbers in the newsreel were applauded"—an assessment supported by reviews of that program.[91] In fact, *Happy Days* incorporated what *Variety* described as "scenic and atmospheric shots from the first Grandeur newsreel woven in as story accessories that stand on their own as stunning pictorial panoramas."[92]

The Grandeur films, however, also harnessed the new frame to display large groups of bodies. Indeed, publicity for the system emphasized that it "is impossible to crowd more than four or five chorus girls into the old size frame and portray them in a close enough shot for clear detail, and when so arranged the picture is crowded and without freedom of motion," but "Grandeur gives a field that permits of natural action and makes it possible to put such a picture across in proper style."[93] In addition to its focus on landscape, *The Big Trail* also employed the wide scope of the new frame to this end, filling it with throngs of human and animal bodies.[94] *Happy Days* pursued this possibility as well, using the format to emulate a large theatrical stage playing host to a spectacular musical revue, thus capitalizing on the capacity to pack in chorus girls. As *Variety* reported, "Three-quarters of the footage is devoted to a series of spectacular stage ensembles that are eye-openers in screen pageantry, numbers involving masses of people and bigness of backgrounds," including "several dance ensembles, one of them with a leader and 32 girls in intricate maneuvers, and each separate dancing girl visible in what would otherwise be semi-closeup."[95]

The *Fox Grandeur Movietone News* items also featured masses of regimented bodies. As Mordaunt Hall described the item "Tiller Girls" in the *New York Times*, it depicted sixteen Tiller dancers "going through one of their precision terpsichorean performances," a spectacle that was, in Hall's opinion, "impressive because of the size of the images of the girls."[96] "West Point," an item that Richard Watts Jr. of the *New York Herald Tribune* also considered "particularly impressive," revealed "the maneuvers of a great body of men with a magnitude and, at the same time, a sort of close-up intimacy impossible on the smaller film."[97] An earlier feature in the same paper suggested the social ramifications of such spectacles, proposing that with the Grandeur camera, "which can deal with large masses of people as easily as present day cameras deal with individuals, new types of story will be permissible. It is conceivable that a good story could be written to deal with masses rather than with individuals in scenarios for Grandeur."[98] On the heels of standard-screen experiments with depicting the masses, such as *The Crowd* (King Vidor, 1928) and *Lonesome* (Paul Fejos, 1928), Grandeur was thus conceived as a particularly apt format for depicting what Siegfried Kracauer had theorized in 1927, via a discussion of the Tiller Girls, as the "mass ornament"—a spectacular display of the masses' experience of urban-industrial modernity, wherein humanity had been subordinated to rationality.[99] In line with Kracauer's argument, we can view the masses of performers collected on the new immense screens as holding up a mirror to the mass audiences convened in the immense auditoriums of the picture palaces in which these screens were installed.

In addition to capitalizing on the wide scope of the new frame, Grandeur films also emphasized deep space. Because wide-gauge film also increases depth of field, as Ralph Fear explained, it "furnishes a clear background: you will see close-ups and yet miles away there will be clearly defined results."[100] As a result, contended Arthur Edeson, cinematographer for the Grandeur version of *The Big Trail*, the director "must in a Grandeur picture pay considerably more attention to his background action than is usually the case, for, even in close-ups, the depth of focus demanded by Grandeur makes the background an important part of the picture."[101] Reviews of *The Big Trail* in particular took notice. The *Los Angeles Times* began its review by exclaiming, "Dramatizing the background!"[102] Another review described the film as having "a background so stupendous and unremitting and powerful as to be wearisome and a real bore."[103] The background in the Grandeur version of *The Big Trail* is indeed notable—not only in its display of the wide landscapes of the West but also in its depiction of a remarkable amount of human and animal movement. Shots depicting characters conversing in the foreground tend also to feature backgrounds that are, in William Paul's words, loaded "with competing activity."[104]

A scene depicting a minstrel show in the Grandeur version of *Happy Days* brings together the emphases on width and depth that I have been discussing and shows how they could be put in service of particular ideological organizations of space. As with "Tiller Girls" and "West Point," this scene employs the wide frame to display the orderly array of a mass of human bodies, in this case filling the frame with a large group of (apparently) white performers in blackface (figure 2.9). These performers are arranged in tiers so that they appear as a wall of minstrels, forming a kind of backdrop to the represented stage. In certain shots, they constitute the entire background of the Grandeur frame. Whereas "Tiller Girls" and "West Point" worked to spectacularize and foreground idealized notions of white femininity and masculinity, this scene from *Happy Days* thus harnessed the new format to proliferate a caricature of Blackness and render it as background.[105] This presentation not only perpetuated the ideological work accomplished by blackface minstrelsy—as a construction and objectification of Blackness usually (though not always) undertaken by and for whites—but also mapped that work onto the spatial dynamics of the image as it materialized onscreen.[106]

**2.9** *Happy Days* (Fox, 1930) used the Grandeur format to depict a minstrel show.

*Source*: Courtesy of the Academy Film Archive, Academy of Motion Picture Arts and Sciences, Beverly Hills, CA, and the Seaver Center for Western History Research, Los Angeles County Museum of Natural History.

In the silent era, Hollywood films had often employed African American performers, in Miriam Petty's words, for "the creation of 'scenery' or 'atmosphere'" and (until the mid-1920s) reserved more prominent Black roles for white actors in blackface.[107] The sound era saw the reinvigoration of white blackface as well as a rise in more high-profile though stereotyped roles for African American actors.[108] In its use of the Grandeur format, *Happy Days* combined the vogue for white blackface with the atmospheric role that had been attributed to Blackness, using the new frame's scope and depth to construct Blackness as both environment and background. In this regard, the use of Grandeur in *Happy Days* functioned similarly to the deployment of rear projection for the depiction of Native Americans in the Westerns produced a decade later (though in the latter case, the new screen practice was associated with the casting of Native American performers), as discussed in chapter 1. Indeed, the wall of minstrel masks functioned as a kind of screen within the screen, not only reflecting an ideological construction of Blackness but also simultaneously obscuring and—in a distorted, exploitative, and deeply demeaning way—making present African Americans' cultural contributions.[109] If this spectacle of regimented bodies held up a mirror to its audience, it did not reflect the "homogeneous cosmopolitan audience" that Kracauer, writing in Weimar Germany, conceptualized as the masses, "arranged by the stands in tier upon ordered tier" in the theater.[110] In the segregated American context in which the film was made and exhibited, the use of a hierarchical construction of racial difference to organize onscreen space also reflected practices within many theaters, wherein—especially but not exclusively in the South—racial difference was also often delineated and enforced spatially, including through the use of hierarchized tiers in the auditorium (with white customers permitted into the orchestra section and African American patrons confined to gallery or balcony levels).[111]

After the abandonment of widescreen in 1930–1931, the forms of spectacle associated with its scope in width and depth were taken up and adapted in association with other screen practices. Hollywood's embrace of rear projection in 1931–1932 came right on the heels of widescreen's demise, and this special-effects technique adopted (and significantly transformed) the emphasis on backgrounds. Despite major differences in the depth of focus and definition characterizing the backgrounds of films made with widescreen and rear-projection processes, these processes were united in employing screen technologies in the service of the creation of virtual environments—which, as we have seen, were aligned with particular kinds of spaces (such as vast land- and waterscapes) and in many cases were tied to constructions of social difference. The capacity to fill the viewer's field of vision with wide spectacles would, of course, be threatened by the abandonment of widescreen. As I argue in chapter 3, however, this challenge was taken up by the architects and engineers designing theatrical spaces in the 1930s. That effort extended into the built space of the auditorium the type of spectacle that widescreen practice had contained within the frame of its enlarged screen.

## Flexibility

Beyond transforming the scale and scope of projected images, widescreen cinema practices involved less widely recognized forms of experimentation as well. Notably, widescreen installations made use of screen modifiers, moveable screen maskings that had been employed in connection with both sound-on-film (to address changes in aspect ratio) and Magnascope.[112] For instance, although the 42-foot-wide screen installed at the Roxy for the debut of *Happy Days* replaced that theater's previous screen, the new screen featured a modifier to allow the projection of standard films. The *New York Herald Tribune* explained that the "screen can thus be transformed in size so that a change from week to week, or even one on the same program, can be executed with a minimum of inconvenience."[113] This emphasis on flexibility, as we will see, would also be carried over into—and transformed with—exhibition practices in the 1930s.

As with Magnascope, the use of screen modifiers in conjunction with widescreen emphasized the display of the large screen surface through comparison with smaller apertures. As *Motion Picture Projectionist* reported of the installation of Grandeur at the Fox Carthay Circle Theatre in 1930, "The present standard-size picture is first shown and is followed by Grandeur," enabling "the audience to draw a contrast between the respective sizes."[114] Indeed, advice on the use of screen modifiers in the early 1930s emphasized the need for a significant difference (of at least 10 feet) between the standard and large picture sizes "to make the effect worth while."[115] With Grandeur and other contemporaneous widescreen processes, however, such spectacles of contrast did not occur during the course of the film as they did with Magnascope.[116] Rather, the moveable maskings were employed to alter the size of the visible screen area over the course of a program that entailed both standard and widescreen fare. Thus, for instance, an advertisement for the Vallen Automatic Screen Modifier in April 1931 informed exhibitors, "When you're showing a mixed program of Wide and Narrow Film features, quick, silent changes in screen area are as important as good projection itself" (figure 2.10).[117]

Screen modifiers were employed for a range of purposes, including providing what the *Motion Picture Herald* in 1931 deemed "ingenious opportunities" for titles, announcements, and special trailers, and they remained in use even after widescreen had been abandoned.[118] The screen area would often be enlarged for the newsreel.[119] When the RKO Roxy Theatre opened in December 1932, it featured an improved screen modifier, adaptable to four distinct picture sizes,

**2.10** The Vallen Automatic Screen Modifier promised to facilitate altering the size of the visible screen area.

*Source*: *Motion Picture Herald*, Better Theatres sec., April 11, 1931, 85.

whereas previous screen modifiers had permitted only two or three possibilities.[120] At the RKO Roxy, these sizes included the massive 60-foot-wide "full sheet"; the 40-foot-wide "Magnascope" screen; the 25-foot-wide regular screen; and a fourth size that, as described by F. H. Richardson, was still "somewhat larger than the regular motion picture sheet size" and used for "the projection of colored border effects around the motion picture image upon occasion" (this use recalled similar effects that, as I discuss later, were employed at the opening of the Roxy Theatre in 1927).[121] In the RKO Roxy's opening program, the full sheet was used for a short film documenting the construction of Radio City; the large image in this case was achieved by using three projectors.[122] Throughout the 1930s, the *Motion Picture Herald*'s Better Theatres section continued to list "screen masks and modifiers" as a product category of interest to exhibitors.[123]

The employment of screen modifiers—whether during the film or over the course of the program—evidences an emphasis on screens' flexibility as well as their transparency, scale, and scope in this period. Most famously, Sergei Eisenstein delivered a speech at the Academy of Motion Picture Arts and Sciences' symposium on the wide screen in September 1930 in which he advocated what he called a "'dynamic' square screen" that would provide "in its dimensions the opportunity of impressing, in projection, with absolute grandeur every geometrically conceivable form of the picture limit," creating a "dynamism of changeable proportion of the projected picture" by "masking a part of the shape of the film square."[124] Eisenstein's call for a flexible relationship between horizontal and vertical screen shapes echoed ideas Henri Chrétien had laid out vis-à-vis his Hypergonar lens. Specifically, Chrétien had conceived of a cross-shaped screen upon which images could span either the horizontal or the vertical dimension.[125] These ideas were actualized in the legitimate theater several months after Eisenstein's speech, when the Theatre Guild production of *Miracle at Verdun* at the Martin Beck Theatre in New York, directed by Herbert Biberman, employed a huge screen in the shape of a cross. This production used three electrically interlocked projectors to throw contiguous images onto the screen, creating both continuous panoramic images and discontinuous juxtapositions, thus also seemingly emulating Abel Gance's deployment of Polyvision in *Napoléon* (1927).[126]

Alongside such modernist experimentation, flexible screens were also envisioned as a possibility for mainstream Hollywood filmmaking, whether to enhance narrative storytelling or to balance it with spectacle. Eisenstein did deliver his speech to the Hollywood industry. Deeming the speech the "spice" of the symposium on wide screens, *Motion Picture Projectionist* reported that the "colorful Sovkino director" was "in rare form and his frequent coups were roundly applauded."[127] But Eisenstein's ideas were also presented in a context in which Hollywood practitioners were already experimenting with screen modifiers, and he was hardly alone in calling for flexible screen dimensions. At the same symposium, the art director Max Ree proposed a screen with a constant height but variable width in order to maintain balance between scenes "meant to be merely impressive and those which demanded concentration of interest."[128] The SMPE Projection Screens Committee continued to discuss such possibilities when it convened in April 1931. At that time, Benjamin Schlanger noted that "it is possible to change the shape of the screen throughout a picture so as to present different geometrical forms—triangular, rectangular, circular, etc.," and suggested that such work was already under way.[129] In June 1932, an *Architectural Record* article on the design of movie theaters, noting that widescreen process such as Grandeur and Realife had "already been tried and either abandoned or postponed," nevertheless advocated for large movie screens and contended that it "is also considered desirable to be able to alter the proportion and shape of images at will for dramatic effect."[130] Although these ideas did not manifest in widespread practice, their connection to what was a more common practice—the use of screen modifiers across the course of a program—points to a collective and multifaceted interest in the newly flexible screen frame.

Despite the pervasiveness of experiments with widescreen across the studios in 1929 and 1930, the project of developing and standardizing a wide-gauge system in Hollywood had been abandoned by May 1931. Although the trade press blamed this abandonment on "present economic conditions within the industry," there was at play, as John Belton has argued, a more complex matrix of economic, industrial, social, and cultural factors, including the freeze on experimentation prompted by the studios' effort to agree on a wide-film standard, the ousting of William Fox from his studio in 1930, and the failure of the first widescreen releases at the box office, indicating a general lack of interest on the part of the movie-going public.[131] Even in the absence of wide-gauge film, however, many of the concerns elaborated in conjunction with widescreen informed approaches to screen and theater design throughout the 1930s. In particular, engineers sought alternative ways to maintain the reconfigured relationship between auditorium and image space that had been achieved through the expanded and flexible frame of the wide screens employed in 1929–1930. As the next chapter elaborates, this occurred through continuing experimentation with screen masking and attempts to rethink the relationship between the illumination of the screen and the lighting of the auditorium.

## Screens at the Roxy Theatre

The opening of the Roxy Theatre in New York City on March 11, 1927, provides a useful segue since it exemplifies how approaches to screen and theater design that would become prominent in the 1930s emerged in conjunction with synchronized sound and in the context of large-screen exhibition. This theater's debut two and a half years before the stock-market crash of 1929, as Ross Melnick argues, represented "the pinnacle of the movie palace era." Designated the "cathedral of the motion picture," this massive theater was luxuriously decorated, and each component of its technical apparatus was, as Melnick puts it, "intended to be the newest, largest, or some other superlative that proclaimed it the 'last word' in exhibition" (figure 2.11).[132] The Roxy was wired for synchronized sound and equipped to show films using both Vitaphone and Movietone systems.[133] When Roxy's Vitaphone contract was formally announced in February 1927, *Motion Picture News* surmised that the Roxy was "probably the first" theater "to have this equipment built into the structure as a permanent fixture."[134] The spectacular display of Vitaphone, which had debuted the previous August, was touted as one of the highlights of the Roxy's opening-night program.[135] At the opening, while the Roxy Symphony gave a live performance of the overture to Bizet's *Carmen*, a movie screen descended; as the *New York Times* described it, the "efforts of the musicians were greeted with hearty applause and then Giovanni Martinelli and Jeanne Gordon were heard and seen

**2.11** Upon its opening in March 1927, the Roxy Theatre was promoted as "the Cathedral of the Motion Picture."

*Source*: From the Tom B'hend and Preston Kaufmann Collection, Margaret Herrick Library, Academy of Motion Picture Arts and Sciences, Beverly Hills, CA.

(on the screen)" in a scene from the second act of the opera, the tenor's voice "[bursting] from the screen with splendid synchronization with the movement of his lips."[136]

The Roxy, however, also featured several other notable experiments with screens. In January 1927, Samuel "Roxy" Rothafel was quoted expressing his enthusiasm for Spoor's wide-gauge Natural Vision system after he saw a demonstration in Chicago.[137] By February, at the same time that Rothafel's contract with Vitaphone was publicized, it was also announced that the theater would be equipped to show films made in Natural Vision.[138] It was reported (probably inaccurately) that the Roxy's screen would be 44 feet wide, making it the "largest in the world" at "twice the size" of the screen at the newly opened Times Square Paramount Theatre and thus "of ample size for the Spoor invention."[139] The screen was also reportedly equipped with a modifier, touted at the time as a "new device whereby the projectionist can change at will the size or shape of the image flashed on the screen."[140] As the screen at the Roxy was portrayed in the press, it thus emulated the scale and flexibility recently popularized in conjunction with Magnascope. (Having expanded to Los Angeles in January 1927, *Old Ironsides* and, with it, Magnascope were still in the news in the days leading up to the Roxy's opening, when the *Los Angeles Times* was still enthusiastically describing the 40-foot-wide screen.[141])

Although discussions surrounding the Roxy's opening still mentioned that the theater featured a projector "specially constructed" for Natural Vision, that projector was apparently not used at the opening.[142] The Roxy, however, would become an important venue for widescreen exhibition. William Fox acquired a controlling interest in the theater on March 24, 1927, just weeks after its opening night, and although the Roxy, as Donald Crafton suggests, provided Fox a "showcase theater" in which to promote Movietone, it would also supply what proved to be an enduring site for presenting that studio's experiments with widescreen.[143] When *Happy Days* premiered at this venue in 1930, the *New York Times* recalled: "Three years ago when the Roxy Theatre was under construction, its director visualized the enlarged screen of the future, and so great was his confidence in its ultimate perfection that provision was made for it in the original plans."[144] More than twenty years later, when Fox debuted its CinemaScope system with *The Robe* (Henry Koster, 1953), it again employed the Roxy Theatre for the premiere.[145]

In addition to the Roxy's employment of Vitaphone and equipment for (if not actual use of) Natural Vision, its opening-night program featured other screen-related experiments that would find echoes throughout the decade to follow. As I discuss in the next chapter, rear-projection screens would become an important part of the landscape of theatrical exhibition in the United States in the 1930s, when the Trans-Lux Corporation opened a prominent chain of newsreel theaters. These screens facilitated showing movies in a lighted theater, thus taking up the pursuit of "daylight" projection that Rothafel had undertaken as early as 1909, though in that case using front projection.[146] Preceding the Vitaphone presentation on the Roxy's opening-night program was the first-ever theatrical use of a Trans-Lux screen, upon which rear-projected images were visible in a lighted space, an arrangement that was reportedly intended for "panoramas and scenic novelties."[147]

At the theater's opening, the Trans-Lux screen was used to display prefilmed imagery as the background to a live performance of "A Fantasy of the South" by the Roxy Chorus and Ensemble. In the first number, the African American baritone Jules Bledsoe sang "Swanee River" in front of rear-projected images of the Manhattan river front with "boats plying up and down the stream." The second number, "Southern Rhapsody," had performers in blackface backed by rear-projected "scenic effects" portraying a plantation.[148] This act thus appears to have employed the new screen technology to construct a nostalgic picture of the antebellum South, falling in line with minstrel shows. It should be noted that the role played by Bledsoe—already an accomplished classical singer and, as a regular member of the Roxy's music staff, considered to have been the first African American continuously employed "on Broadway"—although not necessarily dampening the demeaning nature of such a spectacle, could have lent itself to the kind of "double reading" that Arthur Knight attributes to conjunctions of Black performance and white blackface in films such as *The Singing Fool* (Lloyd Bacon, 1928), wherein the proximity of an African American actor to a white performer in blackface would have cemented the authenticity of the blackface figure for some audiences while delegitimating it for others.[149] Insofar as the use of rear projection facilitated the simultaneous visibility of live performers and projected backgrounds, it thus created an onstage spectacle—like the onscreen spectacles created with rear projection—that amalgamated virtual and actual spaces and in doing so simultaneously contributed to a particular social formation, with rear projection transporting a plantation scene to a New York stage. Although rear projection had not yet been embraced as a special-effects technique in Hollywood at the time of the Roxy's opening, it had been used to depict the television apparatus in Fritz Lang's *Metropolis* (1927), a film for which Rothafel had persistently but unsuccessfully sought exhibition rights and which ended up having its New York debut at Paramount's Rialto Theatre the weekend prior to the Roxy's opening.[150]

The feature film ultimately chosen for the Roxy's opening night, the Gloria Swanson vehicle *The Love of Sunya* (Albert Parker, 1927), which followed the Vitaphone presentation, was also presented in a manner that would find echoes throughout the decade to follow. The screen lacked the usual black masking and instead, in Ben Hall's words, "seemed to be floating in a luminous mist."[151] As *Film Daily* reported, the screen was the "exact size of the picture. Instead of the usual black masking, a delicate gauze of silver cloth with various colored lamps playing on it gives a soft, diffused light to the picture."[152] Such plays of colored light were another of Rothafel's enduring interests, dating to experiments he undertook at the Family Theatre around 1909.[153] Contending that "gorgeous ensembles of color simulate the effect created by a magician's wand," he had in 1925 also envisioned projection "of light from various parts of the house upon sensitized screens" employed "to create effects."[154] As I elaborate in the next chapter, the interest in eliminating black masking would, as the 1930s wore on, be articulated in relation to the failure of widescreen cinema. In addition, the particular approach to framing the screen taken at the Roxy's opening—replacing black masking with a luminous field—anticipates the role that auditorium lighting would be called upon to play in the 1930s as a mediator between the bright screen and the darkened theater.

The Roxy put such innovations in screen practice in close proximity to experiments with broadcasting in a manner that would also find echoes in the decade to follow—including, most famously, in the two theaters that Rothafel would help to conceive for Rockefeller Center, the RKO Roxy (later redubbed the Center Theatre) and Radio City Music Hall. The Roxy Theatre was designed as a site not only for exhibiting films but also for producing radio broadcasts; the auditorium was littered with microphones to pick up performances as well as audience sounds, and studios upstairs were specially designed for broadcasting. The Roxy was not the first theater from which radio was broadcast. This had been done as early as 1922 at the Kinema Theatre in Los Angeles and subsequently, thanks to Rothafel, at the Capitol Theatre in New York.[155] But the Roxy's high-profile opening, advertised by Rothafel via his radio show, was credited with precipitating the film industry's expansion into radio broadcasting, which, as Michele Hilmes contends, reached its "high point" in 1927.[156] In the 1930s, such practices were reconceived in conjunction with television. Indeed, news of AT&T's successful demonstration of television with the transmission of a speech by Herbert Hoover from Washington, DC, to New York would grace the front page of the *New York Times* just a few weeks after the Roxy's opening. The fact that this news story acknowledges the actual television apparatus's resonance with the imaginary device that the just-released *Metropolis* had conjured with rear projection provides an indication of the relationships that would emerge among the screens used for the production of special effects and those employed for the exhibition of moving images both within and outside the theater.[157]

In the anticipated and actual use of cutting-edge screen technologies for its opening in 1927—ranging from Vitaphone, Natural Vision, and Trans-Lux to the luminous frame of the feature-film screen—the Roxy experimented with merging image and theater space through an expanded and flexible screen frame, the equalization of screen and auditorium lighting, and an illuminated screen border. These practices, together with the interest in employing the theater as a site for broadcasting, would continue to resonate in the years to follow. In the 1930s, at the same time that screens were becoming ubiquitous on the set, creating nested images like those featured in *Metropolis*, they were also morphing and proliferating in the theater. In the wake of the experiments with large-screen and widescreen cinema that altered the screen's relation to its theatrical environment, the approaches to screen and theater design that emerged in the 1930s explored alternative means of integrating the projected image with the built space of the theater. As special-effects screens, film exhibition screens (both theatrical and extratheatrical), and television screens (also both theatrical and extratheatrical) were designed and employed in conversation and conjunction with one another, the forms of spatial synthesis achieved on the set were extended and reduplicated in the auditorium, throughout the theater, and beyond.

## CHAPTER 3

# THEATRICAL SCREENS, 1931–1940

*Integrating the Screen*

The early 1930s were a period of intense change within film exhibition in the United States as theaters newly wired for sound faced the realities of the Depression. Although the massive picture palaces built in the 1910s and 1920s continued to operate (usually with a pared-down service staff and without the stage show), the theaters built or renovated in the 1930s increasingly forewent lavish ornamentation and revival themes in favor of functionalist modern design. In this context, as several scholars have shown, the screen was reconceived as the central focal point and structuring architectural component of the auditorium.[1] Acknowledging the role played in this reconceptualization by the enlarged screens that had been embraced throughout the previous five years, the theatrical architect Benjamin Schlanger proclaimed in 1931, "The very art of the theater should demand more intangible surroundings than those which are obtained by garden walls and other finite unchangeable forms. The larger screen will to a great extent determine the treatment of the interior, its very size making it an integral part of the auditorium."[2] As this chapter elaborates, the effort to integrate the screen into auditorium space—thus rendering that space more dynamic than the old picture-palace interiors—would fuel work on screen and theater design throughout the remainder of the decade, even in the absence of large screens.

This work on screen and theater design should be situated in relation to several of the forces shaping American cinema of this period. Many significant experiments took place in 1931–1933 as the film industry's fortunes reached their nadir. Not only did the embrace of modernist theater design advance an ethos of economy and efficiency appropriate to this period of financial hardship, but it also allowed exhibition spaces to echo the modern aesthetic emerging simultaneously in set design, as seen in films such as *Trouble in Paradise* (Ernst Lubitsch, 1932) and *Grand Hotel* (Edmund Goulding, 1932), wherein modern style signaled wealth and glamour.[3] The popularization of functionalist screen and theater design over the course of the 1930s can, moreover, be viewed as part of American cinema's continuing movement toward industrial consolidation, representational integration, and technological standardization. Indeed, insofar as this approach to screen and theater design more closely meshed representation with exhibition, it echoed both the vertical integration of the industry and the form of spatiotemporal integration achieved through the continuity editing that marked classical style. At the same time, however, certain deployments of screens collaborated with other exhibition practices, such as dish and bank nights, to shape theatrical cinema's address in more heterogeneous and less standardized ways.[4]

Scholarly work has tended to portray the screen's integration into auditorium space in this period as having offered viewers an experience of immersion, replacing the form of distraction associated with older exhibition models.[5] In the wake of the experiments with large-screen and widescreen exhibition that took place in 1926–1931, engineers and architects pursuing screen and theater design did advocate eliminating distractions in favor of facilitating a more immersive cinematic experience. Attention to the range of ways in which theatrical screens were employed and envisioned in the 1930s, however, reveals that screens' integration into theatrical space took a variety of forms. In this context, screens were incorporated into theatrical space not only by virtue of their scale but also through the qualities of scope, flexibility, transparency, and multiplicity that I have been mapping throughout the book. As we will see, this incorporation served not only to immerse and transfix viewers but also to encourage and direct their mobility within the theater. Recognizing the multifaceted nature of theatrical screen practices in the 1930s makes it possible to map connections to concurrent screen practices in the domains of production and extratheatrical exhibition. In particular, the proliferation of screening surfaces within and beyond the auditorium, which emerged alongside and in connection with functionalist theater design, aligned the theatrical presentation of cinema—and its address to women especially—with the contemporaneous exploitation of moving images in sites such as homes and stores.

I have argued that special-effects techniques such as rear projection marshaled filmed fragments into synthetic diegetic spaces in much the same way that sound-mixing practices were beginning to blend heterogeneous recordings into layered soundtracks. Film sound, however, also provides a useful model for addressing how this form of synthesis extended to the context of the movie theater. Emily Thompson has argued that movie theaters of the late 1920s and early 1930s manifested a form of synthetic sound space similar to that found on the soundtrack. In doing so, these theaters culminated what she identifies as a broader change in notions of sound and listening associated with modernity, as sound was gradually divorced from space during the first three decades of the twentieth century. With the use of electroacoustic technologies such as microphones, loudspeakers, and radios, sound was no longer imbricated with its environment through reverberation. Instead, as Thompson puts it, "engineers learned to create electrically a spatialized sound that we would call 'virtual.'"[6] The movie theaters designed for synchronized sound also minimized reverberation, making it possible to treat the synthetically spatialized sound to be reproduced as a signal divorced from the space of reception. In such an environment, Thompson argues, architectural space merged with the sound-reproduction equipment, and the movie theater auditorium was reconceived, in Electrical Research Products Incorporated (ERPI) engineer G. T. Stanton's words, as "a system for transmission of sound."[7] The apparatus of sound reproduction thus collaborated closely with the apparatus of recording to render sound space profoundly virtual. In this chapter, I argue that theatrical screen technologies—and specifically the relationship of screens to auditorium space—were reconceived along similar lines. Engineering discourses framed theatrical screens and their architectural environs as components of a unified system for constructing synthetic cinematic spatialities visually as well as aurally. As in Thompson's account of sound, moreover, the synthetic spatialities thus arising through exhibition worked in conjunction with those achieved in production.

Perhaps surprisingly, synchronization emerged as a key concept in thinking not only about sound but also about screen technologies in the long 1930s. As we have seen, the special-effects technique of rear projection was made newly feasible in the early 1930s in part because electrical hookups that had been developed for synchronized sound also made it newly possible to interlock the camera with a background projector. In fact, rear projection was also known at the time as "synchro-projection."[8] The notion of synchronization (and the prefix *sync-*) were also applied to exhibition screens in this period. With use of screen modifiers, for instance, the images projected on the screen were synchronized not only with the sounds emanating from speakers but also with the moveable horizontal and vertical maskings that framed the screen. One of several companies selling these systems, the Vallen Electrical Company, advertised a screen modifier it called the "Syncontrol" and promoted the "perfect synchronization" that such equipment would provide to an exhibitor's program.[9] Throughout the 1930s, as I discuss in this chapter, engineers also focused on means of aligning the lighting of the auditorium with the illumination of the screen, a project that was bound up with approaches to screen masking. Notably, Benjamin Schlanger introduced a form of masking that he called the "screen synchrofield," which sought to synchronize a play of light in the field around the screen with the play of light upon it. Multiple-camera, multiple-projector systems like the one Fred Waller devised toward the end of the decade also relied on several forms of synchronization, including the synchronization of up to eleven cameras during production and of the same number of projectors during exhibition.

Beyond capitalizing on the popularity of synchronized sound (and on electrical technologies' association with instantaneity and simultaneity more broadly), this persistent emphasis on synchronization evidences a particular temporal construction accompanying the application of screen technologies to spatial syntheses in the long 1930s.[10] By emphasizing the synchronicity of disparate components of cinema—including image(s), sound, screen frame, and auditorium lighting—technical discourses indicated that the projected image and the reproduced sound were not the only elements of the cinematic arrangement that existed in regimented time. As Michel Chion has argued, synchronism was "a marvelous phenomenon" for the first spectators of sound film, and its impact lay in the way sound was inscribed in "real time—time chronographically counted, measured, divided, and fixed."[11] Just as synchronization transformed sound from a contingent element of exhibition to a standardized and technologically integrated part of the film, this concept also provided engineers a model for thinking about the built space of exhibition as another component of the apparatus that existed in regimented time alongside the film. By suggesting an alignment of otherwise divergent temporalities, in other words, the notion of synchronization insisted on the dynamic temporality of the auditorium space itself, a temporality that, as Schlanger intimated in the quotation given earlier, had been obscured by the "finite unchangeable forms" characterizing the decoration of the picture palace. The auditorium was reimagined as a space that through elements such as lighting and screen masking could be technologically linked to the protean quality of the films unfolding within it.

In her work on cinematic time, Mary Ann Doane argues that theatrical exhibition subordinates the temporality of reception to the temporality of the apparatus. The emphasis on synchronization that I am describing highlights the role that components of exhibition such as lighting and screen masking played in this apparatus, not only directing attention to the screen but also in their own right fostering the interplay between temporal regimentation and contingency that, as Doane argues, imbricated cinematic time with modernity. Indeed, I would contend that with the pervasive emphasis on the temporal orchestration of technologies—spectacularizing a form of coordination that was always potentially fallible—film exhibition in this period proffered the experience of time as what Doane calls "presence" or "immersion" that was being threatened by its rationalization.[12]

With the effort to synchronize screen and theater space, screens and their theatrical surroundings collaborated to form dynamic audiovisual syntheses.

Borrowing a term from contemporaneous discourses on special effects, I would like to suggest that the practitioners designing theatrical screens in the 1930s, like those simultaneously developing screens for rear projection on the set, worked to produce new forms of "processed" space by experimenting with diverse means of conjoining virtual and actual domains. The notion of processing emphasizes not only the synthetic nature of the resultant composites—their construction from a process of de- and recomposition—but also the dynamic, processual nature of this activity. The idea of processing, moreover, encompasses a wide variety of spatial syntheses, revealing continuities among the screen practices associated with diverse cinematic modes, such as commercial narrative film, newsreels, and the avant-garde.

As this chapter shows, the screens installed in the theater worked to process heterogeneous spaces in some of the same ways as the screens employed on the set. Here we can also map how screens, together with related surfaces such as windows and mirrors, worked to divide and reassemble spaces into new composites. And here, as well, we can assess these new synthetic spaces by charting how the proliferating surfaces were arrayed and layered. Screened images multiplied laterally not only through the kind of multiple-projector systems devised by Fred Waller and pursued by David O. Selznick but also through the dispersal of smaller screens and views throughout the theater. And these images were stacked in deep space not only through a repositioning of the auditorium screen closer to the audience but also through its situation in relation to observation windows of various kinds. While the arraying of screened images worked to divide and reassemble laterally contiguous spaces through the assertion and transgression of the frame, their layering functioned to divide and reassemble deep space by harnessing the ways in which screens and windows as surfaces could function as barriers and membranes.

Focusing on the various ways in which architects and engineers worked to synchronize the image on the screen with the built space surrounding it reveals that midcentury American cinema, even at its most hegemonic, proffered a much more complex and multifaceted experience than is usually remembered. Discourses on screen and theater design from this period provide insight into the historical construction of familiar notions of the cinematic apparatus by outlining material strategies for rendering the spectator attentive, motionless, and relaxed.[13] However, these discourses also reveal an array of less-well-known and even surprising practices, underscoring the heterogeneity in cinema's arrangement and address even within theatrical exhibition at the height of the classical era. Commercial theaters of the 1930s included configurations that are usually aligned with "new" digital media and the historical avant-garde, including the use of lighted viewing environments, light-emitting surfaces, and multiple screens of various sizes. As I hope to make clear, such configurations were not opposed to the more familiar arrangement entailing a single large reflective screen in a dark auditorium; rather, they show how this arrangement, too, took part in a field of screen practice marked by flexibility and fluidity.

## Integrating the Screen, 1931–1940

In the early 1930s, as theaters' wiring for sound was nearing completion, widescreen was being abandoned, and the effects of the stock-market crash of 1929 spread, there was a widespread effort to rethink the design and operation of movie theaters.[14] Bound up with several converging trends in film exhibition—including the diminishing role of live performance, the recent "little cinema" movement, and the rise of the neighborhood theater in addition to the demand for economy and embrace of functionalist modern design—discourses on theater design emphasized what they identified as the need to reconceive these theaters as structures devoted primarily to film.[15] Contending that previous approaches to movie theater design had simply inherited forms from the "legitimate" theater, architects and critics proclaimed the need for new designs that addressed the particular demands of film projection. Specifically, they advocated eliminating distractions within the auditorium and encouraging concentration and comfort by removing decorations and constructing interiors that would guide the eye toward the screen—an approach that resulted in what Jocelyn Szczepaniak-Gillece terms the "neutralized theater" (figures 3.1 and 3.2).[16] As Maggie Valentine puts it, "By the 1930s the idea of architecture as an escape was redirected into architecture that did not distract from the escape of the movie."[17]

Recent scholarship has emphasized the relationship between the new approach to theater design emerging in the early 1930s and the ways in which synchronized sound had transformed viewers' experience within the auditorium.[18] As William Paul has pointed out, the screen's new position downstage (accommodating the sound-projection apparatus's placement behind the screen) had exacerbated problems related to image contrast, screen size, and distortion, rendering a reconception of auditorium space particularly urgent.[19] In addition, insofar as the novelty of synchronized sound had renewed viewers' attunement to the materiality of cinema, as Robert Spadoni argues, it "reasserted the presence of the screen within the space of exhibition," evoking the uncanny presence of ghostly cinematic bodies within real space.[20] Amir Ameri, in particular, has read the new theaters' effort to integrate the screen with the auditorium space as an attempt to mitigate the potentially unsettling effect of that cohabitation.[21] As the cases of Magnascope and Natural Vision make clear, however, the effort to have the screen dominate the auditorium, blurring the boundary between represented and built space, was already under way before

**3.1** The Valentine Theatre in Toledo, Ohio, pictured here in 1929, exemplifies the use of ornate proscenia and picture settings in the 1920s.

*Source*: From the Tom B'hend and Preston Kaufmann Collection, Margaret Herrick Library, Academy of Motion Picture Arts and Sciences, Beverly Hills, CA.

**3.2** After the Valentine Theatre was redesigned by Rapp & Rapp in 1942, it featured a simplified interior dominated by the screen.

*Source*: From the Tom B'hend and Preston Kaufmann Collection, Margaret Herrick Library, Academy of Motion Picture Arts and Sciences, Beverly Hills, CA.

the popularization of synchronized sound. As I hope to show, considering the changes in theater design in relation to the history and synchronic plurality of screens and their uses can deepen our understanding of the exhibition spaces emerging in the 1930s. Exploring the ways in which architects and engineers addressed the point of transition between the screen and the auditorium reveals relationships to a range of previous and contemporaneous methods of synthesizing cinematic space, including but not limited to those associated with synchronized sound and widescreen.

The effort to integrate the screen into auditorium space was guided by a series of assumptions about the kinds of experience that movie theaters (and, more broadly, cinema) should offer moviegoers. Engineers voiced the belief that theater design should encourage viewers' relaxation and comfort, eliminate distractions and the need for bodily movement, and ensure that each seat afforded an equally uninhibited rapport with the screen. In a *Motion Picture Projectionist* article published in January 1931, for instance, L. M. Dieterich advised exhibitors that in the selection of the proper size screen, the "underlying condition is always the limitation of effortless natural vision without head movement," adding, "If such condition is covered, then the picture is 'easy to look at'—it can be viewed with the maximum of comfort, and the public is pleased (in this respect)."[22] In the absence of the large screens that had previously promised to fill the viewer's field of vision, as we will see, engineers voiced particular concern that viewers' comfort was compromised by the fact that they perceived too great a contrast between the screen and its surroundings within the theater. Attempting to enhance spectator comfort by mitigating such contrast, the engineers sought to calibrate the screen with other components of theatrical space.

As this section elaborates, work on screen and theater design pursued several methods for achieving such calibration over the course on the 1930s. Insofar as such work emphasized scale, scope, and flexibility as means of integrating the screen with the auditorium space, it adapted certain emphases from the earlier work on widescreen cinema and dovetailed with concurrent approaches to production screens and extratheatrical screens. There was an effort, first of all, to integrate the screen within the auditorium space by enabling the screen spectacle to better fill viewers' field of vision. In conjunction with this project, recommendations were made with regard to the size and position of the screen, the situation of seats within the auditorium, and even the composition of films. This effort continued after the abandonment of large-screen and widescreen cinema as engineers concentrated on improving audience members' view of smaller screens, and it contributed to a variegated discourse on screen scale in the 1930s. Another method of calibrating the screen with the auditorium entailed better integrating the screen with its surroundings. Addressing the fact that elements of theater space did indeed enter viewers' field of vision, engineers in this case focused on methods of blending the screen image with the visible components of the auditorium to mitigate the perception of contrast. This yielded recommendations for screen masking as well for achieving balance between screen brightness and auditorium lighting. Such work reiterated the investment in scope also evident in work on widescreen cinema, in this case extending the notion of scope to encompass both onscreen and auditorium space. Finally, the effort to calibrate screen and theater space entailed an attempt to have the flexibility and motion usually confined to the projected film blend into the auditorium. This effort also involved approaches to screen masking as well as auditorium construction.

The disparate components of screen and theater design—from the materials used to form the screen surface to the positioning of seats and the lighting of the auditorium—were treated as mutual participants in a unified system for film projection. Benjamin Schlanger in particular conceived the modern theater as "a machine in itself working with the necessary precision required" and proposed in 1931 that the "problem remains as to how the different functions of the theatre structure are to be coordinated into a unit of mechanical efficiency" offering "comforts and emotional effects" for the moviegoer.[23] Throughout the decade, discourses on screen and theater design generally portrayed strategies for such coordination and especially for diminishing contrast between screen and theater space as means of preventing an awareness of the auditorium (or of their own bodies) from distracting viewers. This was accomplished not by fully obscuring the auditorium space but rather by encouraging components of the film spectacle to blend into it. Indeed, some engineers advocated such blending for the sake of spectator comfort even when they believed it detracted from the viewers' ability to concentrate on the film. Attention to these discourses thus reveals rifts within familiar sets of ideas about exhibition and spectatorship, and it will ultimately highlight connections to a wide range of approaches to composite space.

In the pages to follow, I discuss broad trends in technical discourses on screen and theater design in the 1930s, exploring the ways in which engineers and architects advocated harnessing the qualities of scale, scope, and flexibility to integrate the screen into auditorium space. In the second half of the chapter, I examine cases in which particular theaters also experimented with screens' qualities of transparency and multiplicity. In examining discourses on screen and theater design, I pay special attention to the writings and designs produced in the 1930s by Benjamin Schlanger, one of the leading figures in the effort to reconceive screens and theaters to better suit film exhibition—and easily the most vocal in articulating the ideas and aims underlying that project. Schlanger has recently emerged as an object of academic interest, with scholars exploring his role in advocating for more immersive theatrical spaces as well as his insistence on the relationship between film style and modes of exhibition.[24] Schlanger is, to be sure, a unique figure. He brought to the mainstream film industry notions more clearly at home in the avant-garde—such as the prospect of using light as a building material, which was central to the way modern architects and artists such as Le Corbusier and László Moholy-Nagy were incorporating cinematic concepts into their work around the same time—and many of the ideas he proposed were not widely adopted.[25] But even so, Schlanger was an important figure in the industry, active in the SMPE throughout the 1930s and beyond and widely published in venues such as the *Journal of the Society of Motion Picture Engineers* and the *Motion Picture Herald*. Indeed, Schlanger's articles, as Paul argues, "became the strongest and ultimately most influential voice for a modernist aesthetic in theater design."[26] Even Schlanger's most fanciful ideas, moreover, were rooted in broadly shared concerns and practices. By situating Schlanger's work within larger discourses on screen and theater design, I aim to elucidate the multiple and particular ways in which architects and engineers conceived the relationship between built and represented spaces within the movie theater in this period. Although Schlanger and his colleagues did generally pursue such integration in the service of immersion, exploring their work within an archaeology of screen practice also reveals its relationship to less-immersive forms of spatial synthesis and to a proliferation of projection surfaces.

## Scale

Large theatrical screens remained in use after the abandonment of wide-gauge film. Proposed screen sizes continued to range up to the 40-foot width popularized with Magnascope, and prominent theaters featured even larger screens, such as the 60-foot-wide screen installed at the RKO Roxy Theatre and the 70-foot-wide screen at Radio City Music Hall, which were, as of January 1933, reported to be the largest in the world.[27] The effort to enhance the cinematic illusion and promote spectatorial immersion by having the screen fill the viewers' field of vision, however, came into conflict with the concomitant goals of maintaining image quality (another means of achieving illusion) and spectator comfort (another means of promoting immersion). Engineers insisted that massive screens such as those installed at Rockefeller Center, together with the large projecting distances, projection angles, and viewing distances often associated with them—and in light of the return to a 35mm standard—threatened to produce grainy, dim, and distorted images while still failing to fill the viewers' field of vision.[28] Indeed, many put the upper limit for acceptable screen widths at 22 to 25 feet, beyond which point, they claimed, image quality and sightlines would suffer due to factors such as the resolution of 35mm film, the limited strength of existing light sources, and balcony obstructions in existing theaters.[29] In this context, the attempt to have the film spectacle fill the viewers' field of vision as completely as possible without diminishing image quality or spectator comfort entailed adjustments to the theater space, focusing in particular on the positioning of both the screen and the viewers within the auditorium in order to make the fullest use of a screen of limited size.

Discussions about the positioning of the screen also balanced notions of how best to achieve the desired illusion with the goal of maximizing sightlines and spectator comfort. General advice in the early 1930s held that the screen should be positioned as close to the stage floor as possible, with a maximum clearance of 18 inches. As of June 1930, this vertical positioning was said to proffer a "legitimate stage illusion"—simulating the position of a live performance on the stage, in other words, and presumably anchoring the virtual space onscreen.[30] Later formulations of the 18-inch maximum clearance simply contended that it preserved the "illusion of realism."[31] As F. H. Richardson put it in November 1931, "The placing of a picture higher than is necessary only serves to impress upon the minds of the audiences [*sic*] the fact that it is a picture."[32] Contradicting the assumption that the screen should be situated to preserve the illusion of live performance, however, some commentators highlighted the difference between screen and stage and suggested that the screen be raised in the service of sightlines. The *Architectural Record*, for instance, distinguished "the problem of cinema sight lines from those of the stage theater, since the screen, which confines the dramatic action to a limited area, can be adjusted to any desired position" and could in particular "be raised."[33] Given the limitation on screen height in existing theaters, screens in those spaces could be raised only so much. But the 1930s saw significant approaches to redesigning the theater in such a way as to position the screen surface for maximum visibility and comfort, including raising it higher.

This project was carried out most prominently by Benjamin Schlanger as part of a broad effort to design theaters specifically for cinema and, in doing so, to calibrate the relationship of viewer and view in a more scientific manner than had previously been achieved. Born in 1904 and trained as an architect, Schlanger was twenty-three years old when his first complete theater project, the St. George Playhouse in Brooklyn, opened in December 1927.[34] Barely three years later, his articles on theater design began appearing in trade journals. As early as February 1931, Schlanger was articulating several of the concerns that would dominate his and his colleagues' attention throughout the ensuing decade, including the benefits of the "smaller local theatre" over the large downtown palace (especially with the move to sound, which enabled small theaters to show "on the screen, almost everything that the larger ones previously presented in the personal performance"); the interest in eliminating the clear division between auditorium and stage (which was "usually marked off by an architecturally burdened proscenium frame, flanked by unused elaborate box treatments and obsolete ornamental organ grilles"); and a focus on addressing how that point of "transition" could instead "enable the viewer to feel as little conscious of the surrounding walls and ceiling as possible, so that he can completely envelop himself in that which he is viewing."[35]

Schlanger's writings reveal how screen scale and the positioning of both screen and viewers could work together to achieve that effect. Although Schlanger would eventually advocate for limiting screens to no more than about 22 feet wide (due in large part to the 35mm standard), in his earliest writing—done at a time when wide-gauge filmmaking still seemed viable—he promoted the use of an enlarged screen, identifying 22-by-33 feet as the optimal screen size.[36] Notably, Schlanger insisted on the need to increase the height as well as the width of the screen, criticizing recent experimentation with wider shapes as giving an "unbalanced picture" and insisting that the "vertical accent is needed to allow for good picture composition."[37] Whereas proponents of widescreen had assumed that existing theaters placed limitations on the height of the screen,[38] Schlanger suggested that the prospect of a larger—and especially a taller—screen should instead transform the design of the theater. Observing that large overhanging balconies would obstruct the view of the "complete height" of an enlarged screen for patrons toward the rear of the orchestra, he contended that only shallow balconies should be employed. And he suggested that as many seats as possible be "placed on a level with a point about one-third of the height of the screen, measured from the bottom upward," in order to ensure a "comfortable view of the screen, thus remedying the necessity of craning one's neck to look up, or of pitching one's body forward to look down at the screen."[39] Throughout the 1930s, Schlanger would continue to identify the enlarged screen, as popularized in 1926–1931, as an ideal means of integrating the virtual and actual spaces of cinema. Although he worked at first to design theatrical spaces that would be better suited to large screens than were the old picture-palace auditoriums, he eventually turned his attention to pursuing similar forms of integration through alternative means.

At the SMPE meeting in the spring of 1931, Schlanger presented the design for a theater that would accommodate a larger screen.[40] This proposal included a screen that at 22-by-36 feet approximated the size Schlanger had previously recommended; however, he later suggested using the design for smaller theaters, which would accommodate a screen that at 15-by-25 feet was as large as was generally recommended for 35mm projection.[41] Key to this design was a reverse-sloped floor meant to provide better and more comfortable sightlines of a higher screen. With the new design, the orchestra floor no longer sloped down toward the screen but rather up toward it, so that the floor at the front of the auditorium was higher than the floor at the rear. This slope would tilt the bodies of viewers in the orchestra back slightly, so they would no longer have to crane their necks to see even a raised screen. The angle of the tilt was designed to align viewers' sightlines with the point on the screen that was, at about one-third its height (thus echoing his earlier suggestion), "where most of the action takes place and where the spectator's eye is chiefly focused."[42] Since the reverse-sloped orchestra floor facilitated a lower balcony level, the design also promised to align viewers in the balcony more closely with the level of the screen, eliminating the need to pitch forward in their seats. In addition to permitting a more comfortable view of a higher screen in these ways, the design promised to facilitate the use of large screens (by overcoming the problem of balcony obstructions) and reduce image distortion (by limiting the angle of projection to no greater than ten degrees). As of 1947, Schlanger estimated that about sixty theaters had been constructed according to his plans for this floor design, with others applying similar principles.[43] Among Schlanger's own buildings, notable early implementations of the reverse-sloped floor include the Thalia and Sutton Theatres in New York, which Schlanger converted in 1931 and 1933, respectively, and the Pix Theatre in White Plains, which opened in 1935 (figure 3.3).[44]

Although Schlanger maintained in 1933 that enlarging and changing the shape of the screen was "undoubtedly an inevitable and necessary step" toward improving cinema, by this time he admitted that "this step would be quite a costly one, and perhaps its realization will be deferred by present economic conditions." Thus, he proposed an "in-between step," which entailed more properly orchestrating production and exhibition to encourage the use of a larger field of action within existing screens. As Schlanger observed, "An appreciable area of the screen is being ineffectively used and even wasted by present

**3.3** After its conversion by Benjamin Schlanger in 1931, the Thalia Theatre in New York featured a reverse-sloped floor as well as recessed screen masking.

*Source*: "Thalia Theater," *Architectural Forum*, September 1932, 201.

practices of placing the main focal action or interest only within a limited portion of the screen," specifically the areas at the center and top of the frame. The current technique, he suggested, resulted in static compositions wherein the camera constantly changed position in order to keep the action confined to those areas. He opined, by contrast, that films "would be far more effective if the action moved and revealed the background instead of covering and dividing it as it commonly does now." Schlanger suggested, moreover, that there was a correlation between the current approach to composition and limitations in the design of contemporary theaters. As he pointed out, the unused portions of the screen at the bottom, extreme sides, and corners of the frame, "it so happens, are not within the range of even fairly comfortable or unobstructed vision in present day theatre structures. Vision of the bottom portion of the screen is obstructed in a theatre by preceding heads due to poorly arranged floor slopes. Images on the sides of the screen appear distorted to those spectators sitting in the opposite extreme side seats too commonly found in most present theatres." He thus suggested that changes in theater design (to eliminate obstructions and distortion) could encourage the production of films that would make fuller use of the actual screen area.[45]

In addition to evidencing Schlanger's focus on the relationship between architecture and film style, his insistence on designing exhibition spaces that would facilitate movement across the full span of the screen also exemplifies a particular approach to integrating theater and screen space. As with his design for a reverse-sloped floor to accommodate a larger and higher screen, the interest in employing theater design to extend the usable portion of the frame attests to an ongoing investment—in line with broader technical discourses on the size and positioning of screens—in calibrating the spaces within and surrounding the screen to expand the cinematic spectacle within the viewers' field of vision. In envisaging a form of diegetic movement that also entailed movement across the physical space at the front of the theater (rather than staying centered on the screen), moreover, Schlanger's call for a fuller use of the frame exemplifies how discourses on screen and theater design in the 1930s pursued the kind of dynamism and flexibility at the juncture of built and represented space that was previously accomplished through the use of large screens, wide screens, and screen modifiers.

## Scope

Despite the effort to have the screen dominate the viewers' field of vision, in the absence of large screens—especially in the massive auditoriums still in use—elements of theater space threatened to intrude on the viewers' overall visual experience. As indicated earlier, engineers were particularly concerned that viewers' perception of contrast between the illuminated screen and its darkened surroundings would result in discomfort. Throughout the decade, there was thus a broad effort to minimize contrast without thereby detracting from the visibility of the screen image. This work involved overlapping practices related to the brightness of screens, the borders framing them, and the lighting of the auditoriums in which they were installed. The experiments undertaken to those ends reveal how the sense of scope associated with widescreen cinema carried over into later approaches to screen design. Widescreen had endowed the film image with a sense of scope in its own right, allowing a broad and busy mise-en-scène to fill the viewers' visual field. In their attempt to blend adjacent spaces within and beyond the film image, engineers in the mid- to late 1930s attributed such scope to a visual field now taken to incorporate both screen and auditorium space.

There was massive variation in screen brightness across theaters in the United States in the 1930s (with one estimate suggesting a range from 2 to 25 foot-lamberts), resulting in the complaint that a single film print looked significantly different from one theater to another and inspiring a concerted effort among engineers to study, measure, and make recommendations about screen brightness in the hope of standardization.[46] In the spring of 1936, the SMPE Projection Screen Brightness Committee proposed a standard range of 7 to 14 foot-lamberts, a recommendation that was upheld throughout the remainder of the decade.[47] This brightness level closely corresponded to the level of illumination recommended for interior spaces such as offices, schools, and libraries at the time (which was 8 to 12 foot-lamberts), and it was framed as a range that balanced the goal of visual comfort with the capabilities of projection equipment.[48]

Several factors are involved in producing and evaluating such brightness. In the front-projection arrangement dominant in theatrical exhibition, screen brightness is a combined product of the illumination provided by the projector and the reflectivity of the screen.[49] The reflectivity of screens, moreover, varies with their material construction and installation. In the 1930s, three kinds of screen surface were in widespread use for theatrical exhibition: diffusive or matte, reflective or metallic, and directive or beaded. All three types were perforated for sound. Diffusive screens were constructed of fabrics (which could be treated, coated, or woven with materials such as glass or metallic fibers) and coated metals. These surfaces were efficient at reflecting light and capable of redirecting it at wide angles—and thus could be used in wide auditoriums and theaters with steep projection angles—but their apparent brilliance diminished with distance, so that they were often too dim for viewers at the rear of the auditorium and too bright for those at the front. Metallic screens were made of polished and coated materials such as aluminum. These surfaces were much brighter, but only within a narrow angle of view, and they did not accommodate projection angles greater than 10 degrees, which means they were suitable only for some long, narrow theaters with very small projection angles. Diffusive screens were embedded with glass globules. These surfaces produced bright images for viewers in rear seats, redirected light in a manner that provided a good view to viewers in the balcony, and reduced glare for viewers at the front of the auditorium. But they were not appropriate for wide auditoriums or for theaters with projection angles greater than 20 degrees.[50] In practice, then, the screen's brightness depended not only on the power of the projector but also on the screen surface and projection angle employed in a particular theater as well as on the layout of the auditorium and the position within that space at which the brightness was evaluated.[51] As of 1935, a large manufacturer of all three types of screen reported that its sales broke down as follows: 55 percent diffusing screens, 30 percent metallic, and 15 percent beaded.[52]

Although engineers generally advocated for bright screens, many, drawing on vision research, held that too dramatic a contrast between a bright screen and dark auditorium produced physical discomfort. B. O'Brien of the University of Rochester Institute of Optics and C. M. Tuttle of the Eastman Kodak Research Laboratories, for instance, explained, "Too great a contrast (too great a range of brightness) results in an unpleasant sensation that is not well understood but is usually reported as 'glare,' although the term may be misapplied in this connection."[53] One suggested means of mitigating contrast had to do with the way screens were framed within the auditorium. General practice framed the screen with a 3- to 4-inch-wide strip of black velvet, which absorbed any light spilling from its edges and thus minimized what the SMPE Screens Committee described as "the effect of 'jumping' of the picture caused by the film or projecting equipment."[54] That mask was in turn framed by a 2-foot to 3-foot border, which could be black or another color. For dimly illuminated screens, F. H. Richardson recommended a completely black border, extending from 1 inch inside the picture to anywhere between 1 and 3 feet outside it, in order to improve picture quality by enhancing contrast within the image.[55] Several others, however, contended that such wide black borders produced too great a contrast with the illuminated screen image, "subconsciously attracting the eyes to the frame rather than to the picture image" and resulting in "eye fatigue" or "even injury."[56] In order to diminish this contrast, many advocated either a soft gray color or a gradual grading from black (next to the picture) to the "predominant surrounding color."[57] Richardson recommended such practices in conjunction with brightly illuminated screens, so long as the thin black mask edging the screen remained in use.[58]

By 1934, Eugene Clute, writing in the *Motion Picture Herald*, suggested extending such grading into the design of the theater. He proposed that, in "the auditorium, graded tones in paint might well be used in broad vertical bands on the walls at either side of the picture screen, especially if these walls are splayed or flared out from the screen," and he contended that this practice "would make an excellent way of providing a dark area of wall next to the screen to absorb the greater part of whatever light may be thrown upon the walls at this point from the screen, preventing it from being reflected back to the screen to dull the picture."[59] Here, the theater space was treated not only as an extension of the screen border but also—in absorbing reflected light and ensuring the brightness of the picture—as an integral component of the projection apparatus itself.

The idea of illuminating the auditorium to a certain degree was offered as another strategy for mitigating the contrast between the bright screen and the dark space surrounding it. This practice had been advocated during the nickelodeon era, as part of the endeavor to uplift cinema, and we might note that similar efforts were occurring in the 1930s as well (in *Our Movie Made Children* [1933], based on the Payne Fund Studies, for instance, Henry James Forman quoted concerns about the "dimly lighted" nature of movie theaters to support claims about cinema's dangers for children).[60] As might be expected, engineers contended that darkened auditoriums enhanced the visibility of the projected image as well as the concentration of viewers' attention.[61] However, illuminating the auditorium was touted as offering several benefits, including not only the "visual comfort" associated with reduced contrast but also enhancements to convenience (helping moviegoers find their seats easily, without the greater visual distraction of an usher's flashlight) and safety (reducing accidents and assuaging fears about untoward behavior). Many engineers suggested that these benefits outweighed the improvements in visibility and concentration associated with fully darkening the auditorium.[62] And they recommended, in the words of M. Luckiesh and F. K. Moss of General Electric's Lighting Research Laboratory, "adding as much brightness to the immediate environs of the screen as is practicable without seriously reducing the visibility and satisfactoriness of the picture."[63]

Discussions of both screen brightness and theater lighting thus focused on calibrating the illumination of the theater with that of the screen in order to enhance the viewer's comfort, convenience, and safety.[64] In 1931, the SMPE Theater Lighting Committee had identified an auditorium lighting level of 0.1 foot-candle as "satisfactory for taking seats easily, provided the eyes had gradually accommodated themselves to that intensity in passing from the high intensities existing at the entrances."[65] Later recommendations, however, allowed that "somewhat greater general illumination be provided at the rear of the theater than at the front" since "the intensity of the reflected light is greater near the screen," and they proposed an illumination of 0.1 foot-candle at the front of the auditorium with a gradual increase toward the back of the auditorium up to a level of 0.2 foot-candle.[66] These levels were comparable to the 0.18 foot-candle value proposed for theaters in Japan, where it was held that auditorium illumination should permit patrons to read programs printed in an eight-point type. By contrast, German theaters customarily did not provide any auditorium lighting aside from what was reflected from the screen.[67]

Some engineers effectively combined the effort to mitigate contrast through graded screen borders with the work on auditorium lighting by exploring illuminated screen borders. In a report to the SMPE in 1935, for instance, O'Brien and Tuttle contended that it "is well known among opthalmologists and others concerned with the care of the eyes that reading or other close work with high central but low peripheral illumination produces eye-strain which may be relieved simply by increasing the peripheral illumination without changing the brightness of central objects." And they urged: "While there is no satisfactory explanation for this need for border illumination, we must accept it as an empirical fact, and be prepared for changes in projection practice that may involve a border brightness that is low, but not negligible, as compared to the brightness of the screen."[68] In experiments with screen borders illuminated at various levels, they found that most observers preferred moderately illuminated borders, with only a minority favoring dark borders.[69] In the symposium on lighting held at the SMPE convention in the spring of 1936, Luckiesh and Moss also suggested "that experiments be made by surrounding the screen with uniform brightnesses of different magnitudes less than the screen brightness," and they diagramed an auditorium layout in which the screen was "surrounded by a luminous intermediate field" (figure 3.4). This layout included reflective curtains in the space immediately

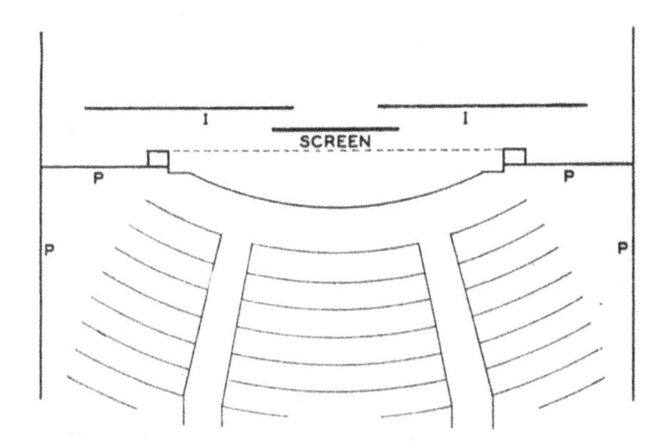


3.4 M. Luckiesh and F. K. Moss of General Electric diagramed an auditorium layout in which the screen was surrounded by fields of light, including luminous intermediate (I) and peripheral (P) fields.

*Source*: M. Luckiesh and F. K. Moss, "The Motion Picture Screen as a Lighting Problem," *Journal of the Society of Motion Picture Engineers* 26, no. 5 (May 1936): 589.

surrounding the screen as well as front and side walls of low brightness, which, Luckiesh and Moss contended, was "better for easy, comfortable and safe seeing than 'darkness,' which is so prevalent now."[70] Like Eugene Clute's suggestion of graded painting for the auditorium in 1934, the idea of creating a luminous field around the screen by framing it with reflective materials would work to extend elements of the screen into the theater space.

Schlanger actively participated in the effort to blend screen and theater space through creative alternatives to black masking, including illuminated screen borders. In 1932, he echoed an idea shared by many engineers in contending that, with black-velvet screen masking, "because of contrast, the eye is always somewhat conscious of the frame, instead of the picture image only," resulting in distraction, eye fatigue, and destruction of the "illusion of space realism so much desired." In place of black masking, he proposed making the screen image "blend into the side walls and ceiling" by having the "illuminated screen and its surrounding surfaces . . . appear as an even tone of light stretching from side wall to side wall." Schlanger put this idea, together with the reverse-sloped floor principle, into practice at the Thalia, which used recessed masking that allowed a "haze of light" emanating from the side wings to surround the screen.[71]

In 1934, Schlanger introduced a related strategy for blending screen and theater space that addressed the way film images entered viewers' field of vision. This proposal involved a new shape for motion pictures that, in Schlanger's words, would make the viewer "least conscious of an obviously committed outline or shape" by more accurately reflecting what he identified as "the shape of the natural field of vision." The shape he proposed was evocative of widescreen insofar as it took the form of a wide rectangle and would require an enlargement of the screen and thus wide-gauge film. However, Schlanger claimed that "there must appear within this shape areas of both central and peripheral vision." As he explained, "Peripheral vision in real life serves as a transition for blending the sharp contrasting details directly in front of the eyes into the complete obscurity that exists behind the head of the viewer." Contending that movies should create this effect, he faulted the Grandeur system for failing to do so and for instead "placing distinct central-vision images in sharp contrasting outlines out at the extreme sides."[72] Schlanger's proposal thus employed areas of central and peripheral vision within an enlarged screen to ease the transition between the edges of the screen and the theatrical spaces surrounding it, which extended from the viewers' peripheral vision to the area of "complete obscurity" behind their heads (figure 3.5).

Schlanger suggested employing adapted illuminated screen borders in conjunction with this approach to onscreen representation. As he put it, in "place of the dead black masking now used to frame the screen, a supplementary border should be used, having a shape conforming to the natural vision contour,

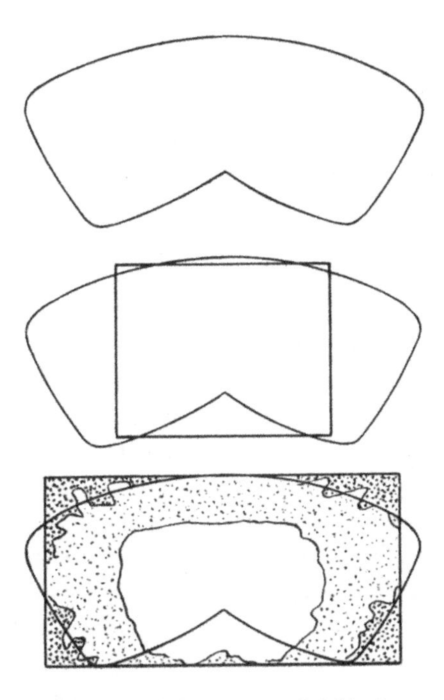

3.5 Benjamin Schlanger diagrammed the contoured field of vision (*top image*), its relation to the existing screen (*center image*), and his proposed image shape, which included areas of central and peripheral vision (*bottom image*).

*Source*: B. Schlanger, "On the Relation Between the Shape of the Projected Picture, the Areas of Vision, and Cinematographic Technic," *Journal of the Society of Motion Picture Engineers* 24, no. 5 (May 1935): 406.

which could be lighted to an intensity that would blend with the lighting of the auditorium and the screen. This screen border lighting would then serve as a transitional blending between the walls of the auditorium and the illuminated screen surface."[73] This "transitional blending" between the screen and the walls would thus echo and extend the blending between central and peripheral vision that Schlanger suggested including within the screen image itself.

## Flexibility

The foregoing has shown how in the wake of large-screen and widescreen cinema practitioner discourses advocated endowing the cinematic spectacle with scale and scope through the calibration of screen and theater design. Specifically, as we have seen, there was a widespread effort in the 1930s to integrate image and auditorium space by addressing the size, position, and brightness of screens, the borders framing them, and the ambient illumination surrounding them. Some practitioners, however, sought to employ light not only to blend screen and theater space but also to endow the point of transition between them with flexibility and dynamism. Benjamin Schlanger in particular was a prominent advocate of this practice as well. His work in this domain, however, again participated in broader trends, particularly related to theater lighting.

Schlanger's pursuit of flexibility at the juncture of screen and auditorium space echoed his early interest in enlarged screens, which, he claimed, provided a "flexible play of the scale of images, objects and backgrounds in relation to the size of the spectator."[74] By extending such luminous flexibility from onscreen images to the space of the physical screen, Schlanger harnessed and combined the functions more widely associated with screen modifiers and illuminated screen borders. Already in 1931, he contended that the front portion of the auditorium, where "the transition from the auditorium to the presentation takes place," should be "flexible, and there is no reason why its form cannot be changeable and its lighting effects varied, to suit the tempo of what is being presented." Moreover, he argued that while "the viewer should not be conscious of the different walls and ceiling that enclose him, he should by all means be conscious of the effect of the unified surroundings, which should assist rather than compete with the presentation." Proposing that this cohesiveness be achieved through "the use and control of light," he advocated employing bands or areas of light to give "an effect, by varying the intensity of the different parts at different times which could be synchronized with the presentation, as well as the musical and sound accompaniments."[75] In short, the flexible use of light within the auditorium, synchronized with the dynamic play of light onscreen, would allow theatrical space to collaborate with the projected film as well as with the reproduced sound as part of a unified system producing an overall "effect."

Schlanger pursued this goal in the second half of the 1930s and beyond. The concept of dynamic screen-border lighting that would integrate the screen with the theatrical space surrounding it was realized with the "screen synchrofield," which Schlanger developed with the electrical engineer Jacob Gilston and first demonstrated in 1937.[76] In the place of black masking, this system framed the screen with diffusive and reflective surfaces positioned at an angle behind and beyond its edges (figure 3.6). As in the other experiments with illuminated screen borders taking place around this time—and as in Schlanger's design for the screen-surround at the Thalia—the screen-synchrofield system thus allowed light from the projector to spill beyond the edges of the frame, creating what Schlanger described as an "illuminated field contiguous to the screen proper." Schlanger, however, now opined that "screen-border illumination of a fixed intensity and color," like that at the Thalia, "can prove to be just as artificial and frame-creating as the present black border when the marginal areas of the

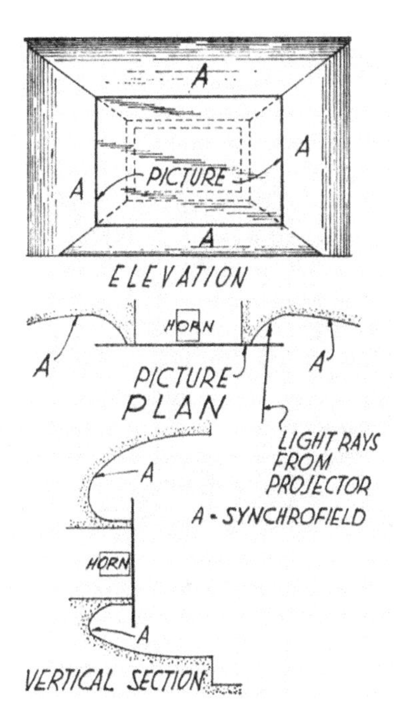


**3.6** Benjamin Schlanger's screen-synchrofield system framed the screen with diffusive and reflective surfaces positioned at an angle behind and beyond its edges.

*Source*: B. Schlanger, "A Method of Enlarging the Visual Field of the Motion Picture," *Journal of the Society of Motion Picture Engineers* 30, no. 5 (May 1938): 507.

picture are dark compared to the contiguous illuminated border." The new synchrofield system ameliorated that problem by endowing the illuminated border with "a constantly changing intensity of light and color" that varied "along the four sides of the screen to match and blend the various edge conditions of the picture into the surrounding field" (figure 3.7).[77]

Schlanger's plan for the synchrofield was situated within a wider context in which, as one observer put it in 1931, trends in theater illumination were "most determinedly" moving in the direction of the "use of light aesthetically."[78] Indeed, some high-profile theaters featured color organs (such as Thomas Wilfred's Clavilux) or elaborate dimmer systems, which could create dynamic lighting effects within the auditorium as well as onstage, facilitating the "harmonization of illumination to the mood of those who come to the theatre for entertainment"—a practice carrying over from the silent era.[79] As an ad for one lighting-control system put it, "People Like Changing Lights—Changing Color."[80] Moreover, certain theaters, such as those run by Samuel "Roxy" Rothafel, as we saw in chapter 2, had for years been experimenting with plays

3.7 With the screen synchrofield, light from the screen radiated outward so that the edges of the projected image matched and blended with the surrounding area.

*Source*: Schlanger, "A Method of Enlarging the Visual Field of the Motion Picture," 505.

of light on or around the screen, which contributed to what Ross Melnick describes as Rothafel's effort to marshal disparate components of exhibition, including lighting and films, into a "unitary text."[81] As Rothafel put it in 1925, the "ballets, musical presentations, stage settings and lighting effects" surrounding the film in his programs were "calculated to form a series of pictures sometimes contrasting and sometimes gracefully merging into one another."[82]

Schlanger, however, maintained that before the advent of the screen synchrofield, it "was practically impossible to create automatically colors and light intensities that would match the ever-changing colors and intensities occurring in the marginal areas of the picture."[83] Although he admitted in 1939 that some "experimental work" had recently been done "with the idea of causing the intensity and color of the secondary sources of light to change in conformance with the screen picture," he contended that this scheme had "all of the disadvantages of the music cue sheets that had to be supplied to each exhibitor in connection with the silent pictures; or of requiring each exhibitor to evolve a new light-synchronization scheme with each picture involved."[84] In other words, whereas the lighted screen border at the Thalia was static and other moving-light surrounds were created as live performances, the screen synchrofield produced an illuminated field around the screen that was not only dynamic but also "automatically" synchronized with the film.

Schlanger continued to develop the screen-synchrofield system in the 1940s. His design for the Island Theatre in Bermuda in 1946, for instance, completely eliminated black masking around the screen, making the screen seem to float on the light-colored wall at the front of the auditorium. This arrangement allowed light from the screen to spill into the surrounding area, creating the kind of synchronous field he had called for in 1937.[85] In the early 1950s, Schlanger would further develop this system into the RCA Synchro Screen or "maskless screen" that, as William Paul has contended, preceded and anticipated the widescreen revolution of that decade.[86] The Cinema—Schlanger's contribution to the Shopper's World complex in Framingham, Massachusetts—featured a 25-by-30-foot RCA Synchro Screen when it opened in October 1951.[87] By the spring of 1952, such screens had been installed in several other theaters, including the Plaza Theatre in Manhattan, the Plaza Theatre in Scarsdale, the Palace Theatre in Chicago, and the Alhambra Theatre in Cleveland.[88]

Already by the late 1930s, however, Schlanger was extending his ideas about the synchrofield beyond the screen border and into the design of the auditorium space. With the use of diffusive screens and increased screen illumination, he pointed out in 1939, a good amount of light was generally reflected from the screen onto the surfaces of the auditorium. As a result, he suggested, auditorium lighting should involve not only secondary light sources but also the absorption and rereflection of light from the screen.[89] He had already experimented with such rereflection in his design for the United Artists Preview Theatre, which used wall and ceiling surfaces to control the light reflected from the screen, including directing it downward to illuminate the auditorium during the film.[90] In 1939, he recommended that all of the interior surfaces of the auditorium seen in combination with the screen, including ceiling and walls, be made white or nearly white in color, and he urged that these surfaces "be broken up into angularized planes or be of such surface textures that their brightness to the audience, under the light reflected from the screen, may be appropriately controlled." As he put it, "The apparent brightness at any point is controlled by re-reflecting the light from the screen toward either the cheeks or the eyes of the patrons in the desired proportions." He suggested, moreover, employing this design in conjunction with his screen-synchrofield system to create a chain of synchronous light effects moving from the screen outward (and eventually disappearing) into the auditorium. With the synchrofield system, as he put it, the "lighting in the area adjacent to the picture is synchronous with the lighting that happens to occur at the edges of the picture." With the new approach to auditorium design, the "surfaces beyond this adjacent area" would in turn "pick up their light in small amounts from the entire picture; but as the distance from the screen increases the apparent brightness of the interior surfaces becomes lower, leaving them in darkness in areas outside the range of vision of the spectators."[91]

In causing light from the projector to seep beyond the edges of the screen, the diffusive and reflective surfaces constituting the screen-synchrofield frame—and potentially also the walls and ceiling of the theater—thus promised to do more than blur the distinction between screen and theater space. By making the screen border and the auditorium interior into secondary surfaces for reflecting light from the projector, Schlanger's synchrofield system and approach to auditorium design promised to unite heterogeneous components of theatrical space (screen, border, walls, and ceiling) into a single synchronized "field," marked, like widescreen, by flexibility. This synchronized theatrical space, moreover, mirrored the composite diegetic spaces that were appearing onscreen by virtue of special-effects techniques such as rear projection. In particular, it echoed the way in which multiple-screen rear-projection arrangements created such composite space through the arraying of image planes. Echoing his earlier discussion of peripheral vision, Schlanger described the work of the screen synchrofield as a "blending of the picture edge into the surrounding field," which would in turn "blend" with the side walls of the auditorium.[92] Like the dual-screen rear-projection apparatus developed at Paramount around the same time, the screen synchrofield thus effected spatial synthesis through the "blending" of laterally contiguous surfaces.

The series of ideas about screen and theater design laid out by Schlanger in the 1930s, though particularly colorful, were, as I have shown, nevertheless rooted in a more widespread effort to blend the main theatrical screen with the auditorium space surrounding it. That decade, however, also saw several alternative approaches to integrating screens into the viewing space. The next sections explore such alternatives, focusing on the use of screens and screenlike views at particular theaters. These case studies include the experimentation with rear-projection screens undertaken at the Trans-Lux chain of theaters, the implementation of auxiliary screens at the Los Angeles Theatre, and the use of windows in conjunction with the screen at the RKO Roxy Theatre in Rockefeller Center. Although these screens and views did not dominate the viewing space in the manner envisaged by Schlanger and his colleagues, they worked in different ways to unite virtual and actual realms. Exploiting screens' qualities of transparency and multiplicity, they offered viewers novel, synthetic experiences of cinematic space that, I argue, remained resonant with Schlanger's notion of the screen synchrofield.

## Trans-Lux Theatres

When the first Trans-Lux theaters opened in New York City in 1931, they, like Schlanger's plan for a reverse-sloped orchestra floor that same year, were presented as an apogee of modern industrial efficiency. These theaters were devoted to newsreels and other short subjects, which were compiled into shows that lasted less than an hour, ran continuously from morning until midnight, and thus catered to what has been described as the "'drop-in' trade."[93] Facilitating the smooth flow of patrons in and out—with maximum efficiency and minimum cost—was a key concern underlying the theaters' design. Turnstiles installed at the entrance allowed customers with exact change to let themselves in; constantly high lighting levels in the auditorium enabled moviegoers to find their seats without an usher; and wide spacing between the rows of seats left ample room for patrons to enter and exit (figure 3.8). Moreover, the Trans-Lux system could be installed in small retail spaces, even those with low ceilings, thus also enabling the theaters to move into and out of existing buildings with ease (figure 3.9).[94]

The technological apparatus that made possible the exhibition of film in illuminated, low-ceilinged rooms consisted of a specially designed projection lens and screen that together facilitated rear projection. With the use of rear projection in exhibition, the auditorium could remain much more brightly lit than usual because the screen transmitted light rather than reflecting it, as was standard—a setup that distinguished the Trans-Lux system from most earlier experiments with "daylight" projection in the United States.[95] Rear-projection theaters reportedly had auditoriums illuminated to about 30 percent of the full lighting (i.e., between screenings) at an average theater.[96] The technical discourse on theater illumination that I discussed earlier treated the Trans-Lux theaters as an extreme example, although Schlanger boasted in 1939 that his

**3.8** With the opening of its first theater in 1931, Trans-Lux promised an illuminated auditorium and "unusual projection."

*Source*: From the Tom B'hend and Preston Kaufmann Collection, Margaret Herrick Library, Academy of Motion Picture Arts and Sciences, Beverly Hills, CA.

**3.9** Trans-Lux Theatre, Manhattan, 1942.

*Source*: From the Tom B'hend and Preston Kaufmann Collection, Margaret Herrick Library, Academy of Motion Picture Arts and Sciences, Beverly Hills, CA.

own front-projection arrangements had matched and even exceeded the light levels "found in the rear-projection theaters."[97]

Expanding an account of screen-based experimentation in this period to include the use of rear projection at the Trans-Lux theaters provides an indication of how the restructuring of cinematic space that I have been discussing bridged fiction and nonfiction cinema, theatrical and extratheatrical exhibition, and the realms of production and exhibition. As Haidee Wasson has shown and as I discuss in greater detail in the next chapter, Trans-Lux screens emerged within and contributed to the wider context of a proliferation of small, portable film screens, many of which employed rear projection.[98] Here I highlight Trans-Lux's concomitant connection to theatrical cinema. Like Schlanger's design for the screen synchrofield, the use of rear projection at the Trans-Lux theaters promised to blend virtual and actual realms within the auditorium through the equalization of illumination. Despite emphasizing newsreels and employing an unconventional exhibition arrangement, the Trans-Lux theaters thus collaborated with the more mainstream exhibition practices discussed earlier insofar as they worked to reshape theatrical space in a way that aligned it with the diegetic spaces that were being reformulated around the same time thanks in large part to the use of rear projection on the set.

When the Trans-Lux Corporation was established in the early 1920s, it was not in the movie business. The company's early work focused on developing a screen upon which slides could be projected in a lighted room. This was made possible through the combination of rear projection and a translucent screen made of silk. When Trans-Lux's first successful "daylight" screen was introduced in 1923, it was sold first for educational lectures in schools and churches and subsequently to financial institutions as a means of displaying stock information. Trans-Lux provided its first "Movie Ticker" to the New York Stock Exchange in 1923. By 1929, the New York Stock Exchange boasted six of them, and Trans-Lux tickers had been widely disseminated among brokerage houses.[99] In fact, as *Time* magazine later recounted, "early Trans-Lux machines in brokerage houses illumined the ticker-tape quotations of the 1929 crash."[100]

Though the Trans-Lux screen was most well known at this time for such financial uses, the corporation also developed several products for other contexts. It continued to promote the educational employment of its screen, which had by 1929 come into use at institutions ranging from Harvard and Yale to public high schools in New York and Philadelphia. The company also developed a cabinet that housed a small screen and rear projector for private use in the home. And it introduced an advertising apparatus, the Daylograph, for use in retail stores, featuring a miniature Trans-Lux screen for the display of a succession of promotional slides.[101] Indeed, Trans-Lux screens contributed to the prominence of extratheatrical exhibition arrangements employing rear projection for the display of both still and moving images in the long 1930s. Such arrangements had become sufficiently pervasive by 1932 that the SMPE Projection Screens Committee claimed in a report on rear projection that in the United States "we are all familiar with the small projectors used in public places for advertising, demonstration, and stock quotations."[102]

Trans-Lux also worked during this time to develop a rear-projection apparatus appropriate for theatrical cinema. This effort involved creating a seamless silk screen of a size comparable to average theatrical screens, about 20 feet wide. Since the projector would now be placed behind the screen, it also necessitated the design of a projection lens that would provide a throw of no greater than 18–20 feet, which was the depth of the average stage.[103] As we have seen, the first theatrical use of the Trans-Lux rear-projection apparatus took place at the opening of the Roxy Theatre in 1927. The establishment of Trans-Lux's own theater chain was the product of its partnership with RKO in 1930.[104] The president of the new Trans-Lux Movies Corporation, Courtland Smith, had been the chief editor at Fox-Movietone news (not to mention a chief figure in Fox's adoption of Grandeur), and the opening of the first Trans-Lux theaters followed the success of Fox's Embassy Theatre, which had opened as a newsreel theater in late 1929.[105]

Like Schlanger's theaters, the Trans-Lux chain took part in the Depression-era movement in movie theater design away from the large scale and ornamental décor of picture palaces and toward the smaller size and functionalist modern design of neighborhood houses. The first Trans-Lux theater to open, located at Madison Avenue and Fifty-Eighth Street in New York City and designed by Thomas Lamb, featured a modern exterior in black and silver, with violet neon lettering on the marquee. The auditorium was about 45 feet long, 30 feet wide, and 14 feet high, had a gray-blue color scheme, and, in the words of the *Architectural Record*, was "designed simply in order to concentrate attention on the screen."[106] More than one commentator noted that such miniscule theaters seemed a throwback to the "store-room" cinemas of years past.[107] However, Trans-Lux presented itself as "the modern theatre" and was treated as such in industrial discourse.[108]

In these theaters, the use of rear projection, translucent screens, and relatively brightly lit auditoriums transformed the relationship between virtual and actual space in several ways. Insofar as the relatively high level of illumination called attention to the space of the auditorium (despite the simplicity of its decoration), the Trans-Lux theaters' lighting scheme worked against the general consensus in theater design that attention should be directed to the screen. Indeed, the SMPE Projection Screens Committee suggested that the light levels at the Trans-Lux theaters exceeded what was practical for eliminating glare and permitting audience circulation. The committee warned that "a partially lighted auditorium tends to prevent patrons from 'living' through a feature presentation, since it makes one too conscious of his immediate surroundings," although the committee also allowed that in theaters that showed newsreels and shorts, this type of awareness was "not objectionable."[109] It is notable in this regard that the Trans-Lux installation used on the opening night of the Roxy Theatre was not for the feature film but rather for a musical number for which it was desirable to see the live performers onstage.

Like the permeable surfaces designed for synchronized sound, rear projection also transformed the screen's role in the exhibition space by exploiting its quality of transparency. Instead of reflecting light from a projector located behind the viewers' heads, the screen transmitted light from a projector situated directly in front of them.[110] Jennifer Wild has explored rear projection's role as an important, if minority, exhibition practice in France in the first decades of the twentieth century, associating its *dispositif* with the assaultive address of modern art. Insofar as the rear-projection arrangement places the projection apparatus within spectatorial space and in front of the audience, she suggests, it collapses a traditional sense of distance between spectator and spectacle and functions reflexively. Important for her discussion is that the screen works to mediate space. She contends that "the formal and also conceptual achievement of the transparent display is how the screen acted neither as a barrier between life and its representation, nor as a mirror reflecting nature or the world back to a passive spectatorial gaze. Rather, it functioned as an interspace between two clearly defined spatial zones behind and in front of the screen, and as a visual field where an interspatial status was conferred on representation itself."[111] With the employment of the transparent exhibition screens that I have been discussing—both the transmissive screens used at the Trans-Lux theaters and the perforated sound screens used widely—American cinemas also adopted an arrangement in which screens (and the forms of representation mediated by them) entered three-dimensional audience space. No longer situated toward the rear of a stage, these screens also both demarcated and bridged the spaces of projection (of image or sound) and reception, emphasizing the mediating nature of the screen surface itself.

The use of transparent screens in the United States in the 1930s did not necessarily produce the radical, assaultive address that Wild associates with that arrangement as it was deployed in France in earlier decades, nor does it make sense in the context of American cinema to distinguish this mode of exhibition neatly as an alternative to the traditional cinematic *dispositif*. Indeed, one of the things that is interesting about the emergence of rear projection as an exhibition practice in the United States is that, in addition to its association with newsreels, it also bore connections with concurrent fiction filmmaking practice. As we have seen, the Hollywood studios also embraced rear projection as a means of producing special effects right around 1931, the same time that Trans-Lux began opening its theaters. Contemporaneous industrial discourses, it should be noted, do not make much of this connection. The fact that rear projection could also be used for production did not figure in Trans-Lux's own corporate narrative until the late 1940s, when it designed special "Teleprocess" rear-projection equipment for television studios.[112] Nor did Hollywood effects engineers' standard accounts of their work to develop projectors and screens for rear projection on the set entail Trans-Lux equipment. However, these engineers did experiment with Trans-Lux's silk screens for the creation of special effects in the 1930s.[113] And regardless of the materials used, the rapid and simultaneous embrace of translucent screens in newsreel theaters and on Hollywood soundstages is notable. With this embrace—with screens in cinematic auditoriums and on movie sets alike taking on the mediating function discussed by Wild—the actual and virtual spaces of cinema in the United States were endowed with the interspatial qualities, if not the radical address, of modern art.

The use of rear projection at the Trans-Lux theaters, like Schlanger's synchrofield, achieved a kind of spatiotemporal hybridity within the exhibition space that mirrors the way rear projection simultaneously functioned as a production technique. It was, indeed, rear projection's capacity to work on lighted sets—enabling live-action footage to approximate if not match the projected imagery—that allowed it to synthesize multiple profilmic elements into new composite spaces. With the Trans-Lux chain, the use of rear projection rendered the auditorium itself a similar kind of amalgamated space. Since, as the *Architectural Record* observed in 1932, the high light levels in Trans-Lux auditoriums decreased the "contrast between [the] screen and [its] surroundings," these theaters also united patrons and screened images into a heterogeneous but composite space.[114]

Attending to the types of images gracing these different rear-projection screens makes the comparison between them particularly apt. As Raymond Fielding contends, the American newsreel generally "devoted much of its program to the presentation of manifestly sensational pictures—footage which could only have been secured by daring photographers," including "strikes, wartime battles, floods, hurricanes, explosions, recalcitrant celebrities, earthquakes, fires, and assassinations," all of which, he claims, "appeared on theater screens with impressive regularity."[115] With the exception of strikes, "recalcitrant celebrities," and assassinations, this list also provides an accurate description of the kind of footage that filmmakers and engineers deemed best suited for rear projection as a special-effects technique. As discussed in chapter 1, many of the films identified in industrial discourses as representative of the challenges and achievements of rear projection as a special effect employ it precisely to depict battles, floods, hurricanes, explosions, earthquakes, and fires. This connection is exemplified by two of the most famous images to grace the different kinds of rear-projection screen in the 1930s: the fire in *Gone with the Wind* (Victor Fleming, 1939) and the flaming airship shown in newsreel coverage of the *Hindenburg* disaster in 1937, which reportedly drew so many people to the Washington, DC, Trans-Lux that the theater "turned over 600 people every hour for several weeks," with lines "around the block."[116] Beyond bespeaking a shared cultural interest in such thrilling and dangerous scenarios, these spectacles evidence a pervasive visual experience combining particular imagery with a particular form of projection. By equalizing the lighting in virtual and actual spaces, both on the set and in the theater, rear projection contributed to the creation of an aggregate form of space and time in the U.S. cinema of the long 1930s. The pervasive presence of dangerous scenarios featuring elements such as fire on special-effects screens and newsreel screens alike epitomized the way in which this aggregate spatiotemporal realm proffered the peculiarly modern sensation of engulfment in virtual environments.

While the experimentation with exhibition screens that I have discussed so far—from widescreen to Trans-Lux—saw projection surfaces expand, morph, and change position within the auditorium, another development was also taking place. Screens, together with the screenlike views provided by windows and mirrors, also proliferated throughout the theater building in this period, allowing moving images to pervade built space, taking a surprising diversity of forms. In the next two sections, the Los Angeles Theatre and Rockefeller Center's RKO Roxy Theatre provide case studies through which to map this proliferation. As we will see, the auxiliary views spreading through these and other theaters were connected to the auditorium screens I have been discussing, but they also took on different roles, emphasizing, for instance, the transmission of views from one space to another and promoting the circulation of viewers throughout the theater. In these and other ways, the proliferating auxiliary views linked theatrical screen practices with the contemporaneous extratheatrical screen practices to be discussed in the next chapter.

## The Los Angeles Theatre

When the Los Angeles Theatre opened on January 30, 1931, it was deemed the "ultra of ultras," and it is usually considered, in the words of the Theatre Historical Society of America, "a supreme exemplar of the movie palace style in its fullest flower."[117] At the time of its debut, however, it was also billed as the "theatre unusual," and it was, in fact, an anomaly in many ways. The last picture palace built in downtown Los Angeles, it embraced the grandeur and period styling of palaces past even as other theaters were already moving toward modern design. The architect, S. Charles Lee, was already known for his work with art deco, having designed the Hollywood-Western Building (home of the Motion Picture Producers and Distributors of America [MPPDA] and Central Casting), which opened in 1928, and the Fox Wilshire Theatre, which opened in 1930.[118] Lee would go on to design the ultimate "automatic" theater, the small and efficient Studio Theatre, which opened later in 1931 and not only emulated the Trans-Lux's turnstiles but also added automatic vending machines, change machines, doors, water fountain, and photo booth—all in a modern style, with the facade featuring bands of black and gray structural glass and highly polished metal.[119] By contrast, the Los Angeles Theatre, inspired by Thomas Lamb's San Francisco Fox (1929), emulated Versailles.[120] At a cost of nearly $2 million during the mounting Depression, the Los Angeles was, on a per seat basis, the most expensive theater ever to have been built. The land on which it sat had been purchased by William Fox; however, the theater was developed and run by the independent exhibitor H. L. Gumbiner, who in April 1930 had arranged to lease the site from Fox for a period of fifty years.[121] Lee had already designed Gumbiner's luxurious Tower Theatre (1927), which evoked the Paris Opera House, and with the Los Angeles Gumbiner was reportedly pursuing something even more sumptuous.[122] The Los Angeles opened with the world premiere of Charlie Chaplin's film *City Lights*, also an independent production and an

**3.10** The Los Angeles Theatre opened in January 1931 with the world premiere of Charlie Chaplin's film *City Lights* (United Artists).

*Source*: Courtesy of the Theatre Historical Society of America, Pittsburgh, PA.

anomaly in its eschewal of dialogue (figure 3.10). As Maggie Valentine reports, according to Lee, Gumbiner "thought the studios would be unable to resist public demand for entertainment in such a lavish theatre."[123] But he was ultimately unable to book first-run films. His company was bankrupt after only three months, and the theater closed in December 1931, having been open less than a year. It was eventually reopened by William Fox, and it operated as a second-run house throughout the 1930s.[124]

Despite its revival style, the Los Angeles was also presented, in the *Motion Picture Herald*'s words, as an "ultra-modern facility."[125] Services and amenities included a luxurious cosmetic room for women staffed by manicurists and cosmeticians; a children's playroom designed as a circus tent; and the innovative inclusion of a full-service restaurant and soda fountain.[126] Moreover, the building was designed (in this sense like the Studio Theatre after it) to maximize comfort and efficiency. The main auditorium grouped seats into rows of six so that no patron would have to pass in front of more than two other people to take a seat (figure 3.11).[127] Neon lights, embedded in the floor under one-inch-thick strips of plate glass, illuminated the aisles.[128] A "detector seat panel"

**3.11** The Los Angeles Theatre grouped seats into rows of six in order to maximize comfort and efficiency.

*Source*: From the Tom B'hend and Preston Kaufmann Collection, Margaret Herrick Library, Academy of Motion Picture Arts and Sciences, Beverly Hills, CA.

installed in the administration office kept track of which seats were occupied.[129] Lighting in the theater was controlled by what the *Los Angeles Times* described as a "$35,000 electric switchboard" that "causes lighting effects to flow on and off with the smoothness of water."[130] The theater boasted one of the largest ventilation systems on the West Coast, allowing simultaneous control of seven different temperatures in various parts of the building. A public-address system, controlled from the projection booth, piped sound throughout the building. Electric cigarette lighters were also installed throughout the building as well as on each dressing table in the ladies' cosmetic room. And the men's and women's washrooms featured newfangled electric hand dryers.[131]

The theater's use of screens, together with the screenlike views provided by windows and mirrors, represented perhaps the most innovative component of its design. The main auditorium screen boasted a three-position screen modifier.[132] More novel—and more widely publicized—was a 3-by-4-foot ground-glass screen installed in the basement-level lounge (figure 3.12). The film that was playing in the main auditorium was also rear-projected onto this screen

3.12 The Los Angeles Theatre featured a ground-glass screen installed in the basement-level lounge, which showed the film that was playing in the main auditorium.

*Source*: From the Tom B'hend and Preston Kaufmann Collection, Margaret Herrick Library, Academy of Motion Picture Arts and Sciences, Beverly Hills, CA.

with a mere two-frame lag time. The image was an indirect projection from the main theatrical projector, diverted and carried through the walls of the theater by means of a channel fitted with a system of prisms and mirrors.[133] An article in *Modern Mechanics and Inventions* illustrated this arrangement (figure 3.13).[134] The projector was fitted with two apertures so that it illuminated two frames on the filmstrip simultaneously. One beam from the projector illuminated the main auditorium screen, while the other was fed through a channel to the basement-level screen.

Lee's plans for the building included several other screens connected through such channels, including one in the outside lobby, one in the children's playroom, and one in a music room that was never actually built. The outside lobby screen was supposed to be placed in one of the small poster cases flanking the entrance. Lee's blueprints for the poster cases indicate that the screen was to occupy the case on the south pier, behind a hinged plate-glass door and juxtaposed with a "pierced loud speaker screen." The blueprints indicate that the corresponding space on the north pier should replace the screen space with

# Movie Shown on 2 Screens at Same Time

A remarkable new movie theater in Los Angeles projects films on auxiliary screens in its lounge rooms by use of a light-proof tube and reflecting mirrors.



This diagram shows how the film is projected through two apertures, the lower being reflected to an auxiliary screen through a black tube.

Cutaway view of theater, showing tube through which indirect projection of the film is made.

Ingenious mechanical devices provide for the comfort and convenience of patrons in the new theater. Seats are arranged for perfect visibility, those at the sides being slightly higher. When a seat is vacated, a light is flashed to inform the ushers. Nursery, beauty parlors and barber shops are at the disposal of patrons. Thermostatic controlled ventilators keep theater constantly at an even temperature.

AN ELABORATE new talkie theater now being built in Los Angeles, home of the movies, is able to project its feature film to one or more auxiliary screens in its lounge rooms, simultaneously with the showing on the main auditorium screen.

To the person not versed in optical laws, this may appear as a simple innovation, yet it is the result of years of experimentation by the most noted scientists in the country. While the people in the auditorium are viewing the picture on the big screen, the waiting patrons in the lounge room, and other ante rooms, see the picture simultaneously on a miniature screen 36"x48", accompanied with complete sound effects.

At the outset, this may seem as a novelty only. But it has a distinct commercial value to continuous performance theaters. While a patron is waiting admission to the auditorium, he can see the picture in miniature. After gaining entry to the auditorium, he probably will leave when the picture reaches the point where he first viewed it in miniature. So the seats are vacated sooner, and the box office profits.

To go into the intricate optical laws which make possible the indirect projection, would be about as intelligible to the average reader as the Einstein theory or the fourth dimension. But the rudiments of the system can be gleaned from the photo-diagram.

It must be assumed that the average reader understands how a picture film passes through the projection machine. The "frames" —that is, each individual picture on the strip of film—are exposed momentarily before the aperture in the gate of the projection machine. A powerful light is focused upon the aperture, and projects the image on the film through a lens to the screen. Seemingly, pictures in motion result.

Now, if a second aperture is cut in the gate, two frames can be exposed simultaneously. A special condenser lens floods both apertures with light. The light beam from the second aperture, carrying the picture image, is conveyed through a black-lined tube fitted with special mirrors and lenses, to a

58 *Modern Mechanics and Inventions for April* 59

**3.13** An article illustrates the Los Angeles Theatre's unusual projection arrangement.

*Source*: "Movie Shown on 2 Screens at Same Time," *Modern Mechanics and Inventions*, April 1931, 58–59.

marble but provide a loudspeaker as well as "three openings for telescopic eyepieces."[135] Although I have not seen a playroom screen mentioned in press coverage or publicity documents, Lee's blueprints also anticipate including one (figure 3.14).[136] A longitudinal section of the whole building reveals the system of projection channels planned to service these screens.[137] It shows a channel leading from the projection room to the screen in the basement-level lounge, similar to the one pictured in *Modern Mechanics and Inventions*. But the blueprint also connects that channel to another room in the basement, which, because of its placement, I believe may have been the music room that was ultimately eliminated. And it also includes a channel leading from the projection room to the outer lobby.

The basement-lounge screen was constructed of ground glass, as the other planned screens likely would have been as well. In addition to using sheets of glass for such screens, the theater also employed plate-glass windows and mirrors to transmit views from one space to another. Reiterating a feature also included in the Tower Theatre, the Los Angeles Theatre's mezzanine level

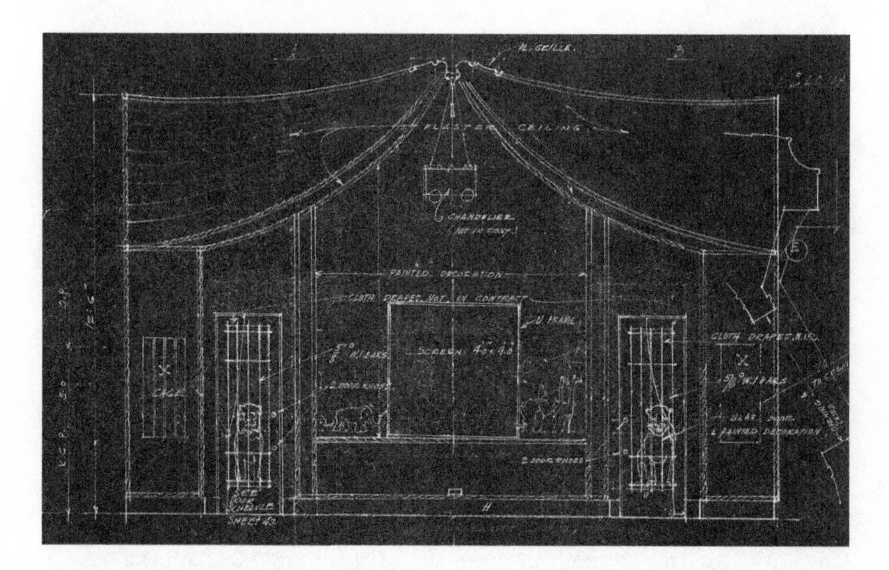

**3.14** S. Charles Lee's blueprints of the Los Angeles Theatre indicated plans to include a screen in the children's playroom.

*Source*: From the S. Charles Lee Papers, Library Special Collections, Charles E. Young Research Library, University of California at Los Angeles.

boasted two sound-proof cry rooms, where mothers could bring their loud infants during the show. The rooms offered these women views of the main auditorium screen through double-paned windows and provided sound through both loudspeakers and earphones.[138] A broadcasting room, capable of sending theater programs into the ether, was also visible to patrons through a thick plate-glass window.[139] One of the exterior poster cases, where the planned outside-lobby screen was supposed to be placed behind glass, ended up instead containing what Maggie Valentine describes as "a recording apparatus that continuously compared the temperature inside and outside." Such views into climate-control systems were also a recurring presence in Lee's theaters of this period. The Tower included a window on the landing between the first floor and the bathrooms that supplied a view of the air-conditioning system.[140] For the Studio Theatre, Lee embedded a similar window in the sidewalk outside.[141]

The foyer of the Los Angeles, lined with massive mitered mirrors, was modeled on the hall of mirrors at Versailles (figure 3.15).[142] The promenade, leading off of the foyer, also featured mitered mirrors, and one of these mirrors was reportedly very special. As the *Motion Picture Herald* described it, "A feature of the promenade is a large mirror on the wall, which, by a special prismatic glass arrangement, shows a full view of the main lounge on the floor beneath. This mirror allows the incoming patron to see who is in the lounge, thereby

**3.15** The foyer of the Los Angeles Theatre, lined with massive mitered mirrors, was modeled on the hall of mirrors at Versailles.

*Source*: Courtesy of the Theatre Historical Society of America, Pittsburgh, PA.

making a trip downstairs unnecessary if the person he is looking for is not there."[143] Although both the *Motion Picture Herald* and the *Architectural Record* describe this contraption as if its existence were an achieved fact, I have found neither photographs of it nor any trace of it in the building as it now stands.[144] Such a device does, however, appear in the blueprints, which show a periscope joining the basement with the foyer above it.[145] Slanted mirrors on either end of the periscope channel are designed to reflect views of the lounge into the foyer, with the foyer-level mirror being marked as adjustable. Mirrors also played an important role in the women's cosmetic room, which not only included individual dressing tables with triplicate mirrors but also featured a three-sided full-length mirror that was at least planned to include a mirrored floor.[146] The *Motion Picture Herald* also reported on this mirrored floor as if it were accomplished fact,[147] but I have not seen it described elsewhere or found a trace of it in the building.

Although the Los Angeles Theatre was, as billed, "unusual," its employment of screens, windows, and mirrors was not entirely unique. Observation windows were a suggested component of theater design at the time. In 1931, for instance, the architect Peter Holsken, writing in the *Motion Picture Herald*, advised exhibitors to place "windows behind the center sections of seats [in their auditoriums] so that the standees [could] see the picture while waiting."[148] Schlanger claimed that it was "desirable to have some part of the lounge command a view of the seating or screen in order that waiting patrons may follow seat availability as well as performance progress,"[149] and he incorporated such windows into his designs for theaters such as the Sutton and the Waldo. As I discuss in relation to the RKO Roxy in the next section, observation windows also played an important role in Rockefeller Center, linking film projection to the project of broadcasting. The Beverly Theatre in Chicago, which opened in 1935, featured observation windows, disguised as mirrors, for the use of the theater manager. As the *Architectural Forum* explained, "Between the photographs hung on the foyer walls are two transparent mirrors about 8 by 30 in. framed to match the photograph frames. By opening a small door behind these the manager can observe the foyer, lobby and ticket booth without being seen."[150]

These theaters, in short, used windows, mirrors, and auxiliary screens to draw together elements that are usually separate (e.g., the auditorium and waiting areas) into a cohesive cinematic space in much the same way that, as Anne Friedberg argues, the main theatrical screen joins built and represented spaces into a larger architecture that unites the actual and the virtual.[151] These examples make it clear that such cinematic space was not limited to the main auditorium but could pervade the theater. The planned use of screens, windows, and mirrors throughout the Los Angeles Theatre in particular operated according to a logic of multiplicity and circulation. According to Lee's design, the same moving image would be multiplied throughout the theater, manifest not only in the main auditorium but also through glass screens and glass windows in various separate spaces, from the cry rooms to the lounges and lobby. Indeed, moving images were piped through theater space in a manner similar to the sounds emanating from the public-address system—and analogous even to the climate-controlled air circulating through the multizone ventilation system. Moreover, the use of mirrors in the Los Angeles resulted in a similar proliferation and transmission of the patrons' own images. Maggie Valentine notes that the lobby's decor "thrust ticket holders into the role of actors, causing them to pose and primp before the mirrors."[152] The ladies' cosmetic room furthered this alignment not only by providing regular women with personal stylists but also by multiplying their images through triplicate mirrors and the curious provision of a simultaneous view up their own skirts. Most assertively, the periscope to the basement-level lounge transmitted distant views of waiting customers in much the same way that the prismatic channels piped views from the main projector to auxiliary screens.

The flow of images and sounds throughout the theater worked most practically to facilitate and encourage the movement of people. Lee suggested, in fact, that the main innovation of his design for the Los Angeles had to do with efficiently managing the flow of customers. Pointing to the seating arrangement, the lounge, and what he called the "general holdout features," he claimed that for "the first time in 20 years an architect dared to try something new in theatre layouts . . . in the methods of handling the public."[153] The basement lounge was unconventional in its sheer size; together with the lobby, it could hold about two thousand waiting moviegoers—a number roughly equivalent to the theater's total seating capacity.[154] The miniature screens and piped sound serving those areas not only functioned to entertain waiting customers but also promised to cut down on the amount of time they would eventually choose to remain seated in the theater, thus enhancing turnover.[155] In his work on advertising film in Weimar Germany, Michael Cowan argues that the small screens that pervaded public spaces in that context functioned as means for "directing the traffic of bodies and attention" in the "newly consumerized arena" of the country. He writes, "Introduced into public spaces, the movement of filmic images was . . . understood as a means of capturing and steering the attention of spectators in motion." Although conventional cinema was, conversely, "seen as an ideal space for advertising precisely because of its ability to *stop* that mobilized attention, monopolizing it in the darkened space of the theatre," Cowan suggests that this moving-image culture framed the movie theater "less as a temple of aesthetic contemplation than as a hub of economic circulation."[156] With its multiple screens and encouragement of spectatorial movement, the Los Angeles integrated such an emphasis on mobilized attention and economic circulation into its very design.

The Los Angeles Theatre's design harnessed screens, windows, and mirrors to mold theatrical space to the contours of cinematic imagery. Windows and mirrors worked in tandem with screens throughout the building to form the kind of synthetic space, mediated by technology, that was associated with the cinematic image itself in this period. The Los Angeles Theatre as a whole operated as an integrated apparatus that employed proliferating screens, windows, and mirrors to incorporate patrons into a particular mode of viewing from the moment they approached the ticket counter. Not limited to the main auditorium, this arrangement bled into all areas of the building, from the restrooms to the spaces put aside for children. The various windows, mirrors, and screens proffered by this expanded apparatus addressed women in particular in specific ways. As mentioned in the introduction, scholarship on cinema's address to women has explored the moving image's alignment with windows and mirrors. Discussing the woman's film of the 1940s, for instance, Mary Ann Doane has argued that the "cinematic image for the woman is both shop window and mirror, the one simply a means of access to the other. The mirror/window takes on then the aspect of a trap whereby her subjectivity becomes synonymous with her objectification."[157] The Los Angeles Theatre literalized such windows and mirrors and employed them to similar effect, privileging women's role as spectators (via the windows provided just for them in the cry rooms) but also encouraging women to view themselves as images (through the mirrors pervading the theater, from the cosmetic room to the lobby and lounge). Beyond echoing the address of Hollywood films (especially the woman's films popular in the 1930s as well), the Los Angeles Theatre's use of screens, windows, and mirrors was also prescient of a range of exhibition practices—within and beyond the theater—that would associate women's role as viewers with consumption and mobility, from chinaware giveaways to the use of moving images in department stores, to be discussed in the next chapter.[158]

Approaching the Los Angeles Theatre through its experimentation with screens, windows, and mirrors highlights other connections as well. Although the Los Angeles might epitomize the larger-than-life spectacle of golden-age cinema, the particular form of synthesis it employed these surfaces to achieve also rested on the notion of seeing at a distance already associated with television. Lee called the system feeding projection to the basement lounge his "'television' system," and its organization is similar to contemporaneous depictions of television, as in the Betty Boop cartoon *The Robot* (Dave Fleischer, 1932), which depicts "Bimbo's Television" as a picturephone connected to Betty's bathroom by makeshift tubing.[159] As discussed in the next chapter, demonstrations of large-screen television were already taking place in movie theaters in 1930–1931. The prospect was also being discussed around this time that television could be used more regularly in the theater, not only in the auditorium but also as a lobby display.[160] The *Motion Picture Herald* reported in November 1931, less than ten months after the Los Angeles opened, that "two important theatre circuits" were already in negotiations to use television sets in their lobbies and possibly also in specially equipped television rooms. The paper explained, "Programs from television stations may be picked up on the 'home' radio-televisor, while regular motion picture film may be broadcast from any part of the theatre over the new small transmitter and picked up by the receiver."[161]

The Los Angeles, moreover, was not alone in employing a device modeled on the idea but not the technology of television, which used mirrors and prismatic channels to transmit moving images from the main projector to auxiliary ground-glass rear-projection screens at the same time as they were being shown on the main auditorium screen. In 1933, the Madrid Theatre in Kansas City employed a similar apparatus, called the "Telescreen," which used a system of mirrors and quartz tubing to transmit images from the main projector to other screens throughout the theater. As with the periscopic projection system at the Los Angeles, this device was envisaged as a lobby display. But like the one-way mirrors at the Beverly, it was also employed for use by the manager, who, as the *Motion Picture Herald* reported, "sits in his office and sees on a small glass panel on the wall all of the action taking place on the theatre screen," while a "speaker attachment completes the projection both of sound and image."[162]

In their use of mirrors, the "television" arrangements at both the Los Angeles and the Madrid emulated Frederick Kiesler's set for the Berlin production of Karel Čapek's play *R.U.R.* in 1923, which not only employed the rear projection of film "instead of a painted backdrop," as Kiesler put it, but also utilized a mirror to bring small images of offstage actors into what was portrayed as a television on the stage.[163] As Kiesler explained, "This *R.U.R.* play was my occasion to use for the first time in a theater . . . television in the sense that I had a big, square panel window in the middle of the stage drop which could be opened by remote control. When the director of the human factory in the play pushed a button at his desk, the panel opened and the audience saw two human beings reflected from a mirror arrangement backstage. The actors appeared in this window as a foot-and-a-half tall, casually moving and talking, heard through a hidden loud-speaker."[164]

Kiesler thus employed mirrors and screens to create a composite space onstage in ways that were reminiscent of special-effects techniques used in the production of films, evidencing what Katharina Loew has identified as the close relationship between mirror-based special effects in cinema and the older tradition of stage illusions from magic shows to Pepper's Ghost.[165] We might note that cinematic depictions of television in the late 1920s and early 1930s, as in Fritz Lang's film *Metropolis* of 1927 (which the *New York Times* described as "a cinema 'R.U.R.' "[166]), were also achieved through special effects—including the rear projection of film onto ground-glass screens—a very similar setup to the Los Angeles and Madrid Theatres' auxiliary screens. There was, in other words, a fluid relationship among these small exhibition screens, the rear-projection screens being embraced for special effects, and the television screens being publicized—all of them roughly concurrently.

Although many of these practices were anomalies in the 1930s, and although the Los Angeles Theatre did not actualize all that Lee had envisioned, the plan for its use of screens, windows, and mirrors thus reveals unexpected relationships among film exhibition, film production, and television. In the Los Angeles, furthermore, the logic of multiplicity and circulation connected materially dissimilar surfaces, from the large, modifiable screen in the auditorium to the diminutive glass-enclosed views provided by the exterior poster cases. The RKO Roxy Theatre embraced a similar logic.

## The RKO Roxy Theatre

Developed amid the deepening Depression, Rockefeller Center in New York City featured two theaters. The better known by far, Radio City Music Hall, opened on December 27, 1932, as a venue for live variety shows. The RKO Roxy Theatre, intended as Rockefeller Center's cinema, opened two days later on December 29. The Music Hall debut was widely considered a flop, and Radio City was converted into a movie house within two weeks. The RKO Roxy continued showing films, but not for long. Renamed the RKO Center in December 1933, the theater was put in the service of live performance in 1934, becoming a venue for ice shows in 1940. It was made a television facility in 1950, only to be demolished four years later.[167] Although Radio City Music Hall's ongoing prominence has long overshadowed the RKO Roxy, the theaters were very much part of the same project in 1932–1933. When *King Kong* (Merian C. Cooper and Ernest B. Schoedsack) premiered on March 2, 1933, it was shown simultaneously in both theaters (figure 3.16).[168]

Both the RKO Roxy and Radio City Music Hall were luxuriously decorated in a modern, art deco style and touted as cutting edge in every respect. Not only did their appointments, from murals to lighting fixtures, accumulate superlatives—such as *biggest*, *tallest*, and *heaviest in the world*—but their technical apparatus was also heralded as the newest and most advanced. In this regard, the Rockefeller Center theaters followed in line with the Roxy Theatre before them. When the RKO Roxy opened in December 1932, it was celebrated as featuring the world's largest screen, which at 60 feet wide purportedly bested

**3.16** *King Kong* (RKO, 1933) premiered simultaneously at the RKO Roxy Theatre and Radio City Music Hall.

*Source*: Courtesy of Photofest.

the previous record holder by almost 50 percent. (The previous record holder was identified as the 42-foot screen used for Fox's Grandeur system; as mentioned in chapter 2, however, a 46-foot screen was reportedly employed in Chicago for the presentation of Natural Vision in 1930.) When Radio City Music Hall was equipped for film projection in January 1933, it got an even bigger, 70-foot-wide screen.[169] It is for such reasons that the Rockefeller Center theaters were considered what the *American Magazine of Art* called, just months after the openings, "perhaps the last monument to the spirit of 1929."[170] Indeed, these theaters, as William Paul argues, "might be fairly described as the last two [picture] palaces," their scale and grandeur already perceived as anachronistic at the time of their debut.[171] At the same time, however, the theaters' participation in Rockefeller Center's multimedial project, during what Ross Melnick identifies as a period of "media divergence" after a flowering of convergence in 1926–1931,[172] situates them, like the Los Angeles Theatre, within a new media landscape in which film's relationship to broadcasting was being negotiated.

The RKO Roxy in particular offers insight into this negotiation through its use of windows in conjunction with the main auditorium screen. In its approach to these surfaces, the RKO Roxy participated in several of the broader trends in screen and theater design that I have already discussed, ranging from widescreen exhibition to the use of observation windows. The massive auditorium screen employed in this theater, like that installed in Radio City Music Hall, failed at achieving the immersive address associated with widescreen cinema and considered a goal by engineers. However, it operated in conjunction with windows to construct a different form of composite space, one marked not so much by scope as by the layering of transparent and translucent surfaces. Indeed, despite the theater's ostensible investment in the large scale of the theatrical spectacle, its use of screens and windows functioned similarly to contemporaneous deployments of smaller screens, such as the various rear-projection screens discussed earlier, insofar as it harnessed these surfaces as selectively permeable membranes. In the next chapter, I argue that the television screens being developed in this period also operated according to a logic of transparency and spatial layering.

With the return to a 35mm standard after the period of experimentation with wide-gauge filmmaking in 1929–1930, as we have seen, engineers throughout the 1930s advised keeping theatrical screens relatively small while continuing to advocate alternative ways to integrate screen and theater space. The massive screens employed at the RKO Roxy and Radio City Music Hall, paired as they were with 35mm projection as well as large projection angles and projecting distances, went against such recommendations. Indeed, Schlanger contended that at Radio City "the picture was magnified to such an extent as to weaken and dissipate the blacks, thereby losing contrast and picture quality."[173] Moreover, even though the screen size at the Rockefeller Center theaters far exceeded recommendations, Schlanger maintained that because of the scale and layout of the theaters, these massive screens nevertheless still proved incapable of predominating the auditorium space in the way he believed they should. At these theaters, the screen, as he put it, appeared "as a tiny focal area made insignificant by the tremendous surface which surrounded it." He claimed that this was "even more the case in the smaller Roxy than in the immense Music Hall, because of the marked vertical expression at the proscenium, which forces the eye upward to the ceiling and away from the screen" (figure 3.17).[174] Although the acoustic engineering in the Rockefeller Center theaters epitomized modern ideals for sound and listening, as Emily Thompson argues (particularly with regard to Radio City Music Hall), their configuration of auditorium and screen space thus failed heartily at approaching similar ideals for image and viewing.[175] In these theaters, the auditorium space intruded upon the viewer's visual field rather than blending with virtual space in the way engineers thought it should.

Despite the Rockefeller Center theaters' inability to integrate screen and theater space in the seamless manner that would seem necessary for the immersive experience the engineers envisioned, their use of screens and related surfaces joined represented and actual spaces in other, no less modern ways. Indeed,

3.17 The vertically oriented proscenium at the RKO Roxy Theatre distracted from the screen by drawing the viewer's eye to the ceiling.

*Source*: From the Tom B'hend and Preston Kaufmann Collection, Margaret Herrick Library, Academy of Motion Picture Arts and Sciences, Beverly Hills, CA.

examining the use of windows at the RKO Roxy challenges assumptions about the kind of views proffered by movie theaters in this era. There were two general ways in which this theater employed windows. First, the auditorium featured "radio observation rooms" to facilitate broadcasting from the theater. These rooms were installed at the sides of the auditorium, with glass-paned portholes providing a view of the stage. Writing in the *Motion Picture Herald*, Eugene Clute explained that, "if desired, a microphone can be placed here for the use of an announcer viewing the performance through the port of one of these rooms" since the rooms were "thoroughly sound-proofed, provisions including use of triple glass with dead air spaces between the panes in the port holes."[176] Like the cry rooms at the Los Angeles Theatre, these "radio observation rooms" thus employed windows to demarcate spaces within the theater by blocking out sound while simultaneously functioning as optical apertures to the featured spectacle. The inclusion of these rooms for broadcasting from the theater also ties in with the larger intermedial project of Rockefeller Center and links the RKO Roxy to the NBC Studios space, where soundproofed, glass-partitioned observation rooms also proliferated. Indeed, NBC Studios featured collective observation galleries as well as private observation rooms for program sponsors and their guests. It also featured a glass-partitioned observation room that provided a view of the studios' main control room so that, as the *Architectural Forum* put it in 1932, "visitors may see the technical apparatus and the staff in operation."[177]

Beyond its auditorium, the RKO Roxy also featured windows of another kind: rear-illuminated windows made of translucent glass. The theater's foyer featured five 24-foot-tall frosted glass windows, creating what Clute called a "great rectangular expanse" broken up by relatively narrow sections of wall (figure 3.18).[178] Each of these windows featured two stacked sheets of glass, with strip lighting sandwiched between them to illuminate both exterior and interior spaces.[179] In addition, the theater had a glass mural in the ladies' lounge, which was illuminated in a similar way. Representing Amelia Earhart's flight across the Atlantic, the mural was 6 feet tall by 18 feet wide and backlit, in Clute's words, to suggest "a wide window" (figure 3.19).[180] The back lighting was provided in a recess behind the mural, where strip reflectors on the sides and bottom of the panel cast light on a serrated mirror on the wall to the rear. As a result, it was reflected and diffused light that passed through the glass. The mural and the light passing through it were further reflected in a large mirror covering much of the opposite wall.[181]

Despite the material differences among these surfaces—the massive though not quite immersive auditorium screen, the transparent but soundproof and bounded glass portholes of the observation rooms, the translucent glass panes in the foyer and lounge, and the mirrors reflecting light and images in the ladies' lounge—the surfaces are connected through their overlapping qualities of

**3.18** The 24-foot-tall frosted windows in the foyer of the RKO Roxy Theatre featured two stacked sheets of glass.

*Source*: From the Tom B'hend and Preston Kaufmann Collection, Margaret Herrick Library, Academy of Motion Picture Arts and Sciences, Beverly Hills, CA.

**3.19** The RKO Roxy Theatre's ladies' lounge included a backlit glass mural representing Amelia Earhart's flight across the Atlantic.

*Source*: From the Tom B'hend and Preston Kaufmann Collection, Margaret Herrick Library, Academy of Motion Picture Arts and Sciences, Beverly Hills, CA.

transparency, reflectivity, and luminescence. These qualities allowed images and light to pass through and volley from one surface to another while simultaneously rendering the surfaces opaque in other ways (e.g., insofar as the portholes transmitted light but blocked sound). Indeed, I would like to suggest that we consider the views provided to audiences at the RKO Roxy on the model of the windows installed in that and other theaters. Such windows functioned not only as apertures but also as barriers, frames, and sources of light. They were not only—or even primarily—large, open windows promising the viewer engulfment in another space but also and more pervasively small, closed ones separating spaces as much as they bridged them.

In these ways, the surfaces structuring the RKO Roxy as an exhibition site functioned similarly to the surfaces structuring the diegesis of *King Kong* (whose successful run has been credited with helping to change the fortunes of both RKO and the RKO Roxy at the depth of the Depression).[182] For viewers watching *King Kong* at this theater, transparent and translucent surfaces proliferated both on the screen (thanks to the film's use of rear projection and glass painting) and surrounding it (with the employment of windows throughout the building). If, as I argued in chapter 1, the use of screens and panes of glass in the making of *King Kong* contributed to the film's emphasis on the separation and bridging of spaces—both within the diegesis and as a mode of address—the deployment of the screen and windows in the theater extended that emphasis into and through the exhibition space. If we follow Paul Young's analysis of *King Kong*, moreover, we can view this spatial construction as a correlate to the changes being brought about by radio, which also prompted the transgression of various spatial and social boundaries (reconfiguring the relationship between the private and the public and bringing African American jazz especially into white American homes) as well as efforts to reassert those boundaries.[183] Although the form of viewing associated with the use of windows at the RKO Roxy departed from the immersive experience championed by contemporaneous approaches to screen and theater design, it thus represented another significant mode of address emerging as cinema accommodated itself to the new media landscape that it shared with radio and television.

## Synchronized Fields

In 1937, at the time Benjamin Schlanger was unveiling his screen-synchrofield apparatus, Fred Waller began working on a system for multiple-projector exhibition.[184] Waller had headed Paramount's special-effects department from 1924 to 1926, where he developed an optical printer that could combine up to seven original images into a single composite.[185] Through that work, he had become interested in what he called "wide-angle photography" and particularly the prospect that such wide angles would supply the impression of depth.[186]

According to his future collaborator, the architect Ralph Walker, Waller's background in special effects had propelled him to approach that "problem from the direction of the wide-angle lens and multiple images on an extra-wide screen."[187]

When Walker's architectural firm was invited to design an exhibit for the petroleum industry at the New York World's Fair to take place in 1939, Walker envisioned "a huge room, spherical in shape, on the walls of which would be thrown a constant stream of moving pictures from an entire battery of projectors," creating "the entire panorama of petroleum" through a "sort of contrapuntal montage."[188] He invited Waller to work out the technicalities of such projection, and Waller set to work on what was quickly named the "Vitarama." The Vitarama system entailed a filming apparatus comprising eleven 16mm cameras collected on a single mount and driven by a shared motor. To show the films, eleven synchronized projectors were employed to throw contiguous images onto a screen shaped as the section of a sphere curving 165 degrees horizontally and 75 degrees vertically.[189]

Waller's work was fueled by the idea that peripheral vision contributes significantly to people's perception of depth, providing a "sense of environment and spacial [*sic*] relationship."[190] Vitarama was thus designed to stimulate peripheral vision by arraying multiple images on the curved screen in what Waller called a "mosaic effect."[191] To "increase the illusion of being in and surrounded by an environment," initial plans for the system also included directional sound, which was to involve a separate soundtrack and speaker for each filmstrip and projector.[192] Although Vitarama was not in the end used for the petroleum exhibit, Waller designed a system to project a similar "mosaic" of still color photographs for the Eastman Kodak exhibit at the New York World's Fair (he also consulted on the film installation inside the Perisphere), and he continued to work on Vitarama outside the context of the fair.[193] By 1939, Waller and Walker had convinced Laurance Rockefeller to invest in the Vitarama Corporation, and they discussed the prospect of installing Vitarama at the RKO Roxy, which was by that time called the Center Theatre.[194]

With the mounting war, that installation did not come to pass. In 1940, Waller instead began working to redesign Vitarama as an apparatus to train machine gunners.[195] The gunnery trainer employed five synchronized 35mm cameras and five synchronized projectors as well as a spherical screen and directional sound. Not only did this system thus offer an immersive environment simulating the audiovisual experience for machine gunners, but it provided responsive dummy gun mounts and immediate auditory and tactile feedback when the guns hit their virtual targets. In order to gauge the gunners' accuracy, the projection system was further synchronized to run with separate "control films" that matched targeting information with each image frame.[196]

The Vitarama system draws together several of the threads weaving through this chapter and chapter 2. Vitarama expanded the scale and scope of the projected image in a manner reminiscent of the large-screen and widescreen experiments of 1926–1931. Since Waller was still at Paramount in 1926, as has been pointed out, he would have known about Magnascope.[197] Vitarama was also, more specifically, one of several multiple-projector systems developed for large-screen and widescreen exhibition throughout the period I have been discussing, ranging from Abel Gance's Polyvision in 1927 to the use of triple projection at the opening of the RKO Roxy in 1932 and Selznick's experiments with multiple-projector exhibition for *Gone with the Wind* in 1937–1939. Notably—also in 1937—Henri Chrétien employed two synchronized projectors outfitted with Hypergonar lenses to fill a slightly concave 60-by-10-meter screen on the facade of the Palace of Light at the International Exposition in Paris (another screen reported to be the largest in the world at the time of its installation, this time far surpassing Radio City Music Hall's 70-foot screen).[198] These experiments with multiple-projector exhibition pursued a form of spatial compositing—the creation of "mosaics" from images shot with different cameras and from different angles—analogous to the processing being achieved concurrently through special effects. Multiple-projector exhibition endowed the auditorium with the kind of synthetic spatiality that optical printers (like the one Waller devised at Paramount) and rear-projection systems (like those I discussed in chapter 1) were creating in production. Indeed, multiple-projector exhibition systems like Waller's and Chrétien's emerged on the heels of Paramount's innovation of a dual-projector, dual-screen apparatus for rear projection on the set.

Like Waller's work on Vitarama in 1937–1939, as we have seen, Schlanger's experiments in screen and theater design in the mid- to late 1930s also emphasized integrating components of central and peripheral vision, expanding screened images so that they might fill the viewer's visual field in a manner reminiscent of widescreen cinema. Both Waller and Schlanger would, in fact, serve as important figures when widescreen reemerged in the 1950s, Waller with the Cinerama system developed from Vitarama and Schlanger with the RCA Synchro Screen developed from the screen-synchrofield system. Although Schlanger, as far as I know, did not explore the use of multiple projectors, his work in the latter part of the 1930s also effectively multiplied screening surfaces within the theater. Both the screen synchrofield and the reflective wall and ceiling surfaces he designed, as I have argued, functioned to carry projected images into theatrical space. In doing so, they "blended" heterogeneous components of that space—screen, screen border, walls, and ceiling—into new dynamic composites in a manner reminiscent of both multiple-projector exhibition and special-effects practice.

If we expand our view beyond the auditorium and into the theater as a whole, the proliferating views provided by auxiliary screens, mirrors, and windows at theaters such as the Los Angeles and RKO Roxy can also be seen to participate in a similar logic of multiple, synchronous projection. With the lounge screen and cry rooms at the Los Angeles and the observation rooms at the RKO Roxy, the same film image appeared simultaneously (or nearly simultaneously) in otherwise separate spaces within the theater. This synchronous projection stitched these spaces together, uniting them through a shared encounter with a single film projection despite their sonic and/or physical differences. This synthesis allowed the film projection to become a structuring component of the theatrical interior as a whole. As with immersive multiple-projection systems such as Vitarama, the pervasion of synchronized images throughout the theater made film not simply something to look at and listen to but rather a feature of the environment.[199] In this case, however, the film was not encountered only by transfixed viewers within the auditorium; rather, it was part of a larger, dispersed space within which the viewers moved. In this regard, the proliferation of synchronized projections and views throughout the theater reflected the emergent logic of television, which was similarly enabling moving images to permeate what was conceptualized as the ether in which people lived and circulated.[200]

As Walker's description of multiple projection as a sort of "montage" suggests, these syntheses also worked in certain analogous ways to editing. Contrary to Walker's advocacy for a contrapuntal approach to this montage, however, the syntheses I have described tended to construct spatial unities evocative of the continuity editing style that was well established by this time. The viewers Selznick envisioned seeing *Gone with the Wind* with multiple projection, for instance, would encounter a diegetic world featuring spatiotemporal unities owing not only to editing between shots but also to syntheses within the shot (from the compositing of principal photography, background photography, and matte paintings). Moreover, these viewers would encounter this unified diegetic world within a theatrical environment designed to "blend" it with the actual auditorium, creating a second-order synthesis. In this way, the approaches to screen practice that I have been mapping functioned differently from the more obviously modernist experiments with multiple screens occurring concurrently in the avant-garde.

Designers associated with the Bauhaus, including Walter Gropius and László Moholy-Nagy, were incorporating multiple projections into the design of spaces for theater and art exhibition by the mid-1920s. In 1925, for instance, Moholy-Nagy called for a "simultaneous or poly-cinema," which would entail multiple films projected onto (and moving across) a large screen shaped as the segment of a sphere, as well as for a "Theatre of Totality," which would include—in the service of "spatial light displays"—a full-size rear-projection screen as well as additional reflective surfaces for onstage projections.[201] Gropius's unrealized plan for a "Total Theatre," commissioned in 1927 by Erwin Piscator, entailed twelve projectors and screens on the walls, ceiling, and stage backgrounds, which would help transform the theater into a "flexible space-machine."[202] Frederick Kiesler applied a similar concept to the design of a movie theater with the Film Guild Cinema, which opened in 1929. This theater featured a "project-o-scope" that Kiesler described as "a gallery of light stations encircling the auditorium and sending rays in all directions."[203] Not only did the accompanying "screen-o-scope" include a kind of rounded screen modifier to adjust the size and shape of the main screen, but the wall (and, though unrealized, ceiling) surfaces were designed to be used as secondary screens, contributing to Kiesler's goal of enabling the spectator "to lose himself in an imaginary, endless space."[204] Such designs thus employed multiple projection in the service of a modernist interest in spatiotemporal reorganization through multiperspectivality, transparency, and simultaneity.[205]

The Bauhaus designers, in particular, were invested in harnessing such multiple perspectives to create what Fred Turner terms a "democratic surround," an immersive but diverse environment that would encourage observers to assimilate ideas in order to better understand—and shape—themselves and society.[206] In emphasizing unity over fragmentation, Hollywood's employment of multiple projection failed to offer viewers the level of psychological independence advocated by members of the Bauhaus. However, the film industry's use of multiple screens in production and exhibition nevertheless proffered an experience of spatial synthesis similarly marked by multiperspectivality, transparency, and simultaneity. As we have seen, such an experience arose with the multiplication of screens on the set, in the auditorium, and throughout the theater; as the next chapter elaborates, it also emerged with the proliferation of these surfaces beyond the exhibition space—and institution—of the cinema.

CHAPTER 4

# EXTRATHEATRICAL SCREENS IN THE LONG 1930s

## *Film and Television at Home and in Transit*

Although multiple projection was ultimately not installed in the theaters showing *Gone with the Wind* (Victor Fleming), that film participated in another form of multiscreen exhibition when its New York premiere on December 19, 1939, was broadcast by television.[1] In this context, multiple screens were not arrayed within the local environment of a theater but rather distributed among the atomized domestic and commercial spaces—scattered across the New York area, though limited in number—that were in possession of early receivers and in range of the transmission.[2] The diffusion of television would be suspended shortly thereafter as programming and the already halting sale of sets further dwindled during World War II.[3] But over the course of the long 1930s, popular and industrial discourses on television technology, together with widely publicized practical experiments, worked to disseminate the idea—if not, for many, the firsthand experience—of a new screen's proliferation far beyond the setting and (often) institution of commercial cinema.

In that project, the television screens designed and promoted in the long 1930s conspired with the extratheatrical film screens that were mushrooming concurrently. In conjunction with the introduction of 16mm and 8mm film gauges and an array of cameras, projectors, and screens suited to them, this period also saw the exploitation of cinema's possibilities within many sites and institutions beyond movie theaters and Hollywood. What Haidee Wasson describes as cinema's "newly expanded apparatus" helped to integrate film projection into homes, schools, churches, stores, and many other places of work, commerce, and education, revealing that cinema, in Wasson's words, "could serve a wider range of purposes, affording a shifting assemblage of production, distribution, display and performance techniques that were highly adaptive to an evolving media ecology and the institutions that shaped them."[4] As this chapter discusses, the screen practices associated with that expanded cinematic apparatus in many ways dovetailed with those applied to the developing televisual apparatus, especially insofar as these practices worked to adapt cinema and television to certain shared social and material contexts.

Examining the design and employment of screens across these domains—especially in conjunction with the exploration of theatrical screen practices undertaken in the previous chapters—reveals intersections among the shifting assemblages of production and display techniques afforded by cinema and television as both media traversed theatrical and extratheatrical sites. As I hope to show, the actual and anticipated proliferation of screens within and across homes, stores, and places of transit allowed the form of spatial synthesis being pursued in the production and theatrical exhibition of Hollywood films also to encompass everyday spaces, meshing cinema's *dispositifs* with the emergent *dispositifs* of television. In achieving spatial synthesis onscreen and in the theater, as we have seen, screen practices in the long 1930s collaborated with concurrent sound practices, including the construction of layered soundtracks and the installation of sound-reproduction equipment. In extending such synthesis into domestic and commercial spaces, extratheatrical cinematic and televisual screen practices also collaborated with radio, which was portrayed popularly as permeating the "ether" even after the idea of this supposedly pervasive and connective substance had been discredited scientifically.[5]

There were, in fact, several points of overlap in the conceptualization of extratheatrical film and television that linked them to conceptions of radio, further evidencing the porousness of the borders demarcating the component parts of the ascending mass-media triumvirate of radio, television, and film.[6] For one thing, all three media (in certain strains and at particular historical junctures) were conceived via notions of amateurism, which had long been attached to practices in both art and engineering and which had taken on complex meanings by the early twentieth century. In this context, as scholars such as Patricia Zimmermann and Charles Tepperman have argued, the notion of amateurism suggested not simply a lack of professional training or compensation but also an investment in forms of human creativity and spontaneity that modern rationalized industry was perceived to threaten.[7] With amateur cinema, extratheatrical exhibition was closely linked to the production of films, thus uniting what were often treated as separate domains in industrial discourses on commercial cinema despite the studios' vertical integration. During the mechanical television boom of 1928–1931, kits were marketed for amateurs to assemble, as they had recently done with radio sets.[8] Although most amateurs in this context did not produce moving images (as with cinema) or transmissions (as with radio), the capacity to intercept moving images and render them legible on these receivers also tethered exhibition to a kind of creation.

In addition, although the form of tinkering associated with radio and early television was often conceptualized as a masculine pursuit, the adaptation of radio, television, and cinema to the parlor between the 1910s and the 1950s was, as scholars such as Lynn Spigel, Susan Douglas, William Boddy, and Haidee Wasson have shown, tightly bound to dynamic conceptions of femininity and the family, conceptions also imbricated with ideologies of class, race, and ethnicity.[9] With cinema and television, that adaptation is made particularly clear by the effort, especially in the late 1920s and early 1930s, to install both projectors and receivers within ornate wooden cabinets similar to those that had recently contributed to the construction of radios as objects of family enjoyment suitable for the living room rather than the garage.[10] Moreover, by bringing moving images—often of distant locations—into domestic spaces, both home movies and television acted as "windows on the world" in a way that simultaneously echoed older parlor amusements such as the stereoscope and contributed to the forms of spatial reconfiguration characterizing modernity into the twentieth century. In doing so, extratheatrical cinema and television not only collaborated with other modes of transportation and communication in seeming to compress or eradicate distance but also worked in connection with radio especially to reframe the relationships between the private and the public and between the local and the global.[11]

The screen practices associated with extratheatrical cinema and television in this period were quite heterogeneous and, indeed, called into question the very nature of the screen as well as its role within moving-image apparatuses. Like contemporaneous screen practices in the areas of production and theatrical exhibition, however, they worked to synthesize space by harnessing the surfaces' qualities of flexibility, multiplicity, transparency, scale, and mobility. With extratheatrical film and television, as with theatrical Hollywood cinema, the exploitation of these qualities contributed to a layering and arraying of image planes, forging separations and unions within and among diverse spaces, both virtual and actual. With television practice, the layering and arraying of surfaces took a broad scope thanks to television's capacity for "space-binding" or what Anna McCarthy terms "scale-shifting," creating spatial conjunctions and divisions across great distances and between local and global registers.[12] The capacity for such spatial amalgamations is not, however, limited to television or to the temporal simultaneity often associated with it. Indeed, both cinema and (often) television screens, as scholars such as Anne Friedberg, Margaret Morse, and Paddy Scannell have shown, conjoin the here and now not only with an elsewhere but also with an elsewhen.[13] The extratheatrical film screens discussed in this chapter thus likewise reconfigured familiar realms—whether homes, stores, or train stations—not only by creating new conjunctions and divisions within these spaces but also by conjoining them with and dividing them from the elsewheres and elsewhens pictured onscreen.

Beyond cinema and television's common capacity—if not, in television's case, the technological necessity—to join the now with an elsewhen, a focus on screen technologies reveals that both media shared another construction of time. The distinction that is usually made between the temporalities of cinema and television is rooted in an opposition between cinema's purported base in inscription and television's in transmission. Whereas (analog mechanical) cinema's reliance on the inscription of film necessitates a temporal lag between registration and exhibition, rendering it a "storage" medium, television's employment of transmission via electrical impulses facilitates the temporal (near) coincidence of sending and receiving operations, providing it with a potential for simultaneity.[14] The employment of transmission in television—as in other modes of communication such as the telegraph, telephone, and radio—has contributed to its association with the concept of liveness and has thus been taken to distinguish its temporality from that of cinema and other storage media such as photography.[15]

It is important to recognize, however, that these notions of storage, simultaneity, transmission, and liveness all address the relationship between the operations of registration (whether filming or scanning and transmitting) and the operations of exhibition (whether projecting or receiving). A focus on screen technologies, by contrast, highlights the temporal dynamics among other components of cinematic and televisual apparatuses as well.[16] In the theatrical cinema of the long 1930s, as I have argued, the screen image entered into a relationship not only with the temporally and spatially removed context of shooting but also with temporally continuous and spatially contiguous elements of the exhibition arrangement, such as the sound apparatus, the lighting system, and even other screen images. Attending to those relationships, I have contended, reveals the pervasiveness of synchronization as a temporal formation marking theatrical cinema in this period. Through the emphasis on synchronization, exhibition practices in their own right accomplished the display of instantaneity and contingency that, as Mary Ann Doane argues, cinema both achieves through its photographic indexicality and—somewhat counterintuitively—shares with television and digital media.[17]

This chapter reveals how a focus on synchronization underlay extratheatrical screen practices as well, especially in the domain of television. As we will see, practitioners employed the concept of synchronization not only to celebrate television's capacity to join moving images and sounds but also to represent the coordination of televisual technologies more broadly. The synchronization of sending and receiving apparatuses was necessary for the functioning of television, endowing it with the capacity for simultaneity or liveness. That same synchronization, moreover, also effected the synchronization of receiving apparatuses—making it possible, in other words, for multiple television sets to receive the same transmission simultaneously. Far from distinguishing television from cinema, such synchronization connected television to the contemporaneous multiscreen practices that, as I have shown, were also being embraced in cinema.

In what follows, I examine screen practices associated first with extratheatrical cinema and then with television in the long 1930s. In both cases, I focus in particular on screens developed for use in the home as well as on those designed to exploit and promote mobility, including installations on moving vehicles and in spaces of commerce and transit. My discussion of television also addresses screens devised for theatrical use. In connection with the previous chapters, this exploration illuminates the multifaceted ways in which practitioners harnessed screens' qualities of flexibility, transparency, multiplicity, scale, and mobility to adapt both cinema and television to a range of shared contexts and purposes. In the domains of extratheatrical cinema and television, I argue, the proliferation of screens—or, to be more precise, the proliferation of diverse surfaces operating as and in conjunction with screens—worked to unite a variety of otherwise disparate realms into new spatiotemporal composites in the long 1930s. The screen's multiplication within and throughout everyday space thus drew the forms of spatial synthesis and temporal synchronization also achieved onscreen and in the theater into a much wider realm.

Like the experiments with production screens and theatrical screens also taking place in the long 1930s, deployments of extratheatrical screens often favored particular—and in many cases overlapping—types of imagery. Travel and mobility represented a recurring focus in extratheatrical screen practice, uniting in a shared project the actual installation of screens and the virtual spaces gracing their surfaces. Attention to that project reveals how the new forms of synthetic mobility proffered by extratheatrical screens promoted particular social and political objectives, from the dissemination of propaganda to the advancement of consumerism and tourism. In connection with the emphasis on travel and mobility was a special focus on the elements of air and water, which served as subjects of representation, sites of exhibition, and media through or across which moving imagery itself could travel. In this focus, the use of screens in extratheatrical cinema and television also recalled the treatment of sound in radio. As Jeffrey Sconce has argued, the idea that radio was permeating the "ether" was bound up with notions of the ocean, extending from the early use of wireless to communicate with ships at sea to the later employment of oceanic metaphors to capture "the seeming omnipresence, unfathomable depths, and invisible mysteries of both radio's ether and its audience."[18] An exploration of screen practices thus shows how extratheatrical cinema as well as television took part in radio's touted capacity to span geographical distance and to saturate the "ether" with omnipresent, if fluidly manifesting, sounds and images.

## Film Screens at Home and in Transit

Portable projectors, designed for use in the home and for small groups, had been in existence since the earliest days of cinema, linking the movies with other domestic entertainment technologies such as the magic lantern, the stereoscope, and—by the first decade of the twentieth century—the phonograph.[19] Cinema's movement into extratheatrical venues accelerated after 1923, when Eastman Kodak introduced its 16mm reversal film and accompanying Cine Kodak camera and Kodascope projector.[20] Both Victor Animatograph and Bell and Howell also introduced cameras and projectors for use with the new film.[21] Like the 28mm and 22mm stock used in older home projectors such as Pathé's Pathéscope and Edison's Home Projecting Kinetoscope, the new 16mm stock was nonflammable, marking what Wasson describes as an "attempt both to assuage the fear of film's dangerous flammability and to render film safe for a range of settings."[22] Moreover, its reversal form and small gauge significantly lowered the cost of making and shipping films and increased the ease of showing them.[23] By 1930, 16mm film, as Wasson contends, "was associated with a feverish enthusiasm for the promise that it would extend the cinematic frontier everywhere."[24] Driven in part by the availability of 16mm film, cameras, and projectors, together with the 8mm gauge introduced in 1932, cinema continued to expand its reach throughout the decade as it was employed for industrial, educational, medical, religious, legal, and military purposes as well as for home movies and other forms of amateur experiment.[25]

As with concurrent approaches to theatrical film screens, practitioner discourses highlighted the effort to integrate extratheatrical film screens into exhibition spaces. Whereas screens increasingly dominated and even structured theatrical spaces in the long 1930s, they were treated as a more fungible and even peripheral component of the extratheatrical exhibition apparatus. As Wasson argues, the portable projectors and small screens often employed for extratheatrical exhibition emphasized adaptability.[26] Indeed, as discussed in this section, these screens were often designed to fit their environment rather than vice versa. Exploring how practitioners harnessed screens' qualities of flexibility, transparency, multiplicity, scale, and mobility to achieve that adaptation within a range of extratheatrical sites—from the home to the department store—illuminates the multifaceted relationship between theatrical and extratheatrical cinematic arrangements in this period. Although I address the design of screens for classrooms and places of business in what follows, I focus primarily on domestic and mobile exhibition because these areas represent the greatest points of overlap within the public and professional discourses on film and television screens that I have examined. It should be emphasized, however, that further research might usefully pursue a fuller exploration of the connections between the screen practices discussed in this book and the significant exploitation of extratheatrical screens for educational and industrial use in (and beyond) the long 1930s.[27]

## Flexibility, Transparency, and Multiplicity

Three different strategies can be discerned in the effort to assimilate the apparatus of film exhibition—including screens, projectors, and, eventually, equipment for sound reproduction—into domestic spaces. There was, first of all, an interest in domesticating these objects by rendering them articles of furniture. In this regard, attitudes toward home movie technologies emulated those already associated with other home media technologies. Like the radios that had recently taken their own place in the parlor, many film projectors were ensconced in wooden cabinets intended to act as pieces of furniture and thus to become permanent fixtures of the interior décor.[28] At the same time, however, thanks in part to those same radios, what J. B. Carrigan and Russell C. Holslag of *Movie Makers* magazine described in 1930 as the "already crowded condition of the living room in the average American home" contraindicated the "introduction of another cabinet to take up floor-space," with the result that in many—and, indeed, most—cases the projector was instead "regarded as a piece of portable equipment to be packed up and stowed away in the closet when not in use."[29] In addition, there was also a countervailing effort to incorporate the projection apparatus even more fundamentally into domestic space by installing it permanently within other components of the interior decoration and construction, from bookshelves to walls. As this section discusses, the film cabinets, portable units, and more permanent installations served, if in different ways, to integrate the new screens into their environments. In achieving that integration, these arrangements often exploited the screens' qualities of flexibility, transparency, and multiplicity. The screens' integration into domestic spaces thus involved processes of layering and arraying reminiscent of those applied to diegetic and theatrical spaces. In this case, though, the layered and arrayed surfaces were employed not simultaneously (as with multiscreen projection) but rather serially.

Some of the cabinet models were self-contained units, incorporating projector and screen within the same device. To do so, certain early models emulated phonograph cabinets, featuring lids that, when raised, revealed a screen on the underside. In 1923, for instance, the inventor C. Francis Jenkins had introduced a "Discrola" machine, which he likened to the Victrola and which used a "picture disc" to project onto a screen embedded in the lid.[30] In 1931, the similarly named Visionola combined a projector, phonograph, and radio in an elaborate console, with a screen on the underside of the lid and a mirror employed to redirect the projection beam back toward it.[31] Other self-contained units incorporated small, translucent screens for rear projection. Eastman Kodak's Library Kodascope of the mid- to late 1920s had such a screen attached to the projector with a metal arm (figure 4.1).[32] Still other cabinets housed the projector but required the use of a separate screen. The cabinet Pathé offered in 1930 for 9.5mm synchronized sound projection, for example, contained portholes through which the projector could send images to a detached screen.[33] By 1931, Pathé offered a 48-inch screen and stand as part of this outfit (figure 4.2).[34] Some such cabinets, such as the console cabinet Bell and Howell introduced in 1928—touted as a "beautiful piece of furniture comparable to the finer radio and phonograph consoles"—were semicontained, featuring compartments to store separable screens.[35]

**4.1** Eastman Kodak's Library Kodascope had a small, translucent screen as part of its construction.

*Source*: *Movie Makers*, January 1930, 47.

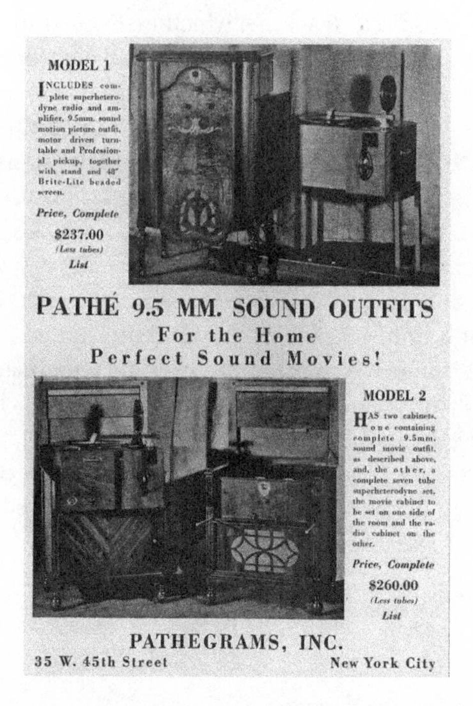


**4.2** In 1931, Pathé's projection cabinets came with a separate 48-inch screen.

*Source*: *Movie Makers*, August 1931, 447.

When screens were not incorporated into wooden consoles, they could be treated as components of the living room and its trappings. Some arrangements simply employed the living-room wall as a screen. Early advertisements for Bell and Howell's Filmo camera and projector, for instance, depict this practice.[36] Commentators and equipment manufacturers, however, argued (often self-interestedly) against it, suggesting that properly reflective and diffusive surfaces—such as the metallic and glass-beaded surfaces provided by the equipment manufacturers' screens—were necessary for bright, clear images.[37] Projection expert F. H. Richardson, by contrast, allowed that fresh, white, nongloss paint and plaster made good projection surfaces, though he insisted that they needed to be renewed every three months in order to maintain their reflectivity.[38] By April 1928, at which time Bell and Howell was selling its own screens, its depiction of living-room projection included what appears to be the New Filmo Framed Motion Picture Screen.[39] That screen featured a rigid frame and, as advertising put it, could be hung "on the wall like a picture."[40] Whereas earlier arrangements had treated the wall as a screen, the new screens were thus likened instead to tastefully framed artwork mounted on the wall.

Other screens were integrated into other components of the living-room furniture and décor. Perhaps most creatively, Kodak introduced a combination screen and bridge table, dubbed the Kodacarte, in 1928. As *Movie Makers* described it, the table top "is hinged and snaps into projection position in a moment's time."[41] Kodak suggested that this piece would enable guests to be entertained through the alternation of cards and movies.[42] In 1929, B. Saubiac & Son offered the Coutard Projection Screen Panel, which combined "Projection Screen and handsome panel for your wall—in One." Advertising explained that the screen was "lowered like a window shade," promising that "when the picture is over, you can simply *raise* your projection screen *out of sight*, and reveal in its place a beautiful Coutard wall panel—exquisitely designed and gracefully blending with the appointments of the room!"[43] With such products, the screen was thus aligned not only with the living-room wall and the framed artwork hanging on it but also with tapestries, window shades, and even the platforms upon which other leisure activities were experienced. As discussed later, screens' association with window shades in particular proved a persistent phenomenon.

Screens designed to be portable and stowable were also quite diverse. These surfaces often came with carrying cases and could be temporarily mounted on the wall, placed on floor stands, or erected on tables (figure 4.3). These portable screens were also likened to existing household objects. Equipment

**4.3** Da-Lite touted its projector stand and Challenger screen as offering flexible and easy home projection.

*Source: Movie Makers*, September 1936, 399.

manufacturers explained that wall-mounted screens were "hung up like a map."[44] And the tabletop models, which often unfolded from and remained based in the boxes in which they were stored, were dubbed "box screens."[45] Here, emphasis was placed on the capacity for adaptability and easy—even "automatic"—setup, collapse, and storage.[46] The Universal Screen Company, for instance, boasted that its aptly named Ray-Flex screen could both raise "to seven feet high" and lower "to one foot from the floor" and that the screen stood "on its own feet—independent of tables, walls, etc.," which, it suggested, meant "no re-arranging of your furniture to show movies."[47] Whereas other screens blended with home décor by becoming pieces of furniture, these flexible screens integrated with domestic space by preserving the arrangement of the existing furniture. In addition, it was emphasized that such portable arrangements could allow the screen to be installed in diverse—and unconventional—spaces "indoors or out," including "any room" as well as the porch or lawn.[48]

The metallic, beaded, and matte surfaces adopted for home projection mirrored those employed in theatrical settings, and instructions for their use emphasized many of the same considerations with regard to screen brightness, projecting distance and angles, and viewing distance and angles.[49] In contrast to theatrical exhibition, however, many extratheatrical projection systems, both self-contained and otherwise, also employed translucent screens and rear projection. The use of such screens as well as—and often interchangeably with—conventional reflective surfaces further enhanced the flexibility of extratheatrical projection arrangements. By 1927, for instance, Bell and Howell was selling Trans-Lux screens as well as reflective surfaces for use with its projector.[50] By 1928, the Truvision Screen Corporation was also offering glass screens for the rear projection of home movies, and F. H. Richardson's advice for amateur projection addressed projection from behind as well as from in front of the screen.[51] In 1930, Willoughbys in Manhattan offered amateurs a "Marshaloptic Screen," made of curved ground glass, for rear projection.[52] About one-third of the 16mm projectors employed at the 1933 World's Fair in Chicago (thirty-one out of ninety-one total) reportedly employed rear projection.[53] In that year as well, Bell and Howell introduced a miniature translucent glass screen for the use of amateur movie makers during editing—thus anticipating the Preview Moviola, which was introduced for professional use in 1937 and also had the film image rear projected onto a ground-glass screen.[54]

The use of translucent screens and rear projection facilitated showing films in brightly illuminated environments (which is one of the reasons they were useful for editing). Proclaiming its Trans-Lux screen a "new, up-to-the-minute way to show *better* home movies" in 1927, Bell and Howell explained, "No darkening of the room is required to use this screen. Excellent results are secured in daylight or the regular evening lighting employed in your home."[55] As discussed in chapter 3, the 1930s saw a debate about ambient illumination within discourses on theatrical exhibition, as many engineers, citing vision research, advocated for slightly illuminated auditoriums such as those found in the Trans-Lux theaters. Around the same time, general practice for 16mm and 8mm projection was to keep the room dark "except for stray light from the projector and light reflected from the screen."[56] There was, however, also an acknowledgment that some of the spaces in which such projection occurred, such as classrooms, did not accommodate complete darkness. Indeed, the use of Trans-Lux surfaces in particular aligned home projection with screens' entry into other spaces beyond the theater—such as the New York Stock Exchange, where, as discussed earlier, Trans-Lux screens had been used to display stock information since 1923.

Reporting on the use of 16mm equipment at the Chicago World's Fair also acknowledged the challenge of projection "where screen brilliance has to compete with a relatively high level of general illumination"—a challenge that the use of rear projection likely helped to meet.[57] One of the installations at the fair evoked the concurrent use of rear projection on Hollywood soundstages. As the SMPE Non-Theatrical Equipment Committee described this installation, it

> involved the use of an automobile operating on rollers. In front of the automobile was a screen on which were projected pictures of other automobiles crossing and recrossing a busy intersection. The driver in the test automobile operated his car, changed gears, put on the brakes, as dictated by the driving conditions shown on the screen. The driver's trials came up thick and fast, requiring quick and correct thinking and acting. After the test the driver was given a certificate rating his ability. As may be appreciated, the test was watched with great interest by large audiences.[58]

A similar installation at the 1939 New York World's Fair had children ride treadmill-mounted ponies while watching rear-projected scenery.[59] As with similar special-effects shots, these installations did not obscure—or, we can assume, fully darken—the actual space in front of the screen but instead worked to integrate it with the virtual space onscreen, creating a spatial and (especially with the emphasis on the drivers' quick responsiveness) temporal synthesis.

Beyond facilitating projection in illuminated as well as dark environments, the availability of both rear and front projection also enabled the projection apparatus to accommodate diverse exhibition arrangements and audience sizes. Eastman Kodak, for instance, explained that for larger audiences the small, translucent screen attached to its Library Kodascope could be "swung aside" to facilitate projection onto a second, larger screen also supplied with the system.[60] Similarly, many portable projectors employed for sales and advertising incorporated small, translucent screens for rear projection while often also allowing for front projection onto larger reflective surfaces.[61] Encompassing several of the qualities mentioned earlier, a home cabinet model presented by the United Research Corporation in 1932 included a pocket in back to store a translucent screen but also allowed for the use of reflective surfaces. It was emphasized that a variation on that model intended for classrooms could be positioned either at the front of the room, projecting upon a translucent screen, or at the rear, projecting upon a reflective screen or wall.[62]

By the early 1930s, there was an interest in exploring how domestic space could not only incorporate the film-projection apparatus but in doing so also serve as a variant of the movie theater.[63] Advice regarding the design of domestic theaters also straddled valuing permanence and flexibility. Commentary in the pages of *Movie Makers* tended to favor carving out a dedicated space in the home, especially in the basement or attic, for the "little theatre" (figure 4.4).[64] In such cases, as in professional movie theaters, it was advised that the size, surface, and positioning of the screen be selected based on the particular conditions of the space. Such discourses evoked the "little cinema" movement of the mid- to late 1920s, wherein small venues had been devoted to showing films produced outside contemporary Hollywood, including amateur movies as well as revivals and especially international cinema.[65] Indeed, as Charles Tepperman has noted, amateur filmmakers often modeled their own movies on the modernist aesthetic of the European films with which they shared billing in these theaters.[66]

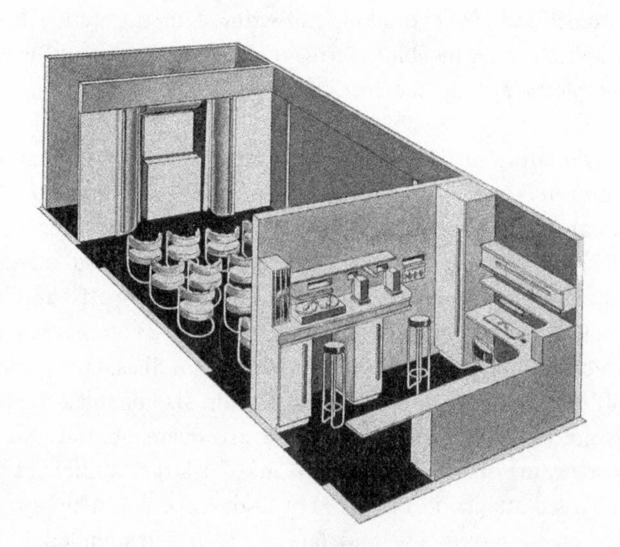

4.4 *Movie Makers* offered ideas for converting domestic spaces into dedicated home theaters.

*Source*: Ralph G. Wildes, "Pointing Up Projection," *Movie Makers*, February 1935, 67.

Despite the efforts to install such "little cinemas" in domestic space, however, the living room often remained the assumed site of home projection, in which case commentators emphasized adaptability. Though some continued to recommend portable equipment, others in the late 1930s persisted in emphasizing how components of the exhibition apparatus could be integrated into the accoutrements of the living room. In an article for *Movie Makers* in March 1937, for instance, James H. Blauvelt diagrammed how living-room furniture could, with shifts in its arrangement, do double duty as theater seating, and he outlined several ways in which the screen could be integrated into the décor. He suggested, for example, mounting a roller screen on the inside of one of the upper shelves on a large bookcase, with the roller hidden by "a false face of book bindings" or a "piece of carved wood" when not in use. Alternately, the screen could be mounted on a window in front of the shade, with the roller hidden by a valence box.[67] In another article later that year, Walter Downs and Aubrey Burnett described how they had adapted their living room for projection, including cutting portholes for the projector beams in the wallboard, to be covered by pictures at other times.[68] In 1938–1940, the idea of integrating the screen into window dressings continued to offer itself as a component of cutting-edge design, and practitioners continued to advocate hiding the screen behind tapestries (figure 4.5).[69]

Such efforts to hide the projection apparatus within and behind furniture and the quotidian trappings of interior decoration marked a shift from the earlier interest in rendering this apparatus *as* furniture. In doing so, they aligned themselves with what Lynn Spigel identifies as midcentury modern design's rejection of ornate media furniture such as radio cabinets in favor of the effort to conceal media technology—hiding, with it, the "entire social and political apparatus that supports the technology," including the systems enforcing gendered, racialized, and class-based forms of labor and leisure within the home.[70] In this regard, the drive to conceal the technology of domestic film projection, in bypassing the elaborate decoration that had marked the earlier media furniture as amenable to women, simultaneously furthered what Haidee Wasson identifies as a tendency in the presentation of home projection in the 1920s—aligned with the modernization of the home—to obscure "the fact that everyone else's leisure likely came at the expense of even more work for Mother."[71]

As with theatrical exhibition, there was an interest in treating the home projection apparatus as a coordinated system. This was especially the case with the move to home sound film, which threatened disjuncture among the proliferating components of the apparatus. In 1930—by which time, as we have seen, there was already consensus vis-à-vis theatrical exhibition that speakers should be positioned behind a permeable screen—Carrigan and Holslag complained that the "home sound projectionist usually makes no effort to place his loud

**4.5** In promoting its Electrol Screen, Da-Lite touted an installation that integrated the screen into the window dressings.

*Source*: *Movie Makers*, February 1940, 56.

speaker in such a position relative to the screen that the illusion of sound actually emanating from the picture is produced."[72] In this case, amateurs did not simply fail to place the speaker behind the screen; rather, they neglected to situate the speaker anywhere near the screen. As Carrigan and Holslag put it, this negligent projectionist "is usually content to leave his loud speaker in a fixed position with relation to the radio set—many times the loud speaker [*sic*] is incorporated in the set—and to erect his screen on a wall or table. The sound volume simply fills the room, with no directional effect whatever."[73] One of Pathé's 9.5mm sound-movie outfits of 1931, for instance, contained two cabinets, one housing the projector and one the sound system, with "the movie cabinet to be set on one side of the room and the radio cabinet on the other."[74] Because the screen accompanying this outfit was to be mounted on a stand, it presumably could be placed near the radio cabinet, but there was no suggestion that it should be, let alone assurance that it would. In this context, Carrigan and Holslag presented the prospect of entirely self-contained and "technically coordinated" home projection equipment—incorporating turntable, projector, pickup, amplifier, and loudspeaker—as the solution to the problem of disjuncture. And they considered the possibility of incorporating the screen into the device so desirable as to justify smaller picture sizes.[75]

Apparently addressing such concerns, later systems accommodated the placement of speakers near the screen—or even behind it, as in theatrical arrangements, although perforated screens were generally not used.[76] The sound-film projectors introduced in 1932 by United Research Corporation, for instance, incorporated the projector and sound equipment, including speakers, into a single cabinet. Because the screens were separated from the cabinet, they threatened disjuncture; however, because the systems accommodated the use of translucent as well as reflective screens, they also permitted the situation of sound projection, together with image projection, behind the screen.[77]

Whereas advice about the configuration of dedicated home theaters tended to echo concurrent discourses on theatrical exhibition—which emphasized functionality and simple design, conceiving the various elements of the projection apparatus as components to be scientifically calibrated and coordinated in a machinelike efficiency—the foregoing discussion shows how the screen occupied a more flexible place within that apparatus in the film cabinets of 1923–1932 and the portable units popular throughout the 1930s. As we have seen, projection outfits often anticipated use with multiple screens of differing size and material construction, accommodating diverse arrangements, including both front and rear projection. Although this equipment did not facilitate simultaneous multiple projection, it did encourage a proliferation of screens to be used interchangeably and serially. As we have also seen, multiple screens were often provided with a single projection unit. At the same time that the theatrical use of large-screen and widescreen systems, together with screen modifiers, was rendering the screen frame flexible in the context of theatrical exhibition, this proliferation of diverse screens in the home endowed domestic moving images with a similar protean quality.

## Scale

In the context of theatrical exhibition, as discussed previously, the period 1926–1931 saw a range of experiments with large screens, often with the goal of enabling the cinematic spectacle to fill the viewer's field of vision and proffer the experience of engulfment. In 1931–1940, with the abandonment of widescreen exhibition and the effort to achieve such visual fullness and immersion through alternate means, practitioners tended to advocate for smaller theatrical screens. In the context of domestic exhibition, the shifting value placed on large screens followed a reverse historical trajectory. From the introduction of 16mm film in 1923 until around 1932, commentators insisted that the quality of images be favored over their size, advocating for what Hiram Percy

Maxim (a.k.a. Dr. Kinema) of *Amateur Movie Makers* described in 1927 as "good, but small pictures, rather than poor but large ones."[78] By 1932, however, with the introduction of 16mm projectors possessing greater illuminating power, some commentators and equipment manufacturers began promoting the capacity for larger screens, for use both at home and in auditoriums. That shift, especially in light of the converse one taking place nearly simultaneously in theatrical exhibition, propelled a significant transformation in home screens' relationship to theatrical screens. Exploring that transformation illuminates the complex nature and heterogeneous functions of screen scale, showing how even screens that appeared large in relation to domestic space nevertheless functioned differently within that space than did large screens in the theater.

In the period 1923–1932, the sizes available and recommended for extratheatrical screens remained stable. When Eastman Kodak introduced its 16mm film in 1923, it offered screens up to 4½ feet (54 inches) wide for home use and suggested that with a special lamp house the Kodascope projector could fill screens up to the 7-foot width it identified as sufficient for classroom use.[79] In the ensuing years, extratheatrical screen makers, including Kodak, Bell and Howell, Arrow, De Vry, Truvision, and Da-Lite, offered certain standard sizes: 22-by-30 inches, 30-by-40 inches, and either 36-by-48 or 39-by-52 inches, sometimes together with less-standard (usually smaller) sizes.[80] In 1932, it was proposed that the SMPE Non-Theatrical Equipment Committee recommend a maximum picture size for 16mm in order to "discourage the use of excessive magnification and inadequate illumination," the same qualities that led to recommended maximums for theatrical screen size around that time. In this case, however, the committee held that such a recommendation was not necessary since a "high degree of standardization" already existed: "For home use the 30 by 40 and the 36 by 48 inch screens are the most popular, and practically all manufacturers make screens of these sizes. For auditoriums the 6 by 6 or the 6 by 8 foot screens are most generally used."[81] In other words, screen size for home use had not increased, and the screens employed in auditoriums had expanded only marginally since Kodak had made its recommendations nine years earlier.

With the use of relatively low-wattage projectors and small screens, advertising in this period touted the detail, clarity, and brilliance with which particular projectors and screens would endow the moving image. Beauty, in particular, was treated as a desirable quality for both projected images and the screen itself. For instance, Eastman Kodak claimed that its Kodascope screens were "specially treated to bring out the maximum beauty and detail of your pictures."[82] And the Arrow Screen Company advertised its product as "the Screen of Beauty," contending that movie stars and directors favored the way it made their films look.[83] Such comments show how the work of screens could be tied to discourses on Hollywood glamour as well as to what Patricia Zimmermann identifies as a "trend toward pictorialism" in amateur filmmaking, which resurrected "residual painterly standards of ideal beauty in nature and pleasing composition."[84] The rhetoric of beauty also served to domesticate the screen as an object in the home by aligning it with attractive domestic furnishings in the late 1920s, thus dovetailing with the descriptions of media cabinets as "beautiful" pieces of furniture. For example, Bell and Howell described its New Filmo Framed Motion Picture Screen as a "beautiful, everlasting screen for the home."[85]

The short-lived Cine-Panor system of 1931 exemplifies how the interest in widescreen cinema—which revolved around the use of very large screens in theatrical exhibition—was transformed in the context of domestic projection and much smaller screens. Announced at the outset of the year, Cine-Panor was a widescreen system for extratheatrical exhibition that arrived at the tail end of the theatrical widescreen cinema boom. Between 1929 and mid-1931, as discussed in chapter 2, discourses on widescreen cinema in Hollywood had identified two primary benefits to the new format. First, like the slightly earlier Magnascope system, it employed much larger screens (up to 40–46 feet wide) than had been standard, promising to better match the scale of large picture palaces and harnessing screen size to distinguish cinema from emergent television technology. Second, unlike Magnascope, most of the new systems also employed wide-gauge film (50mm to 70mm), promising to provide images of higher resolution and broader scope on those larger screens. The Cine-Panor system also achieved wide aspect ratios, but it neither provided larger screens nor offered wider-gauge film. Rather, Cine-Panor, intended for use with 16mm film, consisted of an anamorphic lens that when mounted on the camera compressed a wide image onto the filmstrip and then when mounted on the projector uncompressed it as a wide image onscreen (H. Sidney Newcomer, a former student of French anamorphic-lens pioneer Henri Chrétien, was involved in its development).[86]

Russell Holslag of *Movie Makers* explained that the "result on the screen is a picture fifty percent wider than normal, but of normal height."[87] Discussing Cine-Panor in *American Cinematographer*, Fred Schmid claimed that the "result is a 'Grandeur' effect produced with standard 16mm. equipment and film material," though he also advised that with projection a "larger screen, twice as long as it is high, will have to be provided."[88] Unlike the professional systems, however, Cine-Panor does not appear to have incorporated such a screen. Many advertisements for the system emphasized the wider field of view provided by the new lens without mentioning screen size at all.[89] Those ads that did promise "increased screen width" still did not indicate how the screen would or could be expanded (figure 4.6).[90] When the system was demonstrated for the SMPE in the fall of 1931, the displayed image was approximately 20-by-40 inches in size.[91] Since 30-by-40 inches was a standard size for extratheatrical screens at

**4.6** Advertisements for the Cine-Panor system emphasized the wider field of view provided by the new lens.

*Source*: *Movie Makers*, August 1931, 445.

that time, it appears that in practice the system, rather than achieving a wider image on new screens, simply produced a shorter image on existing screens (thus anticipating letterboxed home videos later in the century).[92]

The reported "normal screen size" for extratheatrical projection remained fairly stable throughout the 1930s, and a 52-inch width was still considered typical in 1941.[93] Over the course of the 1930s, however, some screens for home and auditorium use grew significantly larger. In 1932, the Victor Animatograph Corporation promised that its new projector, with an optical system "more than doubling our former screen illumination," could accommodate projection distances from 5 to 150 feet and screen sizes from 1 foot to 12 feet.[94] And Da-Lite introduced a portable screen, "designed to take advantage of the big picture possibilities of the new super-illumination projectors," available in both 5-foot and 6-foot widths.[95] Whereas early 16mm projectors had employed 100- or 200-watt lamps, the 750-watt bulbs in use by early 1934, as Holslag contended, "indeed are adapted to auditorium projection."[96] In the ensuing months, advertisements in *Movie Makers* touted portable screens up to 8 feet wide and auditorium models up to 12 feet wide.[97] By the fall of 1934, a new 1,000-watt 16mm projector had reportedly achieved a 20-foot-wide picture "of theatrical brilliance."[98] By 1940, Da-Lite offered its Electrol Screen in standard sizes up to

20 feet wide, claiming that this screen was "ideal for use in auditoriums as well as in homes."[99] Although the largest sizes were likely intended for the use of bigger groups, some home screens reached at least 12 feet wide by 1940.[100]

Coincidentally or not, this shift took place around the same time as the move away from large screens in theatrical exhibition.[101] As a result, whereas there had been a significant difference between large (20- to 46-foot-wide) theatrical screens and small (2½- to 4-foot-wide) home screens in 1926–1931, the gap between the two closed significantly in 1932–1940 as theatrical screens were limited to around 24 feet wide and some (though not most) extratheatrical screens reached quite close to that size.[102] Notably, in the mid-1930s, when extratheatrical projection began filling 20-foot-wide screens, 20 feet was, as discussed in the previous chapters, also a common width for both production screens and theatrical screens. Recognizing that confluence recommends against making easy distinctions between theatrical and extratheatrical screens in terms of size. In addition, it points to another transformation in the relationship between theatrical and extratheatrical screens, in this case pertaining to the relative quality of scale.

As we have seen, the cavernous auditoriums of the picture palaces had provided one of the incentives for increasing the size of theatrical screens toward the end of the 1920s. In other words, large-screen and widescreen systems such as Magnascope and Grandeur increased the scale of the screen in relation to the space in which it was installed and especially in relation to the viewing distances that this space imposed. In the wake of the abandonment of widescreen cinema around 1931, there was a call for smaller theatrical auditoriums that would better match the scale of smaller screens.[103] Concurrent discussions of extratheatrical screens echoed the interest in correlating screen size and viewing distance, sometimes employing the ratios of viewing distance to screen size recommended for theatrical contexts to justify the use of even smaller screens in the home. Noting in 1932 that with 16mm film the image could reach up to 5 feet wide, for instance, Russell May of the RCA Victor Company explained, "When sitting in a small room, the angle at which the eye subtends the screen is not greatly different from the similar angle in the theater. In other words, sitting in a living room and watching a picture as large as can be projected from a 16 mm. film, one obtains the same illusion of size as he does in the theater."[104] In 1934, Russell Holslag similarly observed that

> whenever you visit one of the great palaces of the professional cinema, or even a neighborhood movie house, if you sit in the orchestra a reasonable distance back, or in the gallery, the picture seen on the screen actually forms an image on the retina of your eye *no larger* than that given by a twenty by thirty inch movie screen at home. The reason is that, at home, the observer is seldom more than ten or fifteen feet from the screen; whereas, in the movie theatre, he is

> fifty, sixty or a hundred feet away—perhaps more. Hence the smaller screen actually may loom larger because it is viewed at a closer distance.[105]

Both May's phrase "illusion of size" and Holslag's invocation of retinal images emphasize not the screen's linear measurement but rather the viewer's perception of its scale based on the relationship between screen width and viewing distance. In this context, it becomes clear that a 12-foot home screen could actually "loom" much larger than a 20- to 24-foot theatrical screen.

Viewing distance, however, was not the only factor taken to affect the perception of screen scale. Holslag also contended, "When we take our pictures, we deem our subjects important and it may be that our subconscious wish is that, upon reproduction, the picture should not be displayed in smaller proportions, but should appear in a breadth and size that remind us of the scale of the original subject." This appearance of breadth and size, he suggested, was not achieved through close viewing distances but rather emerged through comparison with actual living figures and with the screen's surroundings. As Holslag put it, "In the movie theatre, the picture on the screen assumes an important size solely because of the relation of its dimensions to things around it. Therefore, even though the image of the theatre screen picture within the retina of our eye is actually small, we are impressed."[106] The suggestion was that it would therefore be beneficial to expand the size of the home screen relative to its surroundings despite the viewer's proximity to it—and thus despite its actually "looming" larger than a theatrical screen.

Holslag's observations highlight the home screen's relationship to a range of other objects, from the domestic trappings alongside which it was situated to the other screens proliferating concurrently. As we have seen, the diverse factors affecting the perception of screen scale put home screens into a complex relationship with theatrical screens, allowing the domestic surfaces to seem smaller, equivalent, or even larger depending on the point of comparison. That relationship was further complicated in discourses on television, which used theatrical viewing distance-to-screen size calculations to justify even smaller picture sizes and went so far as to critique home movie screens as being too large for domestic space.[107] In addition, although home screens could duplicate the look of theatrical proscenia, their actual size and installation often aligned them more closely with the other domestic objects upon which their functionality was modeled and with which they were sometimes physically integrated, including walls, pictures, tapestries, maps, boxes, and window shades.[108] Comparison with such objects frames home screens as domestic articles in their own right, like the family albums with which they were compared or the hunting trophies, rifles, and souvenirs of travel by which they were sometimes surrounded.[109] Such a framing not only aligns these surfaces with particular configurations of class, gender, and nation but also portrays them as objects to be admired and manipulated rather than as spaces to be entered, despite their capacity to "loom" large within the home.

## Mobility

Despite the effort to integrate some home screens into permanent installations similar to those found in movie theaters, it should be clear that many of these surfaces were characterized by their mobility. Advertising and practitioner discourses emphasized the way in which such screens could be erected in various configurations and spaces and how they were not only adaptable to different parts of one's own home but also capable of being transported to auditoriums, clubs, churches, and friends' homes. In this regard, domestic screens participated in the wider association of extratheatrical film screens and mobility at midcentury.[110] Examining installations in moving vehicles and spaces of transit such as stores, hotels, and train stations illuminates multifaceted relationships between the movement onscreen, the movement of the screen, and the movement of viewers as they encountered the screen. Attending to these relationships not only highlights the phenomenological heterogeneity of cinema spectatorship in the long 1930s but also reveals its ties to particular social practices, including propaganda, consumerism, and tourism.

Film screens had been installed on ocean liners and trains since the 1910s, and there had been experiments with projecting films in-flight on airplanes since the early 1920s.[111] Such installations recalled even earlier simulations such as Cinéorama and Hale's Tours, which made it seem as though viewers were riding, respectively, an ascending balloon and a moving train.[112] In October 1929, *Variety* announced that "50 big planes of the Transcontinental Air Transport Corp." were to be equipped for 16mm projection, creating an "air theatre chain" or "ether circuit" featuring films from Universal.[113] By the end of the year, the Beaded Screen Corporation was boasting that its Brite-Lite Glass Beaded Movie Screen had been used by Transcontinental Air Transport (later to be renamed TWA) "on the first transcontinental air ship to show 16mm. motion pictures while flying in daylight." The films were projected on a flight from Port Columbus, Ohio, to Los Angeles and back, and as the advertisement enthused, "Three shows a day were given during the flight, at heights varying from 10,000 to 13,000 feet, with the plane traveling at an average speed of 125 miles an hour."[114]

Trains were offering synchronized-sound projection by 1930.[115] Observing in January 1932 that it was also "not unusual in this modern age for the projection room to go to sea," *Motion Picture Projectionist* reported that "a great steamship company has recognized to such a degree the increasing importance of motion picture entertainment as a factor in ocean travel that it has included the motion picture theatre as an integral part of modern ship design." As the article explained, the Furness liner *Monarch of Bermuda*, having had "talking pictures" incorporated in its original design "for the first time in shipbuilding history," offered "four shows on each round trip, one every night that the ship is on the water."[116] The theater was located in a "luxurious lounge room" on the Sun Deck and seated six hundred people. In the daytime, draperies concealed the screen, the speakers were hidden, and a fan-shaped mirror covered the projection ports; for the nighttime shows, the curtains parted to reveal the screen, and the mirror telescoped to uncover the ports.[117]

In addition to better-known examples of mobile cinema in the 1920s and 1930s—such as those in the Soviet Union and across the British Empire—there were also experiments with truck-mounted mobile projection within the United States, likewise in connection with newsreels, education, and propaganda.[118] By 1927, the year Movietone debuted, the Fox-Case Corporation had also developed a portable Movietone projection apparatus installed in a truck and removable for setup in extratheatrical venues; by 1928, Fox-Case had developed a new outfit now arranged for projection "directly from the truck."[119] The screen and loudspeaker were raised through the roof and erected above and toward the rear of the truck, with the projector mounted on its floor, producing a 6-by-7-foot picture (figure 4.7). An alternate arrangement that had been envisaged (but not necessarily actualized) was a rear-projection setup inside the truck, employing a mirror to extend the projection throw and resulting in a 3-foot-high rear-projected image at the back end of the truck.[120] As of 1928, engineer Earl Sponable reported that the outdoor projection trucks had "already found use in the political, commercial and advertising fields."[121] By 1930, Movietone employed a fleet of four such trucks converted from REO Speed Wagon models and employing Simplex projectors.[122]

In December 1930, *Motion Picture Projectionist* announced that the Democratic State Committee of New York had also been utilizing a fleet of trucks "equipped to show sound motion pictures."[123] It is likely that these trucks were borrowed from Movietone since press coverage mentions a "movietone [*sic*] machine on a truck," and Local Union 306, which had purportedly equipped the trucks for the Democratic State Committee, had checked out three of Movietone's trucks that September.[124] The twice-daily screenings featured speeches by the Democratic candidates. Evening shows occurred outdoors, reportedly employing a 6-by-6-foot screen. For afternoon shows, the portable equipment was removed from the truck and assembled in other spaces such as halls, schools, churches, and stores, where the projected images reached 10-by-10 feet. It was estimated that the entire fleet was responsible for more than 350 separate screenings. The president of Local Union 306 contended that the units had "aided in introducing a 'new idea' in the dissemination of propaganda," predicting that in the future such "sound trucks" would "be in demand for many

4.7 The Fox-Case Corporation outfitted trucks for Movietone projection, with the screen and loudspeaker raised through the roof.

*Source*: E. I. Sponable, "Movietone Field Projection Outfit," *Transactions of the Society of Motion Picture Engineers* 12, no. 36 (September 1928): 1178.

purposes other than political work" and explaining that "already there are apparent certain signs which indicate that such equipments [*sic*] will find application to many fields of endeavor, particularly in sales promotional work."[125]

In 1932, it was announced that MGM was introducing a "Trackless Train" equipped for sound and image projection. Deemed a "miniature studio on wheels," the vehicle was purportedly designed "to show the public how talkers are made" and thus included a miniature studio and lab together with the projection facilities; *Variety* identified "exploitation [as] the main angle but with Hearst-Metrotone News tied in and a H-M camera aboard."[126] In this case, a glass screen constituted the rear wall of the trailer car, and rear projection from inside the car made the moving images visible to outside audiences. As *Motion Picture Projectionist* explained, "En route the screen area at the rear of the car is covered by removable panels. During a show these panels project from both sides of the screening window with a canopy across the top, forming an effective shadow box to minimize the reflection of lights from the street."[127] During shows, two speakers were lifted through hatchways in the roof and held in place above the screen by the hatchway lids. In addition, four public-address horns were mounted at the center of the car's roof for use "on occasions where battery current for the screen horns must be conserved or when sound projection must be obtained in all directions."[128]

The same portable screens marketed to amateurs for use in the home were also marketed to businesses for purposes such as employee training, sales, and advertising.[129] Some portable outfits proffered for sales and advertising functioned as self-contained units by employing translucent screens for the rear projection of either still or moving images.[130] For instance, Illustravox, which made portable still-projection equipment, described its "Salesmaker" model thus: "A complete portable theatre in a single case. Put it on desk or table, unfold, plug-in and your sales message goes into action. Projector shows pictures on translucent screen in front of case."[131] Automotion Pictures, Inc., made portable projectors with small, attached translucent screens; in some models, these screens could also be detached for projection onto another surface (figure 4.8).[132]

Both portable and more-permanent screen installations were also being utilized in department stores and expositions. In such cases, the screens

**4.8** The Automotion Pictures, Inc., double-purpose projector allowed for both straight projection onto a large screen and rear projection onto a small, attached screen.

*Source*: *Movie Makers*, October 1935, 416.

themselves could be moveable, and they were also employed to guide the movement of viewers.[133] Not only did department stores offer forms of visual display that aligned in certain ways with cinema, but they served as long-standing sites for the presentation of moving images and the demonstration of moving-image technologies.[134] In 1938, in a section on the use of motion pictures in the department store, the inaugural issue of *Business Screen* suggested that movies be used in this context not simply as attractions in their own right but also as a way to train employees and appeal to customers. In addition to constructing dedicated theaters for this purpose, department stores, it was advised, could integrate portable projection units into the building's walls, install them on department counters, and place them in the store windows. It was emphasized, moreover, that the placement of such displays was important since they should work not only to attract crowds but also to "draw customers through departments featured on the program."[135] A survey of department stores in Chicago indicated that many were employing films to sell goods. By 1938, for example, Marshall Field & Company had already been using film for five years to sell luggage, toys, drapery, and women's clothing.[136]

In a section on the use of movies at recent expositions, the subsequent issue of *Business Screen* similarly outlined the range of installations achieved and modes of crowd management associated with them. As it recalled, at the Century of Progress exposition of 1933 in Chicago, "some projectors were housed in special cabinets, some were built into regular exhibit fixtures, some worked right out in the open with the picture projected across aisles or booth space, while others were concealed behind walls, beneath floors or above ceilings! Projectors may be in front of the screen or behind it, either in a straight line, or, where space limitation forbids this, with the use of mirrors to make possible what otherwise could not be done." While some of these projections worked to arrest attention and "bring people into an exhibit," others offered visitors retreat, comfort, and education.[137] Similarly, in a subsequent report on the New York and San Francisco World's Fairs, the journal explained that the task of moving pictures in such contexts was "that of attracting the largest possible percentage of visitors attending the Fairs and then holding their attentive interest for the longest possible time."[138] As Haidee Wasson argues, such use of moving images at the New York fair enabled a "site-specific targeting of messages, capable of articulating to both attentive and also distracted consumers, wherever a projection device might be set up."[139]

In addition to screens' mobility and their employment to promote the mobility of beholders, these objects were also installed in places of transit in the 1930s. Such installations recalled the positioning of coin-operated Kinetoscopes decades earlier in department stores, hotel lobbies, and barrooms.[140] And they anticipated the placement of Panoram film jukeboxes in department stores, restaurants, and train, bus, and ferry stations in the 1940s.[141] They also intersected with the projected and actual installation of television in similar sites, as I discuss later in the chapter. In early 1932, *Movie Makers* reported that "progressive hotels" were incorporating 16mm projection facilities, which could be used for entertainment, instruction (e.g., in activities such as golf), and the screening of the amateur movies guests were making while on vacation.[142] In 1933, the first drive-in theater opened in Camden, New Jersey. The *Architectural Record* described it as an "outdoor theater to which motorists may drive and sit in their parked cars"; it also offered "an area close to the screen, with space for 600 steamer chairs, for motorists who park their cars at rear of screen."[143] By 1939, Margaret Thorp reported that there were "some dozen" drive-ins "scattered around the country with more to come."[144]

Also in 1933, the small, 181-seat Terminal Theatre opened as part of Cincinnati's new Union Terminal, featuring an 8-by-10-foot screen and catering to passengers waiting for their trains. As the *Motion Picture Herald* reported, "The idea may be said to be Europe's rather than America's. In Paris similar provisions have been made for the travelling public. There, however, the programs are devoted principally to newsreels. That was the original intention in Cincinnati, but the policy has since been modified to include second-run features."[145] Like the Trans-Lux chain of newsreel theaters, the Terminal Theatre featured turnstiles at its entrance. In 1937, a similar theater opened in New York's Grand Central Station. Seating 240 and catering to what the *Architectural Record* described as "between-train idlers," the Grand Central Theatre was devoted to newsreels and, also like the Trans-Lux theaters, featured wide spacing between its rows of seats to accommodate the flow of patrons in and out.[146] By 1938, *Business Screen* observed that railroad terminals "in all principal cities afford the railroad companies a real opportunity to present their own travel films to the audiences created among waiting passengers," and it offered the design for a small terminal theater based on the newsreel theater at Grand Central Station.[147]

In her work on cinema and mobility, Anne Friedberg has argued that cinema took over forms of mobility associated with urban modernity, such as strolling through the city street or shopping at department stores, but also that, in doing so, the new medium rendered mobility virtual and the body static.[148] Recognizing the variety of spaces and contexts in which mobile images graced screens, however, reveals a broad gray area in which actual and virtual mobilities encountered and inflected one another. The flâneur and flâneuse were not necessarily rendered immobile when they encountered mobile cinemas on the street or portable projectors on the department-store counter; rather, the moving imagery onscreen could reduplicate and even direct their own locomotion.

The films that appeared on many of these screens, moreover, also featured various forms of mobility and travel. The screens employed in department stores often advertised either travel or goods associated with it. The Luggage Section at Marshall Field's, for instance, offered films "on travel subjects," furnished by travel agencies; the Fair Store featured film in its Travel Bureau, offering "travel subjects" every Friday morning; and Mandel Brothers also used films in its Travel Bureau, where it employed films "sent by railroads, steamship, and airlines."[149] Such films were likely similar to those that *Business Screen* recommended the railroad companies show in railroad terminals. Although many such "travel subjects" likely focused on the sights one might see while traveling, some focused on the experience of travel itself. Airlines, in particular, employed film to familiarize potential customers with the prospect of flying. As *Business Screen* explained, it is "not an easy job to picture the comfort and delights of air travel to the ground-minded person"; however, "motion pictures can, almost literally, sweep Mr. and Mrs. Prospect off their feet . . . up into 'sky oceans.' And Mr. and Mrs. First-Timer will enjoy the 'trip.' "[150] United Airlines' thirty-one-minute film *Romance of the Mainliner* (ca. 1938) was made for screening in the company's twenty-five branch offices, but it reportedly proved so popular that the airline had additional prints struck, and it was shown by entities ranging from the Chicago schools to the Oil Company of California. As of late 1938, it was reported that at "least one other major airline" also had films in preparation.[151]

The focus on travel aligns the films gracing commercial screens with those projected on home screens. Indeed, much amateur cinema was also devoted to travel. As Charles Tepperman has elucidated, practitioner discourses presented the capacity for travel—and, with it, location shooting—as a significant benefit of amateur filmmaking. Discussing amateur films' reception in *American Cinematographer*, Tepperman explains that amateur "travel films were praised highly for the attractive 'locations' they provided; with these films measured according to their technical achievement, the amateur was at a disadvantage, but in terms of mobility, he had an advantage over the professional." Such amateur travel films, as Tepperman notes, often aligned themselves with "quasi-ethnographic works about exotic foreign spaces" such as Merian C. Cooper and Ernest B. Schoedsack's film *Chang* (1927).[152] Film libraries offering professional films for home viewing also included travel films among their offerings.[153]

In opening up the sales counter or home to views of distant places, advertising films and amateur movies, like earlier travel films, took up functions that had been associated with nineteenth-century amusements such as the magic lantern and stereoscope, which, as Tom Gunning contends, acted as "means of appropriating some distant place through an image, of seeming to be somewhere else by being absorbed in a 'view.' "[154] In doing so, these films aligned the screens populating commercial and domestic spaces in the long 1930s not only with long-standing means of displaying images both publicly and privately but also with emerging practices for creating them, since the screens simultaneously invading studio soundstages often featured similar views, including vistas from moving ships, trains, and planes as well as footage of exotic locations. The pervasive emphasis on moving images of travel highlights the imbrication of these diverse cinematic practices not only with other technologies of transportation and communication promising to eradicate distance but also with the social forces, especially colonialism and the tourist industry, that, as Gunning argues with reference to early cinema, both fueled and were fueled by such images.[155]

## Television Screens at Home, in the Theater, and in Transit

By the end of the 1930s, discourses on extratheatrical cinema were addressing the presence of another (largely) extratheatrical screen. Between the summer and fall of 1939, *Business Screen* reported on the use of television as a point-of-purchase sales tool at Bloomingdale's department store in New York.[156] That fall, in an announcement labeled "From Screen to Screen," *Movie Makers* also revealed that "television made its first appearance on the program of an amateur movie club recently, as the Staten Island (N. Y.) Cinema Club screened 420 feet of 16mm. monochrome of the late Max Baer–Lou Nova prize fight, photographed by Edward W. Wilby, ACL, directly from the screen of his home televisor."[157] Television had indeed garnered a great deal of attention in the wake of NBC's telecasts from the New York World's Fair in April 1939, which launched the network's regular television broadcasting service and, paired with the commercial offering of RCA sets, were heralded as the public introduction of the medium.[158] Television, however, had been in practical existence as a technology if not an institution since the mid-1920s.[159] Experimentation with television between 1925 and 1941, moreover, took place with an eye firmly fixed on cinema of both theatrical and extratheatrical varieties. As I hope to show, exploring the relationships between the film and television screens developed in the long 1930s provides insight into spatiotemporal formations drawing together the range of social and material contexts increasingly structured around these surfaces.

The development of television in this period took place in two related waves, mapping fairly closely onto changes in the theatrical exhibition of film. The years 1925–1931, simultaneous with the embrace of synchronized sound film and the experimentation with widescreen cinema, witnessed the first public demonstrations of mechanical television and what Erik Barnouw describes as the period of "television fever" following in their wake.[160] During this period, color television, long-distance television, and two-way television were realized.[161] By 1931, however, observers were forecasting that television's success would lie in the adoption of electronic systems employing cathode-ray tubes, such as those being developed separately by Philo T. Farnsworth (working independently) and Vladimir Zworykin (at Westinghouse and then at RCA).[162] These systems eliminated the need for moving parts and opened up the prospect of significantly higher-resolution images. By 1936–1939, all-electronic systems had been put into use in several countries, including Germany, England, the United States, and (experimentally) Japan. After a series of regulatory delays and after noncommercial telecasting had long been under way, commercial television broadcasting in the United States commenced on July 1, 1941, employing the standard for "high-definition" electronic television (525-line frames with interlaced scanning at 30 frames per second) newly approved by the National Television System Committee (NTSC).[163] The years 1931 to 1941 thus saw the development and standardization of the television technologies that would gain popularity after the war and hold sway throughout much of the twentieth century.[164]

During this period, television was conceptualized in relation to other technologies and media, especially telephony, radio, and cinema. Television screens were designed and evaluated with an eye to film screens, with comparisons often weighing the surfaces' relative size, resolution, and brightness. Some practitioners explored the prospect of theatrical television, and there were several experiments with projecting television onto screens of a scale closely matching that of concurrent extratheatrical film screens (thus also approximating the size of theatrical film screens by the early 1940s). Work on television screens intended for domestic and mobile use, however, also looked to movie screens as both a model and a competitor. Indeed, television screen practices both experimented with and reconfigured the material qualities also associated with cinema screens—including the qualities of multiplicity, transparency, scale, and mobility—and often within similar sites and social frameworks. The emergent television screen's relationship to film screens thus exceeded contrasts in size, resolution, or brightness to include significant areas of overlap and collaboration. In particular, the introduction of television screens represented another form of multiscreen practice, not only complementing but also expanding the spatiotemporal syntheses achieved simultaneously in cinema.

## Multiplicity and Transparency

The screen has often been treated as one of the elements distinguishing television from classical-era cinema, whether by virtue of distinctions in scale, an opposition between light-emitting and light-reflecting surfaces, or assumptions about the social and material spaces where particular screens have been installed, not to mention the levels of illumination found therein.[165] An exploration of the development of these screens in the interwar period, however, should disabuse us of any such easy distinctions.[166] As we have seen, cinema screens in this era were already heterogeneous. Television screens were as well. For one thing, some early television systems actually used reflective screens. For another, an examination of the range of surfaces employed in these systems raises questions about the very nature of the television screen and its place and function within the televisual apparatus. Indeed, the television systems developed in the long 1930s featured screens of different kinds. In their employment of these varying surfaces, they both manifested and reframed the qualities of multiplicity and transparency that I have been mapping throughout this book. In doing so, these systems simultaneously contributed to and reshaped the forms of spatial synthesis associated with the arraying and layering of image planes.

Mechanical television systems used discs or drums and photoelectric cells to scan profilmic material and convert the resultant patterns of light into sequential electrical impulses. These impulses were transmitted by either wire or radio waves, and in the receiving apparatus a complementary disc or drum recomposed the impulses as an image through a complementary scanning process.[167] As C. Francis Jenkins emphasized in describing his own experiments with what he called "Radio Vision" in 1923, such systems entailed "sending from a flat surface and receiving on a flat surface." He explained: "By flat surface is meant to include the depth of focus of the lens employed . . . while at the receiving station the picture must be caught on a flat surface like a photographic plate, a white wall, or a motion picture screen as employed in the theatres."[168] Cinema, we might note, works similarly insofar as the camera lens produces an image on the flat surface of the filmstrip and the projector recomposes that image on the flat surface of the screen. With television, however, the flat surface in the sending apparatus as well as in the receiver was dubbed a "screen," perhaps due to the ephemerality and dynamism of the imagery it registered. Jenkins's sending apparatus, for example, worked somewhat like an optical printer, in this case employing a movie projector to project photographed material onto "a ground glass screen located in the focus of the radio photo transmitter."[169] Later systems capable of transmitting live action, including the fully electronic systems developed in the 1930s, also incorporated a similar screen in the sending apparatus.

Just as the flat surfaces used in cinema to intercept the projection beam varied significantly (insofar as they were made, as we have seen, from materials such as glass, fabric, plaster, and metal), the surfaces serving that function in mechanical television systems were also diverse. Several of the mechanical systems built in 1925–1931 also employed ground glass as a translucent screen in the receiver. As of 1926, for instance, the "handsome mahogany case" built to house John Logie Baird's television receiving apparatus contained a ground-glass screen about 8-by-8 inches in size.[170] Similarly, an elaborate Jenkins model from 1929 had the "radio-movies and television pictures in action appear

**4.9** This television cabinet from 1929, designed by C. Francis Jenkins, displayed the image in the glass-covered circle at the top.

*Source*: From the George H. Clark Radioana Collection, Archives Center, National Museum of American History, Smithsonian Institution, Washington, DC.

in the glass covered circle at the top" (figure 4.9).[171] These systems also often incorporated magnifying lenses to expand—and sometimes to display—the televised images. As Jenkins put it in 1928, his apparatus at that time produced a 3-inch picture that "appears in the air in a plane tangent to the surface of the drum, and is viewed through a magnifying glass," thus expanding it to about 7-by-8 inches.[172] While Jenkins's drum receiver of 1928 used a mirror to intercept the picture and redirect it to the magnifying lens, his most basic later receiving sets simply mounted a magnifying lens in front of a disc (figure 4.10).[173] The two-way television apparatus that AT&T's Bell Laboratories had constructed by 1930 also displayed moving images on a magnifying lens placed in front of a receiving disc (figure 4.11).[174]

The Jenkins Company's corporate discourse was ambiguous regarding where the screen was located even within such receiving apparatuses. It referred to the magnifying lens as a "magnifying screen" and discussed experimentation with "ground glass used as a translucent screen."[175] But it also described the image constituted by scanning—as opposed to the flat surface that "caught" the image—as a screen: in the instructions for the 300-series Jenkins "Radiovisors" in 1931, the company advised customers that when they looked through the

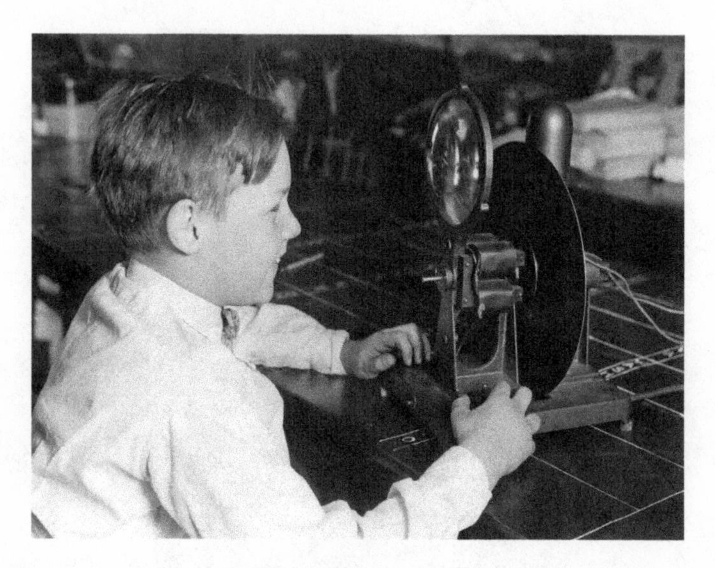

4.10 The Jenkins Company offered kits allowing experimenters to build a basic "Radio-visor" consisting of a sixty-line disc and optional magnifying lens. Ca. 1930.

*Source*: From the George H. Clark Radioana Collection, Archives Center, National Museum of American History, Smithsonian Institution, Washington, DC.

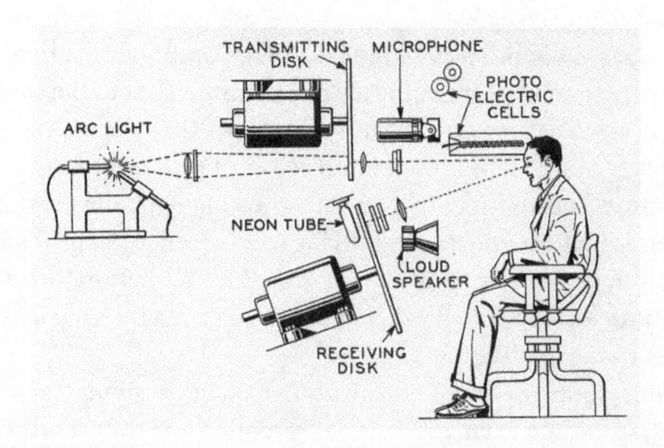


4.11 The two-way television apparatus that Bell Laboratories had constructed by 1930 included a transmitting and receiving disc at each station, displaying the televised image on a magnifying lens.

*Source*: From the George H. Clark Radioana Collection, Archives Center, National Museum of American History, Smithsonian Institution, Washington, DC.

"shadow-box with its magnifying lens," there should "appear a pink spot of light. The bottom switch 'B' is now thrown to the right, starting the motor. The pink spot now becomes a line, a series of lines, and finally"—as the disc attained full speed—"a shimmering screen."[176] Such ambiguity was exacerbated in models employing mirrors. Another Jenkins model from 1928, for example, enclosed the mechanism in "a small cabinet with swinging mirror mounted between uprights resembling nothing so much as our grandmother's dressing table, before which she used to primp," with the televised image appearing in this "magic mirror."[177] In 1931, Jenkins continued to advocate for the use of mirrors in his television apparatus. He proposed, "If a mirror were used out front, similar to the mirror arrangement we use in the home talkie-movie cabinet, you could turn your scanning light-spot backward and make the picture twice as large as it is now."[178] Indeed, as we have seen, mirrors were also used to extend the projection throw and achieve larger images in rear-projection arrangements both on soundstages and in extratheatrical film exhibition around this time. RCA later used a similar configuration to accomplish "large-screen television" with cathode-ray tubes.[179] Such arrangements, in short, incorporated multiple image planes, relaying the television image from one to another.

The electronic television systems developed in the 1930s also incorporated multiple "screen" surfaces, some of which projected onto others. These systems employed cathode-ray tubes in place of the mechanical systems' discs and drums. The cathode-ray tubes used in electronic transmitters, such as the Iconoscope developed by RCA engineer Vladimir Zworykin, featured a mica plate covered with a mosaic of photosensitive silver globules. Zworykin also called this plate a "mosaic screen."[180] A taking lens focused profilmic material onto the plate/screen, to be scanned by an electron gun and transformed into the sequence of current impulses constituting the picture signal (figure 4.12). The receiver, which Zworykin dubbed the "Kinescope," consisted of a cathode-ray tube with an electron gun and a fluorescent screen. In the Kinescope, the fluorescent screen, according to Zworykin, became "luminous under bombardment from the electron gun." As Zworykin explained, the cathode-ray beam in the Kinescope

> is made to sweep across the fluorescent screen in synchronism with the scanning beam in the Iconoscope which is transmitting the picture. Furthermore, the current in the Kinescope cathode-ray beam is controlled by the signal impulses generated at the Iconoscope. This control acts in such a way that the impulse corresponding to a bright area on the Iconoscope causes an increase in current, while a dark region causes a decrease. There will, therefore, be an exact correspondence both in position and intensity between the fluorescent illumination on the Kinescope screen and the light on the mosaic in the

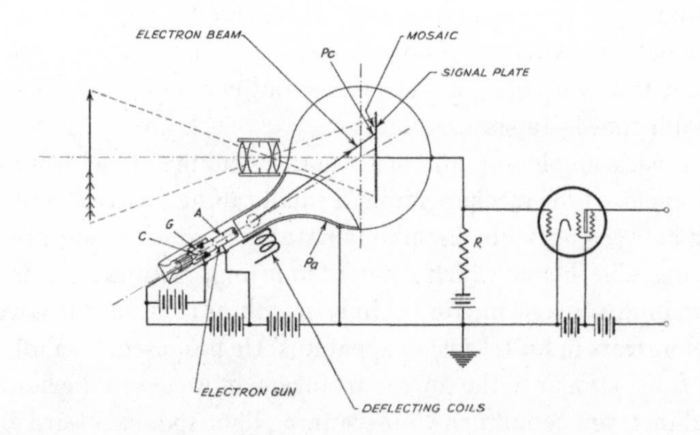


**4.12** As shown in this depiction of the Iconoscope camera, its cathode-ray tube featured a mica plate covered with a mosaic of photosensitive silver globules. NBC, 1937.

*Source*: From the George H. Clark Radioana Collection, Archives Center, National Museum of American History, Smithsonian Institution, Washington, DC.

> Iconoscope. A picture projected on the Iconoscope will therefore be reproduced by the Kinescope.[181]

We can thus also think of such transmission as a relationship between screens, with the Kinescope's fluorescent screen reproducing the image on the Iconoscope's mosaic screen.

Many electronic receivers of the 1930s also involved another surface functioning as a screen. Since the Kinescope's cathode-ray tubes were so large, early console models often mounted the tubes vertically so that the fluorescent screen was visible at the top of the cabinet. The televised image was then viewed through reflection on a mirror installed on the underside of a raised cabinet lid. The models RCA used for its field tests in the mid-1930s and the experimental models Philco produced at that time, for instance, employed this arrangement (figure 4.13).[182] As of 1935, RCA's television sets were described as "like [a] radio set with [a] mirror."[183] Such consoles recalled Jenkins's "dressing table" model of 1928 as well as home-movie consoles such as the Visionola of 1931, which, as discussed earlier, also incorporated a screen on the underside of a raised lid and a mirror to redirect the image onto it. By 1936, *Broadcasting* was reporting on a "trend toward direct viewing," especially in England, where most sets were reported to have viewers look "straight into the cathode tubes instead of seeing the images reflected 90 degrees in a mirror."[184] Mirror models, however, continued to be introduced into the late 1930s (figure 4.14).[185]

This discussion illuminates the way in which television's touted capacity to span or amalgamate spaces intersected with other screen-related practices at

4.13 With this experimental RCA unit, the fluorescent screen was visible through a mirror installed on the underside of a raised cabinet lid. NBC, 1937.

*Source*: From the George H. Clark Radioana Collection, Archives Center, National Museum of American History, Smithsonian Institution, Washington, DC.

this time. The volley of images among a series of planes—the mica plate, the fluorescent receiving screen, and often a mirror—shows one way in which television participated in the proliferation of screening surfaces in the long 1930s. Samuel Weber has observed that television "takes place in at least three places at once: 1. In the place (or places) where the image and sound are 'recorded'; 2. In the place (or places) where those images and sounds are received; and 3. In the place (or places) in between, through which those images and sounds are transmitted." He contends that the "unity of television as a medium of presentation thus involves a simultaneity that is highly ambivalent" since it "overcomes spatial distance but only by splitting the unity of place and with it the unity of everything that defines its identity with respect to place: events, bodies, subjects."[186] Attention to the use of screens and cognate surfaces in early television systems highlights the way in which transmission entered the sites of "recording" and reception themselves, reduplicating the spatial multiplicity underlying television's ambivalent simultaneity. In the television systems I have described, the actual space of the receiving screen was tied to at least two "recording" spaces: the actual space of the transmitting screen and the virtual space focused on it. (The virtual space appearing on the transmitting screen

**4.14** The deluxe RCA Victor Model TRK-12, offered for sale for $600 in 1939, had a 12-inch tube visible through a mirror.

*Source*: From the George H. Clark Radioana Collection, Archives Center, National Museum of American History, Smithsonian Institution, Washington, DC.

could, in turn, exist either in spatiotemporal proximity to the actual space producing it, as with a live broadcast, or at a spatiotemporal remove, as with the transmission of prefilmed material.) This process was reiterated in the space of reception when mirrors were used in the receivers.

Unlike the use of dual-screen rear projection in films or the employment of multiscreen projection in movie theaters, the series of surfaces harnessed in television did not create composite spaces through juxtaposition. They did, however, produce an arraying of images and, as Weber suggests, a form of composite (if fragmented) space. Rather than collaborating through juxtaposition, these arrayed surfaces interacted through relay. Television itself, we might note, had long been conceptualized in relation to mirrors, which volley images from one space to another.[187] Describing television as "a means by which visual representations may be sent broadcase [*sic*] from one point and received at many others," the De Forest engineer C. E. Huffman allowed that it "is conceivable that some combination of lenses or mirrors might be arranged whereby this broadcast could be effected by purely optical methods," and he considered television an alternative "auxiliary system" that was likewise "interposed between the observer and the observed."[188] In other words, he conceptualized television as a kind of expanded periscope, quite like those installed around this time at the Los Angeles and Madrid Theatres and employed by Frederick Kiesler to portray television onstage in the production of *R.U.R.* in 1923, but in this case capable of spanning greater distances. Not only did television's role as a form of multiscreen practice further its aggregate spatiality and link it to the aggregate spatialities arising concurrently through other multiscreen practices, but it thus also expanded multiscreen practice and aggregate spatiality to encompass an extensive domain in their own right. As I have mentioned, the fact that a single transmission could be received on a multitude of television sets represented another kind of multiscreen exhibition, a point to which I will return later.

Conceptualizations of the relationships among television's multiple screen surfaces also participated in and extended the screen's association with transparency as a form of spatiotemporal reorganization. Describing television in 1926 as "seeing what is happening at a distance in the same instant as it happens," *Radio Broadcast* writer A. Dinsdale elaborated: "Over short distances, provided no other objects intervene, this may be accomplished with the naked eye, and, for greater distances, it can be done with the assistance of a telescope or pair of binoculars. But no telescope or binoculars will enable us to see through brick walls."[189] Recounting the history of television from the vantage point of the early 1930s, Orrin Dunlap Jr. claimed that Guglielmo Marconi had in 1915 similarly announced that engineers were working "on a wireless device by which a person can look through a solid wall. It is said to resemble a camera, which, when placed against a wall or floor, makes the wood, stone, bricks, concrete or metal transparent—in this respect resembling the X-ray."[190]

Representations of television in the early 1930s also illustrate this emphasis on transparency. In the animated short film *The Robot* (Dave Fleischer, 1932), for instance, Bimbo's "television" system literally penetrates the walls of Betty Boop's house, enabling him to see into her bathroom. As Lynn Spigel discusses, an episode of the serial *The Naggers*, "The Naggers Go Ritzy" (Roy Mack, 1932), aligns the idea of television with a hole in the wall between two apartments. The occupants of one apartment peer into the adjacent unit through the hole, thinking they are looking into a television set. Not only does this alignment illustrate what Spigel identifies as a "confusion over boundaries between electrical and real space," but it also identifies television with porous boundaries between actual spaces.[191] Similarly, as Doron Galili explains, the novel *Tom Swift and His Television Detector, or Trailing the Secret Plotters* (1933) conceives the eponymous device as "a portable surveillance technology" that makes it possible to see in the dark as well as through "walls and other obstacles."[192] Throughout the 1930s, RCA president David Sarnoff also used the rhetoric of transgressing physical boundaries to describe television. Perhaps most evocatively, he described television in 1935 as a means to lift "the curtain of space from scenes and activities at a distance."[193]

In these formulations, television's capacity to make solid walls transparent or to lift the "curtain" hitherto screening one space from another conjures a form of spatial layering similar to that which I have also mapped to concurrent screen practices in film production and theatrical exhibition. Here, the screens gracing the transmitting and receiving devices are understood not to be arrayed across a wide expanse but rather stacked in (very) deep space. These discourses thus support Samuel Weber's further contention that "the television screen can be said to live up to its name in at least three distinct, contradictory and yet interrelated senses," functioning as "a screen which allows distant vision to be watched"; as something that "screens, in the sense of selecting or filtering, the vision that is watched"; and as "a screen in the sense of standing between the viewer and the viewed, since what is rendered visible covers the separation that distinguishes the other vision from that of the sight of the spectator sitting in front of the set."[194] Like the transparent screens also structuring film sets and movie theaters in this period, in other words, the television screen was conceptualized as putting the actual space in front of its surface into a multidimensional relationship with the actual and virtual spaces beyond that surface. Rear projection, it should be noted, was being used in television production as well by 1939 (though at that point only for still backgrounds), thus reduplicating this dynamic onscreen.[195]

The actual transparency of the many television screens made of glass—like that of the rear-projection screens they resembled and, by the end of the decade, displayed—facilitated conceptualizing them as selectively permeable membranes, inviting viewers to gaze into television consoles as they might into the magnifying lenses dotting the walls of a cosmorama or into glass-paned shop windows.[196] Indeed, the widespread use of the term *looking in* to describe television viewing in this period, while a play on the notion of "listening in" to radio, also suggests this arrangement.[197] John Ellis has employed the concept of "looking in" in distinguishing spectatorship of broadcast television (which he describes as inviting a disinterested "glance") from that of cinema (defined in terms of a concentrated "gaze"), and attention to the concept of transparency bolsters Ellis's concomitant claim that television viewers are "isolated, even insulated" from the events presented onscreen.[198] Indeed, as Spigel shows in analyzing postwar television's analogy to the picture windows also installed in the mass-produced suburbs (which, she claims, embodied a "watered down" architectural modernism), television—like other communication technologies, from the telephone to radio—has long been understood to provide not only contact with but also a barrier from the outside world.[199] As we have seen, however, attention to screen practices in the long 1930s reveals that even cinema did not always or everywhere invite concentration or immersion and that it in fact often employed transparency to establish similar forms of isolation and insulation. The notion of "looking in"—suggesting, as it does, the simultaneous crossing and maintenance of a boundary—should therefore not be taken to distinguish televisual from cinematic spectatorship in this period so much as to identify one modality of viewing among several moving across a range of screen practices.

## Scale

Like the engineers and architects addressing the theatrical and extratheatrical exhibition of film, the engineers developing television in the long 1930s explored the relationships between screen size, image resolution, and viewing distance, in this case often using the work on film as a point of comparison. Home movies, in particular, served as a reference point since they shared with television a domestic viewing environment as well as a lower resolution and (usually) smaller screen size than theatrical cinema. As in the widescreen cinema boom of 1926–1931, a shift in the scale and resolution of images propelled speculation about the representational possibilities of the new screens. With television, commentators often decried the medium's apparent limitation to the close shots that widescreen cinema was simultaneously celebrated as transcending. Demonstrations of television, however, exploited that technology's capacity for different forms of scale and scope. Moreover, experiments with projection television throughout this period aligned television screens particularly closely with the scale as well as the sites of both extratheatrical and theatrical film screens.

Most television images were significantly smaller than home-movie images. Without magnification, the pictures produced by the majority of receivers in the mid- to late 1920s were 1 to 3 inches in size (figure 4.15).[200] Color television images at this time, we might note, were even smaller: the images produced by the color system that Bell Laboratories demonstrated in 1929 were viewed through an eyepiece and compared to the size of a postage stamp.[201] As of 1939, RCA sets offered pictures ranging from roughly 3⅓-by-4⅓ inches (with a 5-inch tube) to 7⅝-by-9¾ inches (with a 12-inch tube), while DuMont's sets yielded 8-by-10-inch pictures (with a 14-inch tube) (figure 4.16).[202] As in the discussions of home-movie exhibition, engineers attempted to justify the small screen sizes by the comparatively short viewing distances associated with domestic spaces. The engineer Alfred N. Goldsmith, for example, explained that "theater pictures may be most conveniently viewed from 45 to 135 feet from the screen, while home television pictures will be viewed from about 4 to 11 feet from the screen," claiming (quite exaggeratedly) that the ratio of viewing distance to screen size was roughly equivalent.[203] Indeed, despite the call from some quarters to make

**4.15** General Electric's experimental home television receiver of 1928 had the viewer look through a small opening in the cabinet. This receiver was installed in the home of E. F. W. Alexanderson (*shown standing*).

*Source*: From the George H. Clark Radioana Collection, Archives Center, National Museum of American History, Smithsonian Institution, Washington, DC.

television screens more closely match the size of home-movie screens, I. J. Kaar of General Electric suggested that television viewers would be stationed closer to the screens than were home-movie viewers, citing even shorter viewing distances than Goldsmith—as close as one foot.[204]

Television also produced images of significantly lower resolution than 35mm and (initially at least) 16mm film. Goldsmith estimated in 1935 that "the theater picture has something of the order of 5,000,000 picture elements, whereas even a good home television picture will probably have something like 150,000 elements."[205] Observing, however, that size and viewing distance also played a role in determining the visibility of grain and line structure, he proposed: "Probably a television picture in the home will be described by most as a 'fair home movie.'"[206] In May 1936, at the time RCA was preparing the first set of receivers for its field tests, employing 343 scanning lines, *Radio Today* similarly opined that the picture quality was "somewhere between the degrees of detail usual with 16-mm movie film and 8-mm film."[207] The following year, by which point a 441-line standard had been adopted, R. R. Beal of RCA contended that such images approximated the detail in 16mm film,

**4.16** This DuMont model, introduced in late 1938, offered an 8-by-10-inch picture.

*Source*: From the Allen Balcom Du Mont Collection, Archives Center, National Museum of American History, Smithsonian Institution, Washington, DC.

and David Sarnoff announced that this "standard of picture-definition is now generally regarded as satisfactory."[208] (By contrast and anticipating discussions of high-definition television decades later, Kaar contended that it would take a 2,000-line picture to approximate the resolution of "high-quality" theatrical film exhibition.)[209]

As with discussions of widescreen cinema, the focus on screen size and image resolution in discourses on television was accompanied by an acute interest in the new technology's impact on representation. Just as discourses on widescreen heralded the new capacity for long shots featuring landscapes and masses of bodies, discourses on television worried that the small, low-resolution images were limiting the new imagery to close shots of a single performer, a limitation that was broadly deemed unsatisfactory. For instance, writing in *Scientific American* in the fall of 1930, coincident with widescreen releases such as *The Big Trail* (Raoul Walsh, in Grandeur) and *Billy the Kid* (King Vidor, in Realife), A. Dinsdale contended that the general public "will have none of the present limited form of television, which can show only a head-and-shoulder view of some distant speaker or singer, and that imperfectly."[210]

During this period, there were efforts to exploit other forms of scope associated with television. AT&T's demonstration in April 1927, for instance, transmitted images of Herbert Hoover's face from Washington, DC, to New York and in doing so allowed the small image to span an immense territory. As the *New York Times* put it, "More than 200 miles of space intervening between the speaker and his audience was annihilated" by the new apparatus.[211] General Electric's demonstration at Proctor's Theatre in Schenectady, New York, in May 1930 spectacularized television's capacity to bridge such distance and to synthesize otherwise separate spaces. As Charles Huffman, who attended the demonstration, reported:

> The usual five acts of RKO vaudeville were presented, after which a Mr. Gleason came out upon the stage and explained briefly what was to follow. He next established voice communication with the studio located about a mile away in Building 37 of the G.E. works and asked that the picture be thrown on the screen. Immediately the head and shoulders of Mr. Trainer at the studio appeared on in a lighted area about six feet square, and after conversing with Mr. Gleason a few moments he introduced several of the artists who had appeared on the stage in person a few minutes previous. First, a girl sang in the studio to the accompaniment of a piano on the stage, her voice being heard thru loud speakers in back of the screen. She could be seen holding a telephone receiver to her ear thru which she heard the piano and other sound from the theatre, which was picked up by a microphone located on the stage. A duet was next sung by two girls, one at the studio and the other on the stage.[212]

Later in the performance, the leader of the theater orchestra appeared onscreen to conduct the live orchestra remotely—although, as Huffman complained, his "head and shoulders were visible in the area included on the screen and he appeared somewhat cramped. His elbow could not be seen, only his hand moving up and down before his face."[213] Here, television was thus employed (if imperfectly) to enact a kind of compositing, joining close-ups from the remote studio together with live musicians in the theater to create a hybrid performance spanning the mile between the two buildings.

Throughout the mechanical television boom, conceptualizations of television's possibilities as a form of entertainment similar to cinema or radio (as opposed to its possibilities as a form of communication like the telephone) also envisioned a much broader scope onscreen, capable of displaying public events and more elaborate spectacles. In an assessment of television in 1931, for example, a writer for *Forbes* predicted that the "broadcasting of news events" would "take a major place in television," imagining that the spectator of the near future would "tune in a baseball game, a movie, a presidential speech, the floor of the stock exchange, in fact, any public place or event with the same abstraction he buys the morning paper."[214] In this context, the capacity to expand the scope of the television image itself was considered an asset. In 1928, Julius Weinberger of RCA insisted that if television "is to become of commercial importance, we must be able to transmit at least the figures of one or more artists, or fairly complex scenes."[215]

The experimental television programming associated with electronic systems in the mid-1930s sought to incorporate a variety of material and, with it, camera distances. In the summer of 1935, Farnsworth Television demonstrated its electronic system by transmitting orchestra and dance performances as well as film subjects including Mickey Mouse.[216] On June 29, 1936, RCA began transmitting television programming from its experimental station, which was picked up by the experimental receivers operated by its engineers, displayed in television demonstrations hosted by RCA, and sometimes also captured by others using home-made receivers or imported British sets. These transmissions also included filmed fare such as newsreels and cartoons as well as performances—ranging from speeches and dramatic scenes to fashion shows, music, and dance—staged in the NBC television studio at Rockefeller Center.[217] As one witness reported, "RCA engineers displayed the flexibility of the television camera" during a xylophone performance "by televising 'close ups' and 'long shots' of the musician in action with the same finesse of the motion picture camera."[218]

Witnesses to the demonstrations in the mid-1930s often complained about how long shots appeared on television. Commenting on a Philco demonstration held in early 1937, for example, Orrin Dunlap Jr. noted that in the fashion shows proving popular at such demonstrations, the "models, in their elegant evening wraps, fur coats, nightgowns and bathing suits, stay far back from the 'eye,' with the result that the television observer gets rather a vague idea of what is modish and chic," and he concluded that television "should be satisfied to present close-ups until the screen is enlarged."[219] Others, however, continued to hold that the "outdoor broadcast of public events," such as sports, parades, and "mass meetings," would "prove to be the best program medium in television," observing that the British had broadcast events such as the coronation and Wimbledon tennis matches.[220] While there is a dearth of existing programming from this period, a fragment from a telecast of Dion Boucicault's play *The Streets of New York* (1857)—which aired on NBC's station W2XBS the evening of August 31, 1939, and was captured from a Kinescope screen with a 16mm camera—is limited primarily to medium close-ups and some medium long shots of between one and three people stationed in front of painted backdrops.[221]

Paralleling the efforts to contend with the small scale of most television receivers, there were throughout the long 1930s also experiments with projecting television images onto larger screens, especially in conjunction with the pursuit of theater television.[222] Such demonstrations often took place in movie theaters, and some of these screens featured reflective surfaces similar to most theatrical screens.[223] At its demonstration in April 1927, AT&T showed television images not only on a screen measuring 2-by-3 inches but also on one that, at 2-by-3 feet, approximated the size of contemporaneous home-movie screens.[224] General Electric's demonstration at Proctor's Theatre in Schenectady in May 1930 featured a 6-by-7 foot "translux" screen.[225] In a series of demonstrations—beginning at the London Coliseum in July 1930 and subsequently touring to the Scala Theatre in Berlin, the Olympic Cinema in Paris, and the Kvarn Cinema in Stockholm—Baird Television utilized a 2-by-5 foot screen consisting of 2,100 metal-filament lamps.[226] In September 1931, Ulises A. Sanabria projected television onto a 10-foot screen at the Radio-Electrical World's Fair in New York, with what Orrin Dunlap Jr. described as "thousands" coming to the exhibition to see the images.[227] The following month, television was shown on a 10-foot screen at the Broadway Theatre in New York City, with a wire link connecting the theater to a "televisor" in the Theatre Guild Playhouse; there were a reported 1,700 audience members at the opening performance.[228] In June 1932, Baird Television's transmission from the Epsom Derby was received at the Metropole Theatre in London before an audience of 4,000.[229] Here zone television was used to fill a (relatively) large screen. As John Logie Baird explained, he "had three pairs of telephone wires from Epsom and sent out three pictures side by side. The three pictures thus formed one big picture on the screen at the Metropole seven feet high and nine feet wide."[230]

In the fall of 1935, it was reported that television "of full movie-screen size and brightness" was being shown in Berlin thanks to a system employing a grid of ten thousand thin-filament lamps, "each 'wired for sight,'" to constitute a 6½-foot-square image.[231] The German government, it should be noted, conceptualized and harnessed television as a vehicle for propaganda (as well as a tool for guiding rockets and torpedoes) between 1935 and 1944 in what William Uricchio describes as an effort at redefining the German public that was intertwined with the employment of both radio and public-address systems.[232] In this context, it is especially notable that the German large-screen projection system of 1935 was reportedly intended "for use in auditoriums or for large mass gatherings."[233] By 1935, the Nazi government was also reportedly "operating six television stations for the benefit of thousands of factory workers, whose workrooms have been equipped with receivers for sound and sight" (an arrangement closely echoed in Charlie Chaplin's film *Modern Times*, released in February 1936).[234] On the heels of Germany's experiments with the use of gigantic outdoor loudspeakers, it was reported in 1938 that the country had also "developed a television public address system in which the picture of the speaker is flashed on a large screen and is synchronized with the loud-speaker horn."[235]

By 1936, images from the fluorescent screens of cathode-ray tubes were also being reprojected onto larger surfaces. In that year, the Berlin Radio Show displayed what *Radio Today* described as a "new form of re-projection television in which the cathode-ray image is actually bright enough to be projected optically upon a screen 3 by 5 ft. in size." The magazine explained that when "the intensely bright cathode-screen image is projected by a simple lens system onto the movie screen, a large group can watch the television picture, as in a theatre," and it reported that "Berlin movie houses are understood to have ordered" the apparatus from developer Telefunken.[236] In England, Baird was also producing large-screen images through the use of cathode-ray tubes and optical projection.[237] As of mid-March 1937, the system was being employed as a "regular program feature" at the Dominion Theatre in London, and it was announced that "a television installation [would] be included in the initial equipment of England's largest movie theater, the Ritz, to be erected in the center of Blackpool."[238]

In 1937, RCA demonstrated optical projection from a Kinescope onto both 3-by-4-foot and 8-by-10-foot screens, with what the Academy of Motion Picture Arts and Sciences Research Council identified as better results than the German experiments.[239] A demonstration in May took place in the ballroom of the Hotel Pennsylvania, a location that was deemed "as large as many small motion-picture theatres," and it was suggested that the device could be used in television receivers for either the home or the theater. On the smaller screen (equivalent to the size of many home-movie screens at that time), the result was reported to "approximate closely the brilliance of the average home movie"; on the larger screen (similar in size to many extratheatrical auditorium screens), the image was considered "clear to persons nearly 100 feet away from the screen."[240] A demonstration in the RCA Building for attendees of the SMPE convention in October 1937 employed the 3-by-4-foot screen but not, apparently, the larger surface (figure 4.17).[241] In 1939, it was reported that Farnsworth had also developed a "cathode ray tube gun" capable of projecting onto "an external receiving screen several feet in width."[242] In a demonstration for the Federal Communications Commission in January 1941, RCA used a theater television projector to fill a 15-by-20-foot screen at the New Yorker Theatre, reportedly retaining "quality comparable with that of a home receiver" (figure 4.18). Notably, that setup was accompanied by a "multisonic sound system," compared to Disney's "Fantasound," employing eighteen loudspeaker systems "scattered throughout the theatre" and thus permitting "movement of sound with action on the screen, rotation of sound around the walls of the auditorium and emanation of sound from any desired point in the theatre."[243] In its screen size, exhibition site, and use of sound, this television installation thus approximated—and, with sound, even surpassed—contemporaneous norms for the theatrical exhibition of film.

4.17 RCA demonstrated projection television, using a 3-by-4-foot screen, in 1937.

*Source*: From the George H. Clark Radioana Collection, Archives Center, National Museum of American History, Smithsonian Institution, Washington, DC.

## Mobility

As scholars such as Margaret Morse and Lynn Spigel have shown, the cultural formation of television has been intimately bound up with the concept of mobility, especially inasmuch as television is constructed as a mode of virtual transportation.[244] This construct is already evident in conceptions of television in the long 1930s, dovetailing with concurrent approaches to cinema. In particular, the notion that television offered a form of distant vision aligned it with the investment in virtual travel also guiding cinematic practices, in this case adding the potential for liveness. In 1936, for instance, Rudolf Arnheim claimed that television was "a relative of the car and the aeroplane" insofar as it functioned as a "means of transmission" that, "like the machines of locomotion that the last century gave us, . . . alters our relation to reality itself, teaches us to know it better, and lets us sense the multiplicity of what is happening everywhere at one moment."[245] Whereas midcentury film and television are usually distinguished vis-à-vis assumptions about their material forms (large versus small) and sites of exhibition (public versus private), I have been arguing that attention to the heterogeneity in their screen practices illuminates long-standing

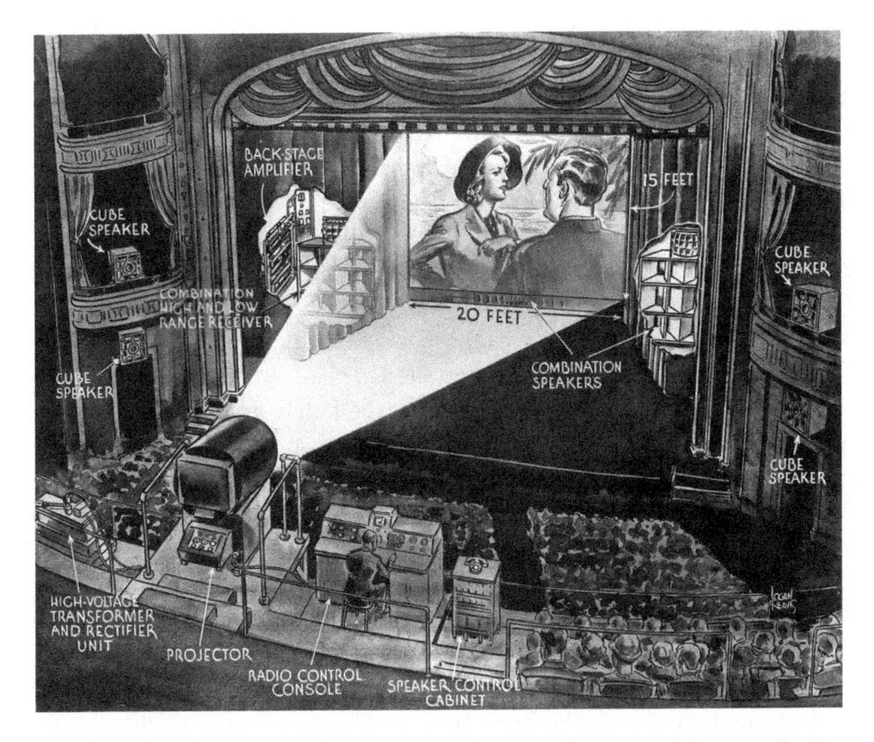


**4.18** RCA employed a system for projection television, diagrammed here, at the New Yorker Theatre for a demonstration for the Federal Communications Commission in January 1941.

*Source*: From the George H. Clark Radioana Collection, Archives Center, National Museum of American History, Smithsonian Institution, Washington, DC.

connections between them. Indeed, beyond and in concert with the promise of virtual transportation, television practices of this period also joined cinematic practices in establishing multifaceted relationships between onscreen movement, the screen's movement, and the viewers' own mobility. In activating these dynamics, television screens joined cinema screens in their contribution to practices such as tourism, sales, and advertising.

Commentators, for instance, lauded the capacity to display moving images on television screens, in addition to movie screens, aboard trains, ships, and airplanes in motion. As early as 1928, Baird transmitted television images to the ocean liner *Berengaria* while it was at sea, recalling the early uses of radio for such communication.[246] On February 1, 1932, television broadcast from London was received on a train "travelling between Sandy and Huntingdon at 60 miles an hour."[247] On May 21, 1932, the Don Lee Broadcasting System demonstrated television reception aboard an airplane using a cathode-ray receiver.[248] In early 1937, David Sarnoff described a four-way conversation in which he sat at his desk in Rockefeller Center and spoke with Guglielmo Marconi, who was on his yacht in the Mediterranean, and two other interlocutors who were flying in two separate planes. Although this unification of land, sea, and sky was achieved by radio, Sarnoff suggested that television would soon be able to add sight to such exchanges.[249] Such actual and anticipated mobile reception, relying as it did (or would) on wireless transmission, aligned television with radio's purported saturation of the "ether." Television screens also collaborated with cinema screens to bring moving images into and through such terrestrial, aquatic, and aerial domains, creating syntheses that not only spanned distance but in doing so also overcame the inhospitable environments of water and sky.

Like cinema, television was also installed in places where potential viewers circulated. Television sets proliferated in stores, hotels, and bars in London in 1937–1938, following the initiation of regular telecasting by the British Broadcasting Corporation (BBC) in November 1936.[250] After returning from a trip to Europe in 1937, Allen B. Du Mont observed that throughout England "one sees television sets on display in radio shops, music stores and department stores," and he reported that in London "people throng the shops and department stores where teleview sets are displayed and demonstrated. The larger stores have equipped little theatres and the receivers operated therein simulate the atmosphere of highly entertaining miniature dramas and puppet shows."[251] In 1937, it was announced that the St. Regis Hotel had become "one of the first hostelries in the world to feature television reception in its lounge."[252] Moreover, by 1938, anticipating the installation of televisions as well as film jukeboxes in taverns in the United States the following decade, it was reported that "nearly every cocktail lounge in the British capital is equipped with a television receiver to serve sight programs with its drinks during the two hours of the day that the B. B. C. transmitter is on the air."[253]

Similar ideas and efforts also emerged in the United States. In 1936, David Dietz of the *World-Telegram* reported on the belief that "department stores will want to use television to stage style shows when the new coats or dresses arrive" as well as the idea that automobile companies would "want to use television to show their new models."[254] Noting the vogue for televised fashion shows, Orrin Dunlap Jr. similarly predicted in 1937 that the "distinctive shops are likely to have television screens so that they may stage a timely Paris showing from the television rods atop Eiffel Tower."[255] In line with the use of film in department stores, a "tele-sales" point-of-purchase television system for such establishments was introduced in 1939. The system was installed at Bloomingdale's that April, where it was used to transmit a "millinery fashion show" from the sixth floor of the store to four 7-by-9-inch viewing screens—dubbed "kinets"—situated in different places on the third floor.[256] The *New York Times* reported on Bloomingdale's plan "to install the 'kinets' at escalator landings and in various departments of the store to encourage multiple sales from customers who come with a fixed idea of what they want to buy."[257] Bloomingdale's sales director, Ira Hirschmann, enthused that "the clarity and incisiveness with which merchandise is televised from our own studio and seen by customers on all floors simultaneously make the new medium a 'must' among selling methods."[258] He claimed that the response from the public was "startling," with crowds "packed around the receiver."[259] By the summer, the system was also being used to display prefilmed material featuring fashions from Bloomingdale's Barbara Lee brand as well as clothing and footwear from the store's children's department.[260] In 1939–1940, both RCA and Farnsworth also toured their television equipment to department stores. The Farnsworth tour alone encompassed stores in eighty-eight cities, beginning with Meier & Frank in Portland, Oregon, in September 1939. Attendance at such demonstrations reportedly averaged ten thousand per day. Both television "studios" and television receivers with 9-by-12-inch screens were set up within the stores, and programs included hairstyling, fashions, merchandise demonstrations, and the televising of the visitors themselves.[261] As Anna McCarthy has discussed, such in-store use of television advertising continued in the period immediately following World War II, though such experiments ceased by 1949.[262]

The employment of television in department stores illustrates William Boddy's observation that early conceptions of the technology "defined television simultaneously as itself a consumer product for the home and as an audio-visual showroom for advertisers' consumer goods."[263] Insofar as the items featured in the demonstrations—dresses, children's clothing, hairstyles—targeted female consumers, moreover, these practices reveal how television's address to women moved outside the confines of the home in the long 1930s. The television screens installed in department stores, as McCarthy points out, enhanced the "gendered visual culture of consumption" confronting the flâneuse.[264] These television screens thus collaborated with the film screens being deployed simultaneously—in department stores; in other public spaces designed with women in mind, including movie theaters; and even on film sets—in harnessing the display of moving images to direct and construct women's movements, both actual and virtual, often in the service of consumption.

As with extratheatrical cinema, the mobility of camera operators was also important for television. Mobile television pickup units were introduced and employed in both Europe and the United States in the mid-1930s. These units emulated the similar mobile outfits used for the filming of sound newsreels, but in this case they incorporated the capacity not only to capture but also to transmit moving images and sounds from the field.[265] In July 1935, it was announced that such a unit had been developed in Germany, which had begun the regular public broadcasting of television in March.[266] In that case, a standard movie camera was mounted on the roof of a car that housed a darkroom containing the apparatus and "extremely fast-working chemicals" to develop the film in

**4.19** Mobile television vans operated by NBC were deployed to pick up outdoor events. NBC, 1937.

*Source*: From the George H. Clark Radioana Collection, Archives Center, National Museum of American History, Smithsonian Institution, Washington, DC.

ninety seconds; the still-wet film was then scanned and transmitted.[267] By 1937, the BBC was operating three television pickup vans. In this case, moving images were shot with television cameras, and the signal was sent to the large transmitter atop Alexandria Palace for rebroadcasting.[268] By the end of the year, mobile television vans operated by NBC were also in service in the United States, intended for the "outdoor pick-ups of sports and other news events" (figure 4.19).[269] Like the British units, these vans operated as relay stations, sending signals to be rebroadcast from NBC's transmitter atop the Empire State Building. In December 1939, NBC introduced new mobile equipment that could be packed into the back of a car, thus opening up the possibility of telecasting from otherwise inaccessible spaces such as nightclubs and Broadway theaters (such as the theater from which the *Gone with the Wind* premiere was transmitted that month).[270]

Practitioners, moreover, advocated for onscreen motion, arguing that it enhanced the interest and clarity of small, low-resolution images. In 1933, for instance, E. W. Engstrom of RCA contended that moving objects necessitated less detail than still pictures since they possessed continuity, facilitated scanning, and captured the attention of viewers, dissuading them from critically analyzing the picture.[271] Echoing these ideas, Harry Lubcke of the Don Lee Broadcasting System explained in 1940 that "scenes of limited detail, as in television, will appear to be of greater clarity when in motion" and thus recommended that, in programs made for television, "the principal characters should move, gesticulate, or talk whenever possible."[272] When this was not possible, he suggested moving either the camera or elements in the background. Camera mobility would become what Lynn Spigel identifies as the "reigning aesthetic choice" for television productions in the postwar period, working together with other techniques, including rear projection, to create an illusion of spaciousness within cramped television studios.[273]

Some of the types of material featured in broadcasts of the long 1930s, such as animated cartoons, dances, and boxing matches, would seem particularly conducive to producing motion-filled imagery. Uses of Mickey Mouse, in particular, recurred in research into, discourses on, and displays of television. RCA employed images of the mouse—together with close-ups of women and long shots of baseball games, two other forms of recurring imagery—in its study of television picture detail in 1933. In this case, 16mm film was utilized to simulate the resolution of television images.[274] In articles for both *Radio News* and *Radio-Craft* in 1935, Wilhelm Schrage included images purporting to show Mickey as received on a television screen (though these images suspiciously reproduced what appears to be the same frame used by RCA, with added curvature suggesting a television screen).[275] As mentioned earlier, Farnsworth's demonstration of electronic television in 1935 also featured Mickey as well as orchestra and dance performances. Although NBC had long employed the figure of Felix the Cat revolving on a phonograph turntable as a test image, RCA's internal publication *Family Circle* misidentified that character as Mickey Mouse upon the figure's "retirement" in 1935.[276] Mickey Mouse's popularity and simple black-and-white form certainly contributed to the character's recurrence in these contexts. But its capacity for what Sergei Eisenstein identified as "plasmaticness"—as a being "which behaves like the primal protoplasm, not yet possessing a 'stable' form, but capable of assuming any form"—made it an especially suitable choice and shaped conceptualizations of the television image in ways that resonated with other screen-related experiments.[277]

Eisenstein associated the plasmatic nature of Disney's animation with fire, and, indeed, a concomitant enthusiasm for using television to capture fires is also notable.[278] In 1936, *Radio Today* cited the spectacle of seeing "the Camden fire department extinguishing a roof blaze" to exemplify the kinds of "outdoor scenes" possible with the new technology.[279] In 1938, the telecast of a fire on Wards Island was harnessed to demonstrate the achievements of NBC's mobile unit. As it was reported, "Tele-observers pronounced the telecasts as 'graphic' and 'amazingly clear.'"[280] Such spectacles, to be sure, capitalized on television's capacity for liveness and association with catastrophe, and in this regard they fell in line with other televised spectacles of the time, such as that of a woman falling to her death from a skyscraper near Rockefeller Center.[281] But, as we have seen, the visual imagery of fire itself was especially pervasive in screen-based experiments of the 1930s. I argue in the book's coda that such imagery—sharing the plasmatic quality of the Disney cartoons with which it also shared the airwaves—conspired with the screens upon which it was displayed to liquefy spatial boundaries and forge new spatial amalgamations.

## Multiplicity Redux

Demonstrations of television throughout this period often employed multiple screens. As mentioned earlier, AT&T's demonstration in April 1927 presented the televised images on both a small screen and a relatively large one, displaying both console and projection-television systems. RCA reiterated that strategy in presentations of its own console and projection-television systems in 1937.[282] Demonstrations of electronic systems, moreover, frequently employed multiple consoles arrayed for gathered audiences. In 1935, Farnsworth Television was displaying its transmissions on two sets, one with reception via wires and the other slightly larger set receiving via radio.[283] When Philco demonstrated its system in 1936, it arranged four consoles in a row. As *Broadcasting* reported, when the images started to come through, "two of the sets were operated, permitting easy view for those in the three tiers of seats on either side of the room."[284] Philco demonstrations in 1937 reportedly employed six receivers.[285] Not only did RCA distribute receivers to its employees in and around New York City as part of its field tests (with these test models numbering about seventy by November 1936), but the demonstrations it staged in the mid- to late 1930s also featured many such receivers arrayed within a single space (figure 4.20).[286] Early in 1937, David Sarnoff bragged that in "various demonstrations we have used fifteen receivers to entertain two or three hundred people."[287] In its demonstration for the SMPE meeting in the fall of 1937, RCA set up twenty such receivers "in a long hall so that groups of ten to fifteen persons could view the image at each receiver."[288] On the occasion of NBC's inaugural broadcast from the New York World's Fair on April 20, 1939, engineers in the RCA building, demonstrating reception for a group of journalists, reportedly "[tuned] in pictures simultaneously on six of the $600 sets" about to go on the market.[289]

As the journalistic coverage of these events makes clear, the use of multiple receivers for such demonstrations served a practical function, enabling large gathered audiences to view televised material on small screens. Indeed, citing the difficulty of group viewing in 1935, the pulp-magazine publisher Hugo Gernsback outlined the idea for a "multiple-image television receiver" boasting three screens, thus anticipating later multiple-image sets such as DuMont's experimental Duoscope of 1954.[290] The exhibition arrangement employed in the demonstrations, however, also displayed television's touted capacity to convey

**4.20** Demonstrations of television throughout the 1930s often employed multiple receivers.

*Source*: From the George H. Clark Radioana Collection, Archives Center, National Museum of American History, Smithsonian Institution, Washington, DC.

a single transmission to multiple receivers simultaneously—a feat that can be viewed as complementary to a single set's potential, spectacularized by the Duoscope, to receive multiple transmissions. In describing the mechanical system he initially used to transmit still images, C. Francis Jenkins explained in 1923, "The whole apparatus is comparable to a camera with a lens in Washington and its photographic plate in Boston, for example; with this difference, that one lens, in Washington, may put its picture on ten, one hundred, or one thousand photographic plates in as many different cities at the same time."[291] With the transmission of moving pictures, the analogy to (and possible employment of) photographic plates would, of course, become less relevant, as Jenkins and others embraced surfaces conducive to the display of motion, such as reflective screens, ground glass, and, later, cathode-ray tubes. In this case, the figure of an expanded still-camera apparatus might be replaced by the figure of an expanded movie-projection apparatus, with a single projector illuminating a multitude of screens—as was actually the case (though on a much smaller scale) in the Los Angeles and Madrid Theatres discussed in chapter 3.

In this context, it is notable that discussions of television technology also frequently emphasized the notion of synchronization. This term came into play in a couple of ways in discourses on television. As Richard Koszarski has noted, the term *synchronization* was used to mark the parallel between television and sound film.[292] The idea here was that, as one observer put it, television would "bring sight to the American radio, so that it may become a *complete* medium of expression, as was accomplished in motion pictures in bringing sound to the silent screen."[293] Indeed, as late as 1937, David Sarnoff touted the "synchronized sound and light effects" offered by television—seemingly borrowing language from the discourses on cinema of a decade earlier.[294]

However, technical discourses in particular also relied heavily on the notion of synchronization to describe the temporal alignment of the apparatuses employed to send and receive television signals. In this case, synchronization referred not to the coordination of sound and image but rather to the mechanism for achieving the kind of simultaneity between transmission and reception described in other contexts as "instantaneity" or "liveness." The notion of "liveness," in particular, tends to emphasize representation, suggesting the temporal coincidence between an event and its mediated apprehension. Thinking about television's temporal dynamic in terms of synchronization, however, shifts focus away from representation and toward the coordination of technologies. What was important to technicians was not so much what was being transmitted as the coordination of the sending and receiving apparatuses. Such coordination was equally crucial for prefilmed programming as it was for live broadcasting.

Achieving this coordination was key to expanding the projection apparatus—extending it, in other words, beyond the local configurations obtained with film projection—and thereby also opening up the possibility of displaying synchronous moving images on dispersed screens. With Jenkins's mechanical system, this synchronization was achieved with tuning forks, which controlled and aligned the motors at the sending and receiving stations.[295] Thus, if, as Jenkins put it, "all the receiving stations are in synchronism with the sending station[,] the light values fall in the proper places on each picture receiving screen, and the motion picture at the sending station is duplicated in a thousand homes."[296] Bell Laboratories' one-way television system of the mid-1920s incorporated a separate channel for the transmission of what Herbert Ives called the "synchronizing pulses" provided by a master tuning fork.[297] In its two-way arrangement, four discs (two at each station) needed to be kept in synchronism, which was now achieved with a vacuum-tube oscillator, whose frequency was also transmitted over a separate pair of wires.[298] With RCA's electronic system, the electron beams in the sending and receiving apparatuses were kept in alignment through the transmission of "synchronizing impulses" after each line scanned and at the end of each frame. As R. R. Beal at RCA contended, "The requirement of accurate synchronization between the scanning beams at the transmitting and receiving ends of the circuit is one of the important factors necessitating a uniform standard for all television systems to be used in broadcasting services in this country."[299]

Such coordination would, as discussed earlier, seem to distinguish the temporality of television from that of (mechanical analog) cinema, which does not allow for the simultaneity of capture and reception. Attending to the concept of synchronization, however, highlights television's connection to the apparatuses employed within—rather than between—the domains of film production and exhibition in the long 1930s. Especially insofar as many television demonstrations arrayed the receivers side by side within the same space, they participated in the emphasis on synchronous multiple projection that was, as we have seen, taking place concurrently on Hollywood soundstages and in movie theaters. Indeed, RCA's television demonstrations employing arrays of fifteen to twenty receivers around 1937 occurred in close historical proximity to several of the multiscreen experiments discussed in the previous chapters. Television's concomitant capacity to disperse its multiple projections among heterogeneous spaces and across wide distances expands the significance of multiscreen exhibition.

This discussion of the screen practices associated with extratheatrical film and television has thus mapped both contractions and expansions of the apparatuses of moving-image exhibition. As film and television screens were designed for domestic use, not only did they take on a smaller scale than most theatrical screens, but they also entered into a different relationship with other components of the technological arrangement for reproducing moving images and sounds. The console models developed for extratheatrical film and television literalized what was in theatrical exhibition a more figurative call to coordinate the elements of the apparatus into a machinelike collaboration. By integrating the technologies for generating images (projector or receiver), reproducing sound (speakers), and—often—displaying the images (the screen), however, these consoles did not merely consolidate several otherwise disparate components of the apparatus. They also seemingly squeezed viewers out from within the technological arrangement, resituating them on the outside and—to borrow a term that was, as I have mentioned, attributed to spectatorship of television at this time—"looking in." At the same time, however, practices associated with extratheatrical cinema and television simultaneously expanded the apparatus, as Wasson suggests, by proliferating a diversified array of technologies for producing and reproducing moving images and sounds, forming shifting assemblages in a range of social, material, and institutional contexts. Insofar as moving images and sounds became increasingly pervasive in spaces such as homes, stores, and places of work and transit, the cinematic and televisual apparatuses alike—heterogeneous and protean though they were—can also be understood to have incorporated a vast realm.

# CODA: MULTIPLICITY, IMMERSION, AND THE NEW SCREENS

I have argued that during the long 1930s screen practices across a variety of domains—including production, theatrical exhibition, extra-theatrical exhibition, and television—shared an emphasis on spatial synthesis and temporal synchronization. Contrary to most conceptions of the Hollywood cinema of that period, this exploration of screens has shown in particular how prevalent multiscreen practices were in shaping such spatiotemporal formations. In addition, the book has revealed that in the achievement of synthesis and synchronization these practices often harnessed particular types of imagery. As we have seen, this included an emphasis on travel and mobility as well as on water, fire, and sky, which John Durham Peters describes as "elemental media" in their own right. Peters defines media as "our infrastructures of being, the habitats and materials through which we act and are," a definition that brings natural elements such as earth, water, and air, which can function as habitats, together with the cultural techniques (including the use of fire) that enable humans to treat them as such.[1] Examining how this imagery manifests on screens illuminates the ways in which these surfaces have also been constructed as agents of mediation, operating as both habitats and means of organizing them. Exploring such constructions provides insight into the social and material functions that screens perform within particular historical contexts.

The onscreen presentation of fire has been especially notable across the multiscreen experiments in particular. In the realm of production, as we have seen, films such as *Gone with the Wind* (Victor Fleming, 1939), *Rebecca* (Alfred

Hitchcock, 1940), *Typhoon* (Louis King, 1940), and *Aloma of the South Seas* (Alfred Santell, 1941) used rear projection to immerse characters in fiery landscapes. *The Forest Rangers* (George Marshall, 1942) devoted the new rear-projection equipment introduced in the early 1940s, including more powerful triple-head rear projectors and dual large screens, to such imagery. Experiments with multiple projection in theatrical exhibition also focused on the depiction of fire. In pursuing the use of multiple projection for the exhibition of *Gone with the Wind*, David O. Selznick anticipated employing that arrangement for the "burning of Atlanta" sequence in particular. Upon the Film Guild Cinema's opening in 1929, its capacity for multiple projection was envisioned as a means of engulfing viewers in virtual fire during screenings of "*Jeanne D'Arc*" (version unspecified but likely evoking Carl Theodor Dreyer's film from 1928).[2] In the experiments with television that occurred in this period, which I have argued should also be considered a form of multiscreen exhibition, the capacity to depict fires was also a celebrated achievement.

Fire, of course, has constituted a long-standing screen spectacle, and the promise of safe proximity to flames aligns the cinema and television screens of the long 1930s with everything from the fire screens used in actual fireplaces to the virtual Yule logs that currently enable any YouTube- or Netflix-enabled screen to simulate a fireplace. The persistent vogue for the depiction of fire is, it should be said, ironic given cinema's troubled history with actual fires.[3] In fact, marketing for the Trans-Lux system in the 1930s emphasized that the rear-projection arrangement reduced the hazard of fire by locating the projection booth—and, with it, "everything that usually causes a fire"—behind the screen, shutting it off "entirely" from the auditorium. When the newsreels shown on this screen depicted fires, as in the coverage of the *Hindenburg* disaster, the screen thus did double duty as a virtual and actual barrier.[4] With the predominant front-projection arrangements, however, fireproofing of the projection booth performed the latter role.[5]

Despite the enduring interest in onscreen fire, attention to its presence in deployments of multiple projection in the long 1930s offers insight into the historical specificity of the spatiotemporal formations this practice fostered. The spread of virtual flames across multiple screens promised to spectacularize the screens' offer of immersion by seemingly engulfing the viewer or actor in an otherwise inaccessible space. But onscreen fire collaborated with multiple projection in other ways as well. Insofar as fire manifests what Sergei Eisenstein describes as "a flowing diversity of forms,"[6] its onscreen presence conspired with multiple projection to mask the edges of each screen's frame. In addition, insofar as fire—like the projector—gives off light, its onscreen presence could also work in connection with ambient illumination (whether set or auditorium lighting) to equalize the actual spaces in front of the screen and the virtual spaces beyond it, thus downplaying the screen's surface. In both of these ways, the presence of fire onscreen worked to melt the boundaries constituted by the screen itself, forging new spaces from the resultant amalgamations. Through the depiction of fire, the multiple-projection experiments thereby achieved some of the functions of actual fire, which operates as an engine of both de- and reformation, acting as what Peters calls "our most radical environmental shaper," in part by promoting liquefaction.[7] The employment of fire thus enhanced the way in which multiscreen practices themselves were marshaled to reconfigure both represented and spectatorial spaces. *Gone with the Wind* reveals the ideological function this collaboration could serve. Whereas this film's represented fire signaled social and political change by razing Atlanta and making way for a new formation of the city after the Civil War, Selznick's exhibition scheme harnessed the multiple projection of virtual flames to resurrect a reactionary nostalgia for the antebellum South.

Insofar as fire indexes a temporal event, its onscreen presentation also asserts the temporality of the cinematic or televisual spectacle. Identifying fire as "physical time," Eisenstein describes it as "an image of coming into being, revealed in a process."[8] Fire's ephemerality, urgency, and fast-moving nature allowed it to display television's heralded capacity for liveness. And it also helped to assert the presence of filmed imagery, whether through the rear projection of conjured flames in narrative features such as *Gone with the Wind* and *The Forest Rangers* or through the presentation of actual conflagrations in newsreel footage, such as that capturing the *Hindenburg*'s demise. Beyond this, the emergent nature of onscreen fire also emphasized the temporal coordination achieved through multiple projection, showing how this practice could orchestrate multiple temporal events into a single synchronized spectacle. As I have discussed, instantaneity and simultaneity are often associated with media employing electrical forms of transmission, such as radio and television, and are distinguished from media rooted in inscription, such as film. I have argued, however, that multiple projection shows the important role that simultaneity played in the exhibition of film, including but exceeding the spectacle of synchronized sound. As in radio and television, such spectacles were often also indebted to electrical means of synchronization. The link between the temporalities of fire and electricity is not coincidental since electricity, as historian of fire Stephen Pyne contends, harbors fire covertly.[9] In short, crossing areas such as production and exhibition, traversing modes such as the commercial and the avant-garde, and uniting cinema and television, the practice of multiple projection contributed to particular spatiotemporal formations in the long 1930s by emphasizing the capacity of screen media to forge new spatial composites marked by temporal synchronicity. The recurrence of fire in the multiple-projection experiments of this period underscored and furthered that project by aligning screen media with a much older medium for effecting spatial change and asserting temporal presence.

Fire has continued to operate as a figure for "new" digital media, invoking qualities such as the capability for what Peters calls "contagious spreads."[10] Virtual Yule logs aside, the onscreen representation of fire has arguably become less prominent. Depictions of water and sky, however, have continued to operate as means of conveying and enhancing new screen technologies' promise of immersion. The prominence of such spectacles on contemporary devices thus provides an opportunity to explore transformations in the ways in which screens shape material and social space. In the remainder of this coda, I explore how contemporary virtual-reality headsets employ aerial spectacles in particular to proffer a form of immersion that departs significantly from those I have discussed thus far. The new screens, to be sure, share material qualities with many of the screens developed and employed in the long 1930s, including mobility, a small scale, and forms of flexibility and transparency. I argue, moreover, that the new headsets should be situated within a media landscape characterized, like the media landscape of the long 1930s, by a multitude of screens. The new screens, however, collaborate with aerial spectacles in different ways than did the screens of the earlier period. Indeed, considering that collaboration in light of the practices discussed in the body of this book reveals the historical contingency of screens' role in mediating virtual and actual spaces.

## Virtual Reality and Aerial Spectacles

The term *virtual reality* was reportedly coined by the computer scientist and entrepreneur Jaron Lanier in 1989,[11] but this concept drew on and drew together an array of more long-standing ideas and achievements. By the late 1980s, key components of the technological assemblage that would come to be associated most strongly with virtual reality—head-mounted displays (HMDs) paired with computers and input devices such as data gloves and controllers—had been in existence for decades.[12] Despite virtual reality's eventual association with forms of commercial entertainment such as video games, much of the development of this technology took place in academic and military research laboratories as well as in commercial laboratories focused on industrial applications, with projects ranging from flight and weapons simulators to scientific visualization, surgical training, and architectural walkthroughs.

For instance, Ivan Sutherland, working at MIT and then the University of Utah with funding from the Advanced Research Projects Agency (ARPA) and the Office of Naval Research, developed and refined a computer-aided HMD in the late 1960s and early 1970s. Nicknamed the "Sword of Damocles," that device employed two small cathode-ray tubes and a series of lenses and half-silvered mirrors to project 3D computer graphics depicting objects such as cubes and molecular models 14 inches in front of the user, hovering within the actual environment (and thus functioning more as augmented than virtual reality). The system tracked the position of the user's head and updated the visual display to correspond with the user's changing perspective.[13] Claiming to have been stimulated by Sutherland's writing—specifically his essay "The Ultimate Display" (1965), which conceptualizes the multisensory encounter of virtual worlds—researchers at the University of North Carolina, led by Frederick Brooks Jr., were by the late 1960s experimenting with the use of haptic feedback in conjunction with visual displays, particularly as a tool for scientific visualization.[14] Since 1966, research at the U.S. Air Force led by Tom Furness had focused on visual displays for cockpits; in 1982, Furness and his colleagues introduced the Visually Coupled Airborne Systems Simulator (VCASS), which featured a helmet that employed miniature cathode-ray tubes and mirrors to display computer-generated maps of the landscape synchronized with radar information. Later iterations, eventually under the aegis of the Super Cockpit program, included eye tracking, voice command, 3D sound, tactile gloves, and new helmets that used half-silvered mirrors to overlay graphics on the actual cockpit.[15] In the mid- to late 1980s, researchers at NASA's Ames Research Center—including Scott Fisher, who had been involved with interactive displays at MIT in the late 1970s and worked at Atari in the early 1980s—developed the Virtual Environment Display (VIVED) and Virtual Interface Environment Workstation (VIEW) systems, which employed HMDs with stereoscopic displays and allowed for input not only through position tracking but also through gesture, thanks to the incorporation of the data glove developed by Fisher's former Atari colleague Thomas Zimmerman (who had teamed up with Lanier, another Atari alumnus, to form the commercial firm VPL Research). The newer VIEW system also provided 3D sound and speech recognition.[16]

In uniting various academic, military, and commercial projects undertaken in the preceding decades, the notion of virtual reality identified what was taken to be an emerging form of mediated experience. Although the term *virtual reality* has long conjured the HMD-centered technological assemblages I have described, by the early to mid-1990s it was conceptualized more broadly in terms of the experience of presence in mediated spaces. In particular, it was taken to denote simulated environments that functioned as if authentic by proffering the experience of presence. In some formulations, the notion of virtual reality could also encompass the mediated perception of temporally or spatially distant actual environments via the concept of "telepresence."[17] As Jonathan Steuer argued at the time, virtual-reality systems sought to evoke the sensation of presence in artificial or distant spaces through a combination of sensory breadth (a multisensory address), sensory depth (resolution), and interactivity (understood as the user's capacity to modify the mediated environment).[18] The concept of virtual reality thus encompassed technological configurations beyond the "goggles and gloves" arrangement, including physical installations such as the "responsive environments" that Myron Krueger developed in the 1970s and the Cave Automatic Virtual Environment (CAVE) that Daniel Sandin, Thomas DeFanti, and Carolina Cruz-Neira created in 1991.[19]

The interest in interactive simulated environments and mediated presence was, to be sure, bound up with developments in computing as well as cultural responses to them, especially William Gibson's science-fiction novel *Neuromancer* (1984), which popularized the term *cyberspace*. But it also drew on other recent trends including immersive film and video formats such as Cinerama and Sensorama as well as practices in art and performance, which harnessed various configurations of multisensory address, high resolution, and spectator engagement (and which can in turn be traced to multiscreen experiments of the long 1930s).[20] In bringing together this range of practices through the alignment of immersion with presence, the notion of virtual reality as it was conceptualized in the 1990s thus downplayed the significant material differences among various technological arrangements, including the use of screens with dramatically divergent sizes and levels of mobility. In doing so, it upheld the emphasis on dematerialization associated with virtuality in general. In line with my effort to parse the *dispositifs* associated with specific screen practices, the analysis I undertake here, by contrast, focuses on the conjunction of a particular technological arrangement (contemporary virtual-reality headsets) and form of representation (aerial spectacles), bracketing the concept of presence and focusing instead on the spatial formations produced at the juncture of virtual and actual space.

In adopting aerial imagery, contemporary commercial applications of virtual reality seem to reiterate the means by which cinema has long exploited and flaunted the immersivity of its screens. Consider, for instance, the virtual-reality video game *The Climb* (Crytek, 2016), which positions the player in a series of exotic mountainous landscapes (figure 5.1). The game has the player attempt to scale the steep edifices only to plummet upon misplacing her grip. As with a film such as *Avatar* (James Cameron, 2009), whose visual style the game recalls, *The Climb* thus harnesses a supposedly (but not actually) new immersive technology, together with digital imaging, both to plunge users into a spectacular space and to provide the visceral experience of plunging through that space.[21] Insofar as *The Climb* engulfs players in an aerial environment, it falls in line with several of the earlier screen practices I have discussed, from the use of rear-projection screens to make it seem as though actors were flying airplanes to the employment of extratheatrical film screens to acclimate viewers to air travel and the installation of film and television screens on actual airplanes. In *The Climb*'s focus on provoking the sensation of movement through the air, its exploitation of virtual reality also aligns with large-screen and widescreen exhibition formats such as Magnascope and Vitarama (as well as previous and subsequent instances such as Cinéorama and Cinerama), which have long

5.1 The virtual-reality video game *The Climb* has players ascend to vertiginous heights.



harnessed the spectacle of aerial motion to display the technologies' capacity not only seemingly to position viewers high above the earth but also to provide a visceral experience of kinesis.[22] In virtual reality, as in Magnascope and Vitarama, immersive screens contribute to the experience of kinesis by provoking the visual sensation of motion despite the user's or viewer's simultaneous felt experience of bodily stasis.

Such aerial spectacles have become a prominent component of contemporary blockbuster movies employing digital visual effects, often in conjunction with immersive exhibition formats such as 3D and IMAX. As Kristen Whissel argues, such spectacles shift emphasis away from the screen's *x* axis and toward its *y* and *z* axes, producing what she identifies as a "new verticality," which exploits the capacity of visual effects to create spectacles that defy the laws of physics. In emphasizing descent and ascent within the frame and foregrounding the pull of gravity and its defiance within the diegesis, such spectacles dramatize a range of polar oppositions relevant to global audiences and mark historical thresholds (social, political, and technological).[23] Especially insofar as many films and games being produced for virtual reality also make use of digital imaging, they are particularly closely aligned with recent films such as *Avatar* and *Gravity* (Alfonso Cuarón, 2013), which employ both immersive exhibition formats and computer-generated imagery to supply the sensation that viewers are defying gravity by flying or hovering in aerial environments alongside the characters.[24] Such works thus exemplify how the visual logic of verticality—if not necessarily its narrative function as what Whissel calls an "effects emblem"—traverses a range of forms, as she argues, aligning virtual reality with cinema, gaming, and comics.[25]

The virtual-reality headsets, however, transform the way screens collaborate with such spectacles to elicit immersion. Film formats of the long 1930s such as Magnascope, Grandeur, and Vitarama achieved their claims to immersivity by virtue of the scale—and, with Vitarama, curvature—of the screen, sometimes together with the employment of high-resolution images (as with Grandeur) or surround-sound systems (as with Vitarama). Not only did a large scale enable the screens to function as environments, but it also collaborated with other components of exhibition and representation to facilitate the perception of continuity between actual and depicted space, suggesting the extensivity of the represented realm as well as the viewer's proximity to it. Contemporary virtual-reality headsets, by contrast, push the boundaries of the screen frame beyond the viewer's field of vision not by virtue of the screen's scale but rather through its proximity to the eyes. Indeed, the small scale of the new screens and the viewer's proximity to them aligns the viewing arrangement they foster with those of the extratheatrical cinema and television screens that I have discussed as well as with those of earlier "peeping" devices such as stereoscopes and Kinetoscopes.[26] The contemporary headsets, we might note, can incorporate either dual screens (one screen for each eye) or a single screen divided into two frames (one frame for each eye). As with 3D cinema, the use of stereoscopy facilitates a sense of continuity between the viewer's bodily space and the represented imagery. Although the screens themselves do not possess the scale or dimensionality to engulf viewers, deployments of virtual reality can proffer a sense of vastness and depth through the representation of environments.

With the new screens, this arrangement transforms the construction of verticality, including the portrayal of aerial spectacles, and alters the forms of experience it elicits. With cinema and television, verticality is conveyed representationally (with relation to the depicted world) and graphically (with relation to the frame of the screen). These two forms of verticality often coincide, as when Kong, falling from the top of the Empire State Building at the end of *King Kong* (Merian C. Cooper and Ernest B. Schoedsack, 1933), also moves down along the *y* axis in the frame. But they can also diverge, as in a shot of the woman Kong plucks from the window of a skyscraper, where a high-angle view of her falling figure exploits *z* axis movement. Moving-camera shots depicting the action of plunging can also exploit the *z* axis, as when the camera is mounted at the back of diving airplanes in *Wings* (William Wellman, 1927). Significantly, however, in most cinematic and televisual viewing arrangements, the vertical orientation of the screen itself remains steadfast, matching the upright orientation of seated viewers' heads and bodies. Even when the onscreen depiction of vertical movement diverges from the screen's vertical orientation, as with the depiction of a descent that moves along the *z* axis, the screen's position persists in grounding that depiction, providing it a particular situation in actual space. Such instances of disjuncture in orientation provoke the form of pleasure that Scott Richmond attributes to conflict between the viewer's visual and vestibular senses—for example, by making it look to the viewer as though she is horizontal to the earth when she also feels herself sitting upright—especially when portrayed on large screens, which allow the spectacle to fill the viewer's field of vision.[27]

Contemporary employments of virtual reality also emphasize verticality representationally, as *The Climb* exemplifies. The animated virtual-reality film *Allumette* (Eugene Chung, 2016) also takes place in an aerial environment and articulates danger and redemption in terms of descent and ascent within that space. In this case, the viewer hovers alongside the characters, capable of looking up into the sky and down into the atmospheric depths. The prospect of catastrophe emerges when a burning ship threatens to fall onto a crowd gathered below. The protagonist's mother averts that disaster, sacrificing herself, by boarding the ship and steering it high into the sky, where it finally explodes, raining embers (figure 5.2). The virtual-reality film *Take Flight* (Daniel Askill, 2015), like *The Climb*, proffers the experience of vertical motion, here through the portrayal of an ascent. The viewer's perspective begins on a city street only

**5.2** In the virtual-reality film *Allumette* (Penrose Studios, 2016), the prospect of catastrophe emerges when a burning ship threatens to fall onto a crowd gathered below.

*Source*: Screen grab by author.

to rise quickly through the skyscrapers to a space above the clouds where that perspective hovers alongside several floating celebrities.

Despite these connections, such spectacles, as presented through contemporary virtual-reality headsets, diverge from cinema and television by divorcing vertical articulation from the frame of the screen and establishing it instead in relation to the user's body as it is oriented and positioned in space. Since the screen is now affixed to the user's face and mounted on the axis of her neck, screen space can appear not only to ring her body panoramically but also to exist above and below her head. Indeed, the capacity to present mediated space above and below the user's head represents a prevalent preoccupation of the films and games produced for the new systems. While the experience is similar to having a screen on the ceiling of a small exhibition space, it is different from having a screen on the floor (as in the CAVE system) since mediated space now rises to the level of the user's face, even in the place where she feels her body to be. Many virtual-reality applications portray the space below the user's head as empty, so that in looking down toward her body, the user instead sees vacant diegetic space. However, some applications, such as the virtual-reality film *Invasion!* (Eric Darnell, 2016), present animated bodies in the space below the user's neck. Others can make it seem as though the user is up to her neck in components of the setting. For example, the virtual-reality film *Dear Angelica* (Saschka Unseld, 2017) seemingly buries the user up to the neck in a represented bed, and the virtual-reality experience *The Night Café* (Mac Cauley, 2015) embeds her up to the neck in a represented bar counter.

The availability of mediated space above and below the head not only further enables represented space to surround the user but imbricates that gesture with the articulation of verticality. With *Allumette*, for instance, the looming catastrophe is visible only if the viewer looks down to below the place where she feels her body to be. *Take Flight* enables the viewer to watch the city recede in that space. And *The Climb* conveys how far the player has ascended—and how far she has to fall—by depicting the depth below her. Other virtual-reality applications emphasize verticality through the depiction of objects descending from above. Both *Invasion!* and the virtual-reality game *Trials of Tatooine* (Lucasfilm, 2016), for instance, feature spaceships that seem poised to land on top of the user. The virtual-reality film *Colosse* (Fire Panda, 2015) depicts a looming creature. Both *Invasion!* and *Colosse* alert the user to these overhead threats by having small characters run around and hide behind her, looking up in fear. Taking a cue from these characters, the user also cranes her neck and looks up to see the ship or creature approaching from above. In such cases, verticality is not relative to the screen or to the user's head but rather to the remainder of her body and her experience of gravity. Although the depicted ship and creature appear to move forward on the *z* axis relative to the user's face, in other words, their motion is perceived as a descent because it travels along the *y* axis relative to her torso and felt position on earth.

Underlying this shift in the articulation of verticality is a transformation in the relationship between the user's body, the screen, and represented space. With the cinema and television screens of the long 1930s and throughout much of the twentieth century, the space that appeared onscreen—whether diegetic or graphic, representational or abstract, static or moving—bore a stable relationship to the screen itself. Neither movement of the screen nor the position of the viewer affected that relationship. The screen could incorporate multiple images, as with split screen, and its dimensions could change, as with Magnascope. But even in such cases the screen continued, if dynamically, to operate as what Stephen Heath describes as both receiver and provider of the frame. As Heath puts it, the screen's alignment with the frame is "the basis of the spatial articulations a film will make, the start of its composition."[28] The fact that twentieth-century audiovisual screens simultaneously anchored represented space and gave it a specific position in actual space served as a condition of their application to spatial synthesis in the long 1930s.

With virtual-reality headsets, screens continue to give support to images, but they no longer anchor spatial articulation. Space is instead articulated through the mapping of orientation, position, and movement across actual and virtual realms.[29] It is this mapping that remains stable (if the system is functioning according to design), enabling the screen's relationship to represented space to become volatile. Hence, movement of the screen, propelled by the user's movement, results in transformation onscreen. Whereas no movement of the film screen or spectator will alter Kong's position in the frame in *King Kong*, when a player of *The Climb* moves her head to look down, the screen image shifts accordingly to provide a vertiginous high-angle view.

In permitting and emphasizing the correspondence of such movements, films and games produced for virtual-reality headsets alter the relationship between represented and actual space. The virtual-reality works that I have discussed map the orientation, position, and movement of the user's head onto the orientation, position, and movement of the visual perspective supplied by a virtual camera. As with the earlier use of HMDs such as the "Sword of Damocles," the new systems achieve this mapping through a process that involves tracking the user within actual space and updating the image display accordingly. Oculus Rift headsets, for instance, contain motion and position sensors (gyroscopes, accelerometers, and magnetometers), which work together to track the orientation of the user's head. The headsets also contain infrared light-emitting diodes (LEDs), which function in connection with external infrared cameras to track the user's position in space. Dynamic information on the orientation and position of the headset in actual space is employed, in turn, to orient, position, and move the field of view within a virtual environment.[30]

Thus, despite the fact that represented space has been untethered from the screen, it has not become unmoored. The employment of tracking systems—which measure movement not only in relation to the user's previous orientation and position but also in relation to external forces (e.g., gravity) and reference points (e.g., the infrared camera)—ties representation even more firmly to actual space.[31]

In response to the notion that virtual reality enables users to escape their bodies, scholars have long emphasized how it activates and indeed relies upon users' bodies.[32] Although my observations about the exploitation of verticality support that argument, they also make a further point. Far from tools for dematerialization, these applications of virtual reality rematerialize representation by anchoring it not only to the user's body as it interacts with virtual environment but also to the user's physical environment.[33] Through the use of tracking systems, onscreen representation is made to index the orientation and position of the user's body in actual space. The user's experience of represented space is also tied to actual space, especially in the exploitation of verticality, since it relies on her capacity to gauge up and down proprioceptively, a capacity that is anchored to the earth in part through the way gravity acts on the musculature and inner ear.[34] Thus, for instance, whereas the act of looking forward to see the woman fall in *King Kong* diverges from the viewer's position in and experience of space, the act of looking down to see the threatened townspeople in *Allumette* is grounded in that position and experience.

The forms of spatial synthesis and temporal synchronization that I have attributed to the screen practices of the long 1930s facilitated the construction of new spaces from the spatiotemporal coordination of fragments. Such practices, as I have shown, enabled the synthesis and synchronization of actual and virtual domains to reshape diegetic spaces, auditorium spaces, theater spaces, domestic spaces, and spaces of transit alike. The onscreen display of elements such as fire, water, and sky framed these new spaces as environments. Such mediated environments could proffer the experience of immersion in a couple ways. They could operate as alternate spaces into which viewers could seemingly enter, providing the experience of being "in the picture" often attributed to illusionistic media.[35] Or they could operate as architectures reshaping actual space, as multiscreen displays have been understood to do.[36] The aerial virtual-reality spectacles I have discussed here, by contrast, neither evoke the experience of movement into different spaces nor operate as architectures forming new spaces. Rather, they immerse users in a familiar worldly space imbricated with representation. This formulation may recall descriptions of "cyberspace" as an immaterial realm that substitutes for the material world, but I am suggesting something different. These applications of virtual reality spectacularize the forms of connection and tracking that enable the user's felt experience of the world to drive representation. In doing so, they are exposing phenomena (forms of connection and tracking) that constantly surround us but often remain invisible. Indeed, the exploitation of verticality in virtual reality echoes—and, perhaps, helps to make tangible—what scholars such as Lisa Parks have identified as a more pervasive vertical orientation associated with contemporary technologies of surveillance and control, such as satellites and drones.[37] In short, far from displacing or transforming the material world, this use of virtual reality penetrates its surface.

This formulation of immersion aligns the experience provided by recent applications of virtual reality with the experience associated with the contemporary proliferation of screens more broadly. As Francesco Casetti argues, the new proliferating screens function as "junctions of a complex circuit, characterized both by a continuous flow and by localized processes of configuration or reconfiguration of circulating images." As a result, we find ourselves immersed, as he contends, in the circulation of information.[38] As our interfaces to technological, social, political, and economic networks, screens operate as interlinked nodes in constantly changing formations. Although the screens may be small, these formations are so boundless and complex that they bear comparison to the sublime.[39] In this context, our proximity to a range of screens (especially those we wear or hold in our hands), in conjunction with the sheer scale of the networks to which they connect us, provokes immersion. Like other spaces of immersion—such as the panoramas of the nineteenth century and the multiscreen displays of the twentieth—the networks of the twenty-first century, as many have argued, are not only sites of apparent agency but also increasingly and pervasively means of control and capture.[40]

At roughly the same time as the resurgence of virtual reality, there has also been a swell of interest in practices that exploit the proliferation of screens by encouraging the simultaneous engagement with multiple screens. Some of these practices map the screens' relationships to actual space and to one another, as in *KL Dartboard* and *Darts* for iPad and iPhone (2010), which enabled players to launch virtual projectiles from phone to tablet.[41] With others, screens' relationship to one another is more informational than explicitly spatial, as in "second-screen" applications tied to television broadcasts.[42] Such multiscreen practices, like the virtual-reality practices I have discussed, manifest the screen's relation to other devices (e.g., through network connections) and to physical space (e.g., through the use of sensors and location tracking). In doing so, they also make visible the often hidden but nevertheless constant connections among our devices as well as the often hidden but materially and geographically situated infrastructures supporting them.[43] Multiscreen practices reveal such connections and infrastructures by materializing particular, though fleeting, configurations, thus providing users a (small) point of access to the vast and dynamic networks that pervade our environment yet, as Patrick Jagoda puts it, remain "accessible only at the edge of our sensibilities."[44]

This discussion of recent virtual-reality and multiscreen applications shows how contemporary illusionistic and nonillusionistic screen practices share overlapping *dispositifs*, as I have argued such practices also did in the long 1930s. Moreover, it distinguishes the contemporary from the historical *dispositifs* despite the commonality of material arrangements involving a multiplicity of small, mobile, light-emitting screens. In making hidden connections and structures visible, the new screen practices do not create new spaces so much as they penetrate existing ones. In this regard, they are aligned less closely with architectural objects such as windows or walls than with devices such as X-ray machines and scanners.[45] Upon their discovery at the end of the nineteenth century, X-rays presented the possibility of rendering the invisible visible by penetrating the body and revealing the skeleton, offering a form of "penetrating vision" that was considered both macabre and erotic.[46] In proffering entry into the bodily interior, X-rays crossed the boundary marked by the skin and concomitantly reconfigured the relationship between the interior and exterior. Insofar as such penetrating vision offered access to invisible realities and conjured spatial reorganizations via transparency, it, like the new mode of vision offered by cinema, was—as several scholars have argued—bound up with the spatial reconfigurations associated with modernism and modernity more broadly.[47] As we have seen, several screen practices associated with both film and television in the long 1930s also activated this form of transparency, and some were in fact compared with X-rays. In the second decade of the twenty-first century, however, new screen practices are conjuring a new form of transparency, and a different kind of penetrating vision is revealing a different kind of skeleton. Whereas X-rays simultaneously plumbed and constructed bodily space as a penetrable depth and film and television screens at once probed and formed geographical space as a bridgeable distance, the new screen practices both expose and actualize the expansiveness of the mediated space surrounding us.

# NOTES

## Introduction

1. Vannevar Bush, "As We May Think" (1945), in *The New Media Reader*, ed. Noah Wardrip-Fruin and Nick Montfort (Cambridge, MA: MIT Press, 2003), 45. The article was published in the *Atlantic Monthly* in July 1945. A condensed version appeared in *Life* on September 10, 1945.
2. See James M. Nyce and Paul Kahn, eds., *From Memex to Hypertext: Vannevar Bush and the Mind's Machines* (San Diego, CA: Academic Press, 1991).
3. See Paul N. Edwards, *The Closed World: Computers and the Politics of Discourse in Cold War America* (Cambridge, MA: MIT Press, 1996), esp. 44–52.
4. James M. Nyce and Paul Kahn, "A Machine for the Mind: Vannevar Bush's Memex," in *From Memex to Hypertext*, ed. Nyce and Kahn, 39–66; Jonathan Auerbach and Lisa Gitelman, "Microfilm, Containment, and the Cold War," *American Literary History* 19, no. 3 (Fall 2007): esp. 748–49.
5. On graphical user interfaces and the metaphor of the desktop, see Anne Friedberg, *The Virtual Window: From Alberti to Microsoft* (Cambridge, MA: MIT Press, 2006), 219–32.
6. For claims about the novelty of contemporary small-screen viewing arrangements, see Gabriele Pedullà, *In Broad Daylight: Movies and Spectators After the Cinema*, trans. Patricia Gaborik (New York: Verso, 2012); and Anne Friedberg, *Window Shopping: Cinema and the Postmodern* (Berkeley: University of California Press, 1993), 132–43. For examples of apparatus theory, see the articles collected in Teresa De Lauretis and Stephen Heath, eds., *The Cinematic Apparatus* (London: Macmillan, 1980), and Philip Rosen, ed., *Narrative, Apparatus, Ideology: A Film Theory Reader* (New York: Columbia University Press, 1986).
7. For this designation, see, for instance, Richard B. Jewell, *The Golden Age of Cinema: Hollywood, 1929–1945* (Malden, MA: Blackwell, 2007).

8. See Thomas Elsaesser, *Film History as Media Archaeology: Tracking Digital Cinema* (Amsterdam: Amsterdam University Press, 2016). Indeed, my project represents one way of taking up Erkki Huhtamo's call for an "archaeology of the screen." See Erkki Huhtamo, "Elements of Screenology: Toward an Archaeology of the Screen," *Iconics* 7 (2004): 31–82. Also see, in particular, Frank Kessler, "The Cinema of Attractions as *Dispositif*," in *The Cinema of Attractions Reloaded*, ed. Wanda Strauven (Amsterdam: Amsterdam University Press, 2006), 57–69.
9. Elsaesser, *Film History as Media Archaeology*, 371. Also see D. N. Rodowick, *The Virtual Life of Film* (Cambridge, MA: Harvard University Press, 2007), 31.
10. See especially John Belton, *Widescreen Cinema* (Cambridge, MA: Harvard University Press, 1992), 183–210, and William Paul, "Screening Space: Architecture, Technology, and the Motion Picture Screen," *Michigan Quarterly Review* 35, no. 1 (Winter 1996): 143–73.
11. For an exemplary representative of such work, see Margaret Morse, *Virtualities: Television, Media Art, and Cyberculture* (Bloomington: Indiana University Press, 1998).
12. On embodiment, see Mark B. N. Hansen, *New Philosophy for New Media* (Cambridge, MA: MIT Press, 2004). On the materiality of screens, see Will Straw, "Proliferating Screens," *Screen* 41, no. 1 (Spring 2000): 115–19; Mary Ann Doane, "The Close-Up: Scale and Detail in the Cinema," *differences* 14, no. 3 (2003): 89–111; and Haidee Wasson, "The Networked Screen: Moving Images, Materiality, and the Aesthetics of Size," in *Fluid Screens, Expanded Cinema*, ed. Janine Marchessault and Susan Lord (Toronto: University of Toronto Press, 2007), 74–95.
13. For the effort to historicize virtuality, see Oliver Grau, *Virtual Art: From Illusion to Immersion* (Cambridge, MA: MIT Press, 2003), and Friedberg, *The Virtual Window*. On immersion, see Erkki Huhtamo, "Encapsulated Bodies in Motion: Simulators and the Quest for Total Immersion," in *Critical Issues in Electronic Media*, ed. Simon Penny (Albany: State University of New York Press, 1995), 159–86, and Alison Griffiths, *Shivers down Your Spine: Cinema, Museums, and the Immersive View* (New York: Columbia University Press, 2008). On mobility, see Lynn Spigel, *Welcome to the Dreamhouse: Popular Media and Postwar Suburbs* (Durham, NC: Duke University Press, 2001), 60–103, and Charles R. Acland, "Curtains, Carts, and the Mobile Screen," *Screen* 50, no. 1 (Spring 2009): 148–66. On multiplicity, see Beatriz Colomina, "Enclosed by Images: The Eames' Multimedia Architecture," *Grey Room* 2 (Winter 2001): 6–29; Janine Marchessault, "Multi-screens and Future Cinema: The Labyrinth Project at Expo 67," in *Fluid Screens, Expanded Cinema*, ed. Marchessault and Lord, 29–51; and Fred Turner, *The Democratic Surround: Multimedia and American Liberalism from World War II to the Psychedelic Sixties* (Chicago: University of Chicago Press, 2013). On small screens, see Huhtamo, "Elements of Screenology," and Haidee Wasson, "The Other Small Screen: Moving Images at New York's World Fair, 1939," *Canadian Journal of Film Studies* 21, no. 1 (Spring 2012): 81–103.
14. Haidee Wasson, introduction to "In Focus: Screen Technologies," ed. Haidee Wasson, *Cinema Journal* 51, no. 2 (Winter 2012): 143.
15. Tom Gunning, "An Aesthetic of Astonishment: Early Film and the (In)Credulous Spectator," *Art and Text* 34 (Spring 1989): 31–45; Charles R. Acland and Haidee Wasson, eds., *Useful Cinema* (Durham, NC: Duke University Press, 2011); Noam M. Elcott, "Rooms of Our Time: László Moholy-Nagy and the Stillbirth of Multi-media Museums," in *Screen/Space: The Projected Image in Contemporary Art*, ed. Tamara Trodd (Manchester: Manchester University Press, 2011), 25–52. Also see Miriam Bratu Hansen, "Early Cinema, Late Cinema: Permutations of the Public Sphere," *Screen* 34, no. 3

(Autumn 1993): 197–210, and Noam M. Elcott, *Artificial Darkness: An Obscure History of Modern Art and Media* (Chicago: University of Chicago Press, 2016), esp. 165–228.

16. Friedberg, *The Virtual Window*, 150–51.
17. Here and throughout the book I follow Haidee Wasson in employing the term *extra-theatrical* rather than *nontheatrical*. See Haidee Wasson, "The Reel of the Month Club: 16mm Projectors, Home Theaters, and Film Libraries in the 1920s," in *Going to the Movies: Hollywood and the Social Experience of Cinema*, ed. Richard Maltby, Melvyn Stokes, and Robert C. Allen (Exeter, UK: University of Exeter Press, 2007), 218–20. On the problems with the designation *nontheatrical*, also see Barbara Klinger, *Beyond the Multiplex: Cinema, New Technologies, and the Home* (Berkeley: University of California Press, 2006), 2–6.
18. Francesco Casetti, "Notes on a Genealogy of the Excessive Screen," in *Screens*, booklet for Mellon Sawyer Seminar on Genealogies of the Excessive Screen, February–December 2017, Yale University (New Haven, CT: Yale University, 2017), http://dev.screens.yale.edu/sites/default/files/files/Screens_Booklet.pdf.
19. Friedberg, *The Virtual Window*, 149–80; Francesco Casetti, *The Lumière Galaxy: Seven Key Words for the Cinema to Come* (New York: Columbia University Press, 2015), 131–33. Also see in particular Giuliana Bruno, *Surface: Matters of Aesthetics, Materiality, and Media* (Chicago: University of Chicago Press, 2014).
20. See especially Nanna Verhoeff, *Mobile Screens: The Visual Regime of Navigation* (Amsterdam: Amsterdam University Press, 2012).
21. See Laura Mulvey, "Rear-Projection and the Paradoxes of Hollywood Realism," in *Theorizing World Cinema*, ed. Lúcia Nagib, Chris Perriam, and Rajinder Kumar Dudrah (London: I. B. Tauris, 2012), 212.
22. James Lastra, *Sound Technology and the American Cinema: Perception, Representation, Modernity* (New York: Columbia University Press, 2000), esp. 203–15.
23. Ben Schlanger, "Motion Picture Theatres of Tomorrow," *Motion Picture Herald*, Better Theatres sec., February 14, 1931, 57, 56.
24. Emily Thompson, *The Soundscape of Modernity: Architectural Acoustics and the Culture of Listening in America, 1900–1933* (Cambridge, MA: MIT Press, 2002), esp. 229–315.
25. Lev Manovich, *The Language of New Media* (Cambridge, MA: MIT Press, 2001), esp. 293–308.
26. Manovich, *The Language of New Media*, 298–99. For the argument that contemporary digital visual effects do not mark a radical break from earlier analog effects, see Stephen Prince, *Digital Visual Effects in Cinema: The Seduction of Reality* (New Brunswick, NJ: Rutgers University Press, 2012), 4–5.
27. See Beatriz Colomina, *Privacy and Publicity: Modern Architecture as Mass Media* (Cambridge, MA: MIT Press, 1994), and Friedberg, *The Virtual Window*, 117–23.
28. See in particular Miriam Bratu Hansen, "The Mass Production of the Senses: Classical Cinema as Vernacular Modernism," *Modernism/modernity* 6, no. 2 (April 1999): 59–77. On the new configurations of space and time arising with modernity, see Stephen Kern, *The Culture of Time and Space, 1880–1918* (Cambridge, MA: Harvard University Press, 2003).
29. My approach to screen technologies here draws on James Lastra's work and in many ways extends the project I began in my previous book. See Lastra, *Sound Technology and the American Cinema*, and Ariel Rogers, *Cinematic Appeals: The Experience of New Movie Technologies* (New York: Columbia University Press, 2013).
30. Arjun Appadurai, *Modernity at Large: Cultural Dimensions of Globalization* (Minneapolis: University of Minnesota Press, 1996), 33–36. For a discussion of the transnational

dynamics of cinema in this period, incorporating but decentering Hollywood, see Miriam Bratu Hansen, "Vernacular Modernism: Tracking Cinema on a Global Scale," in *World Cinemas, Transnational Perspectives*, ed. Nataša Ďurovičová and Kathleen Newman (New York: Routledge, 2010), 287–314.

31. On Hollywood's complex relationship with Germany in the 1930s, see Thomas Doherty, *Hollywood and Hitler, 1933–1939* (New York: Columbia University Press, 2013).
32. See Huhtamo, "Elements of Screenology," 54–59.
33. See especially Charles Musser, *The Emergence of Cinema: The American Screen to 1907* (Berkeley: University of California Press, 1990); Gunning, "An Aesthetic of Astonishment"; and Douglas Gomery, *Shared Pleasures: A History of Movie Presentation in the United States* (Madison: University of Wisconsin Press, 1992). On cinema's relationship with other technologies of modernity, also see in particular Tom Gunning, "Fritz Lang Calling: The Telephone and the Circuits of Modernity," in *Allegories of Communication: Intermedial Concerns from Cinema to the Digital*, ed. John Fullerton and Jan Olsson (Rome: John Libbey, 2004), 19–37.
34. See Ben Singer, "Early Home Cinema and the Edison Home Projecting Kinetoscope," *Film History* 2, no. 1 (Winter 1988): 37–69, and Huhtamo, "Elements of Screenology."
35. See Musser, *The Emergence of Cinema*, 16.
36. Maggie Valentine, *The Show Starts on the Sidewalk: An Architectural History of the Movie Theatre, Starring S. Charles Lee* (New Haven, CT: Yale University Press, 1996), 22; Gomery, *Shared Pleasures*, 19.
37. William Paul, *When Movies Were Theater: Architecture, Exhibition, and the Evolution of American Film* (New York: Columbia University Press, 2016), 86–88; Ross Melnick, *American Showman: Samuel "Roxy" Rothafel and the Birth of the Entertainment Industry* (New York: Columbia University Press, 2012), 62–63.
38. F. H. Richardson, *Motion Picture Handbook: A Guide for Managers and Operators of Motion Picture Theaters*, 2nd ed. (New York: Moving Picture World, 1912), 336–38. See also Richard Koszarski, *An Evening's Entertainment: The Age of the Silent Feature Picture, 1915–1928* (Berkeley: University of California Press, 1990), 10. On the emergence of picture palaces, see Eileen Bowser, *The Transformation of Cinema, 1907–1915* (Berkeley: University of California Press, 1990), 121–36.
39. Paul, *When Movies Were Theater*, 195–96.
40. On rear projection, see George E. Turner, "The Evolution of Special Visual Effects," in *The ASC Treasury of Visual Effects*, ed. George E. Turner (Hollywood, CA: American Society of Cinematographers, 1983), 46. On glass painting, see Dan North, "The Silent Screen, 1895–1927: Special/Visual Effects," in *Editing and Special/Visual Effects*, ed. Charlie Keil and Kristen Whissel (New Brunswick, NJ: Rutgers University Press, 2016), 46. On mirror shots, see Katharina Loew, "Magic Mirrors: The Schüfftan Process," in *Special Effects: New Histories/Theories/Contexts*, ed. Dan North, Bob Rehak, and Michael S. Duffy (London: British Film Institute, 2015), 67.
41. See Tino Balio, *Grand Design: Hollywood as a Modern Business Enterprise, 1930–1939* (Berkeley: University of California Press, 1993), 235–55; Lea Jacobs, *The Wages of Sin: Censorship and the Fallen Woman Film, 1928–1942* (Berkeley: University of California Press, 1995); and Eric Smoodin, *Regarding Frank Capra: Audience, Celebrity, and American Film Studies, 1930–1960* (Durham, NC: Duke University Press, 2004), 23–50. For a discussion from the time that identifies women as Hollywood's target audience while also highlighting the Jim Crow practices limiting the access of African Americans, also see Margaret Farrand Thorp, *America at the Movies* (1939; reprint, London: Faber and Faber, 1946), 17, 19.

42. Friedberg, *Window Shopping*, 66; Mary Ann Doane, *The Desire to Desire: The Woman's Film of the 1940s* (Bloomington: Indiana University Press, 1987), 33.
43. Arthur Knight, *Disintegrating the Musical: Black Performance and American Musical Film* (Durham, NC: Duke University Press, 2002); Miriam J. Petty, *Stealing the Show: African American Performers and Audiences in 1930s Hollywood* (Oakland: University of California Press, 2016).
44. On the imbrication of technology, style, and the politics of representation in *The Birth of a Nation* and *Gone with the Wind*, see Clyde Taylor, "The Re-birth of the Aesthetic in Cinema," in *The Birth of Whiteness: Race and the Emergence of U.S. Cinema*, ed. Daniel Bernardi (New Brunswick, NJ: Rutgers University Press, 1996), 15–37; Michael Rogin, *Blackface, White Noise: Jewish Immigrants in the Hollywood Melting Pot* (Berkeley: University of California Press, 1996), 14–16, 159–65; and Jacqueline Najuma Stewart, *Migrating to the Movies: Cinema and Black Urban Modernity* (Berkeley: University of California Press, 2005), 230–44. On Hattie McDaniel's role in *Gone with the Wind*, see Petty, *Stealing the Show*, 27–71.
45. See Jean-Louis Baudry, "The Apparatus: Metapsychological Approaches to the Impression of Reality in the Cinema," in *Narrative, Apparatus, Ideology*, ed. Rosen, 299–318, and Michel Foucault, "The Confession of the Flesh," in *Power/Knowledge: Selected Interviews and Other Writings, 1972–1977*, ed. Colin Gordon, trans. Colin Gordon, Leo Marshall, John Mepham, and Kate Soper (New York: Pantheon Books, 1980), 194–228. For discussions of this concept, also see Frank Kessler, "Notes on *Dispositif*," May 2010, http://www.frankkessler.nl/wp-content/uploads/2010/05/Dispositif-Notes.pdf, and François Albera and Maria Tortajada, eds., *Cine-Dispositives: Essays in Epistemology Across Media* (Amsterdam: Amsterdam University Press, 2015).
46. Jean-Louis Baudry, "Ideological Effects of the Basic Cinematographic Apparatus," in *Narrative, Apparatus, Ideology*, ed. Rosen, 295. Also see, for instance, Baudry, "The Apparatus," 316–17; Jean-Louis Comolli, "Machines of the Visible," in *The Cinematic Apparatus*, ed. De Lauretis and Heath, 122; Christian Metz, *The Imaginary Signifier: Psychoanalysis and the Cinema*, trans. Celia Britton, Annwyl Williams, Ben Brewster, and Alfred Guzzetti (Bloomington: Indiana University Press, 1982), 8, 49–53; Philip Rosen, introduction to part 3, "Apparatus," in *Narrative, Apparatus, Ideology*, ed. Rosen, 282; and Casetti, *The Lumière Galaxy*, 77.
47. Will Straw, "Pulling Apart the Apparatus," *Recherches Sémiotiques / Semiotic Inquiry* 31 (2011): 60–61. For discussions of these issues, also see, for instance, Linda Williams, introduction to *Viewing Positions: Ways of Seeing Film*, ed. Linda Williams (New Brunswick, NJ: Rutgers University Press, 1995), 1–20, and Judith Mayne, "Paradoxes of Spectatorship," in *Viewing Positions*, ed. Williams, 155–83.
48. See, for instance, Gunning, "An Aesthetic of Astonishment," and Miriam Bratu Hansen, *Babel and Babylon: Spectatorship in American Silent Film* (Cambridge, MA: Harvard University Press, 1991).
49. See Hansen, "Early Cinema, Late Cinema"; William Uricchio, "Historicizing Media in Transition," in *Rethinking Media Change: The Aesthetics of Transition*, ed. David Thorburn and Henry Jenkins (Cambridge, MA: MIT Press, 2003), 23–38; and Elsaesser, *Film History as Media Archaeology*. For reconsiderations of the cinematic apparatus in light of recent technological changes, see also Caetlin Benson-Allott, *Killer Tapes and Shattered Screens: Video Spectatorship from VHS to File Sharing* (Berkeley: University of California Press, 2013), and Rogers, *Cinematic Appeals*.
50. See, for instance, Elsaesser, *Film History as Media Archaeology*, 105–6. For a discussion of this issue, also see Straw, "Pulling Apart the Apparatus," 65–66.

51. Gilles Deleuze, "What Is a *Dispositif*?" in *Two Regimes of Madness: Texts and Interviews 1975–1995*, ed. David Lapoujade, trans. Ames Hodges and Mike Taormina (New York: Semiotext(e), 2006), 338–48; Giorgio Agamben, *What Is an Apparatus? And Other Essays*, trans. David Kishik and Stefan Pedatella (Stanford, CA: Stanford University Press, 2009); Kessler, "The Cinema of Attractions as *Dispositif*," 61–62; Elsaesser, *Film History as Media Archaeology*, 101–36; Casetti, *The Lumière Galaxy*, 67–97.
52. See Bruno Latour, *Reassembling the Social: An Introduction to Actor-Network-Theory* (New York: Oxford University Press, 2005); Jane Bennett, *Vibrant Matter: A Political Ecology of Things* (Durham, NC: Duke University Press, 2010); and Doreen Massey, *For Space* (Thousand Oaks, CA: Sage, 2005).
53. Agamben, *What Is An Apparatus?* 14. On Agamben's account of the apparatus and its relation to the work on film spectatorship that emerged in the wake of apparatus theory, also see Straw, "Pulling Apart the Apparatus," 62–63.
54. Casetti, *The Lumière Galaxy*, 79.
55. Elsaesser, *Film History as Media Archaeology*, 131. For an earlier essay developing these ideas in connection with historical work on early cinema, see Thomas Elsaesser, "The New Film History as Media Archaeology," *Cinémas* 14, nos. 2–3 (2004): 75–117.
56. On the way in which the notion of *dispositif* accommodates a machine's function as both an assembly of internal systems and a component of larger arrangements, see François Albera and Maria Tortajada, "The Dispositive Does Not Exist!" in *Cine-Dispositives*, ed. Albera and Tortajada, 22–23.
57. Bruno, *Surface*, 5.
58. Alexander R. Galloway, *The Interface Effect* (Malden, MA: Polity, 2012), vii.

## 1. Production Screens in the Long 1930s: Rear Projection and Special Effects

1. Orville Goldner and George E. Turner, *The Making of "King Kong": The Story Behind a Film Classic* (South Brunswick, NJ: Barnes, 1975), 58, 88–95; Linwood G. Dunn, "Creating Film Magic for the Original 'King Kong,'" *American Cinematographer* (hereafter *AC*), January 1977, 96–97.
2. See, in particular, Judith Mayne, "'King Kong' and the Ideology of Spectacle," *Quarterly Review of Film and Video* 1, no. 4 (1976): 373–87; James A. Snead, *White Screens/Black Images: Hollywood from the Dark Side*, ed. Colin McCabe and Cornel West (New York: Routledge, 1994), 1–27; Fatimah Tobing Rony, *The Third Eye: Race, Cinema, and Ethnographic Spectacle* (Durham, NC: Duke University Press, 1996), 157–191; Noël Carroll, "*King Kong*: Ape and Essence," in *Interpreting the Moving Image* (New York: Cambridge University Press, 1998), 118–42; Dan North, *Performing Illusions: Cinema, Special Effects, and the Virtual Actor* (New York: Wallflower Press, 2008), 66–93; Stephen Prince, *Digital Visual Effects in Cinema: The Seduction of Reality* (New Brunswick, NJ: Rutgers University Press, 2012), 33–34, 35, 190; and Lisa Purse, *Digital Imaging in Popular Cinema* (Edinburgh: Edinburgh University Press, 2013), 77–102.
3. See Purse, *Digital Imaging in Popular Cinema*, 87–90, and North, *Performing Illusions*, 91–92.
4. Julie Turnock, "The Screen on the Set: The Problem of Classical-Studio Rear Projection," *Cinema Journal* 51, no. 2 (Winter 2012): 157–62.
5. For theoretical explorations of these processes, see Anne Friedberg, *The Virtual Window: From Alberti to Microsoft* (Cambridge, MA: MIT Press, 2006); Nanna

Verhoeff, *Mobile Screens: The Visual Regime of Navigation* (Amsterdam: Amsterdam University Press, 2012); Giuliana Bruno, *Surface: Matters of Aesthetics, Materiality, and Media* (Chicago: University of Chicago Press, 2014); and Francesco Casetti, *The Lumière Galaxy: Seven Key Words for the Cinema to Come* (New York: Columbia University Press, 2015), esp. 157–60.

6. Friedberg, *The Virtual Window*, esp. 150–51.
7. The suggestion that cinema reached a peak at the end of the 1930s perhaps most famously appears in André Bazin, "The Evolution of the Language of Cinema," in *What Is Cinema?* vol. 1, ed. and trans. Hugh Gray (Berkeley: University of California Press, 1967), 30–31. On the "Motion Pictures' Greatest Year" campaign, see Catherine Jurca, *Hollywood 1938: Motion Pictures' Greatest Year* (Berkeley: University of California Press, 2012).
8. See Tino Balio's, Richard Maltby's, and David Bordwell and Kristin Thompson's contributions to Tino Balio, *Grand Design: Hollywood as a Modern Business Enterprise, 1930–1939* (Berkeley: University of California Press, 1993). On Hollywood cinema in this period, see also Ina Rae Hark, ed., *American Cinema of the 1930s: Themes and Variations* (New Brunswick, NJ: Rutgers University Press, 2007).
9. See, for instance, Michele Hilmes, *Hollywood and Broadcasting: From Radio to Cable* (Urbana: University of Illinois Press, 1990), 49–77; Charles Wolfe, "The Poetics and Politics of Nonfiction: Documentary Film," in Balio, *Grand Design*, 351–86; Haidee Wasson, *Museum Movies: The Museum of Modern Art and the Birth of Art Cinema* (Berkeley: University of California Press, 2005); Charles R. Acland and Haidee Wasson, eds., *Useful Cinema* (Durham, NC: Duke University Press, 2011); and Devin Orgeron, Marsha Orgeron, and Dan Streible, eds., *Learning with the Lights Off: Educational Film in the United States* (New York: Oxford University Press, 2012).
10. Dominique Païni, "The Wandering Gaze: Hitchcock's Use of Transparencies," in *Hitchcock and Art: Fatal Coincidences*, ed. Dominique Païni and Guy Cogeval (Montreal: Montreal Museum of Fine Arts, 2000), 69. On *King Kong*'s soundtrack, see Claudia Gorbman, *Unheard Melodies: Narrative Film Music* (Bloomington: Indiana University Press, 1987), 70–91. For a reassessment of *King Kong*'s score, also see Michael Slowik, *After the Silents: Hollywood Film Music in the Early Sound Era, 1926–1934* (New York: Columbia University Press, 2014).
11. Carroll, "*King Kong*," 135. See also Purse, *Digital Imaging in Popular Cinema*, 80–81.
12. Thomas Elsaesser, "The New Film History as Media Archaeology," *Cinémas* 14, nos. 2–3 (2004): 109.
13. See Snead, *White Screens/Black Images*, 17, and Rony, *The Third Eye*, 158–60, 176–77. On Noble M. Johnson and his Lincoln Motion Picture Company, see Jacqueline Najuma Stewart, *Migrating to the Movies: Cinema and Black Urban Modernity* (Berkeley: University of California Press, 2005), 111–13, 202–18.
14. Snead, *White Screens/Black Images*, 26, 27.
15. For a different reading of Kong's association with the sense of touch, see Cynthia Erb, *Tracking King Kong: A Hollywood Icon in World Culture*, 2nd ed. (Detroit: Wayne State University Press, 2009), 109–12. On *King Kong*'s effectiveness at frightening viewers upon its opening in 1933, see Mordaunt Hall, "A Fantastic Film in Which a Monstrous Ape Uses Automobiles for Missiles and Climbs a Skyscraper," *New York Times*, March 3, 1933, C12; Edwin Schallert, "'King Kong' Stirring Film," *Los Angeles Times*, March 27, 1933, A9; and "New Films: Keith Theatre 'King Kong,'" *Boston Daily Globe*, March 27, 1933, 9.

16. Paul Young, *The Cinema Dreams Its Rivals: Media Fantasy Films from Radio to the Internet* (Minneapolis: University of Minnesota Press, 2006), 73–135.
17. See David Bordwell, "The Classical Hollywood Style, 1917–60," in David Bordwell, Janet Staiger, and Kristin Thompson, *The Classical Hollywood Cinema: Film Style and Mode of Production to 1960* (New York: Columbia University Press, 1985), 55–59.
18. See Laura Mulvey, "Rear-Projection and the Paradoxes of Hollywood Realism," in *Theorizing World Cinema*, ed. Lúcia Nagib, Chris Perriam, and Rajinder Kumar Dudrah (London: I. B. Tauris, 2012), 212; and Adrian Danks, "Being in Two Places at the Same Time: The Forgotten Geography of Rear-Projection," in *B Is for Bad Cinema: Aesthetics, Politics, and Cultural Value*, ed. Claire Perkins and Constantine Verevis (Albany: State University of New York Press, 2014), 69.
19. On the relation between the "spatial limit" constituted by the frame and the "temporal limit" represented by the cut, see Mary Ann Doane, *The Emergence of Cinematic Time: Modernity, Contingency, the Archive* (Cambridge, MA: Harvard University Press, 2002), 31.
20. Friedberg, *The Virtual Window*, 157.
21. Vernon L. Walker, "Use of Miniatures in Process Backgrounds," *AC*, August 1934, 162.
22. I am borrowing the concept of spatial montage from Lev Manovich, but my argument challenges his contention that it is primarily a contemporary phenomenon. See Lev Manovich, *The Language of New Media* (Cambridge, MA: MIT Press, 2001), 322–26.
23. See, in particular, Prince, *Digital Visual Effects in Cinema*; Kristen Whissel, *Spectacular Digital Effects: CGI and Contemporary Cinema* (Durham, NC: Duke University Press, 2014); and Julie Turnock, *Plastic Reality: Special Effects, Technology, and the Emergence of 1970s Blockbuster Aesthetics* (New York: Columbia University Press, 2015). New anthologies include Dan North, Bob Rehak, and Michael S. Duffy, eds., *Special Effects: New Histories/Theories/Contexts* (London: British Film Institute, 2015), and Charlie Keil and Kristen Whissel, eds., *Editing and Special/Visual Effects* (New Brunswick, NJ: Rutgers University Press, 2016).
24. My approach to special effects vis-à-vis this nexus of issues has been guided by Miriam Hansen's discussion of the classical Hollywood cinema in "The Mass Production of the Senses: Classical Cinema as Vernacular Modernism," *Modernism/modernity* 6, no. 2 (April 1999): 59–77.
25. Bordwell, "The Classical Hollywood Style, 1917–60," 3–4.
26. David Bordwell and Kristin Thompson, "Technological Change and Classical Film Style," in Balio, *Grand Design*, 130–35.
27. Bordwell, "The Classical Hollywood Style, 1917–60," 59.
28. See Jean-Louis Baudry, "Ideological Effects of the Basic Cinematographic Apparatus," in *Narrative, Apparatus, Ideology: A Film Theory Reader*, ed. Philip Rosen (New York: Columbia University Press, 1986), 286–98, and Stephen Heath, "Narrative Space," in *Narrative, Apparatus, Ideology*, ed. Rosen, 379–420.
29. Païni, "The Wandering Gaze," 55.
30. Laura Mulvey, "A Clumsy Sublime," *Film Quarterly* 60, no. 3 (Spring 2007): 3; Mulvey, "Rear-Projection and the Paradoxes of Hollywood Realism," 207, 208. For a similar emphasis on the reflexivity and distancing effects of Hitchcock's rear projections, also see Christine Sprengler, *Hitchcock and Contemporary Art* (New York: Palgrave Macmillan, 2014), 91–117, and Elisabeth Bronfen, "Screening and Disclosing Fantasy: Rear Projection in Hitchcock," *Screen* 56, no. 1 (Spring 2015): 1–24.
31. Christian Metz, "*Trucage* and the Film," trans. Françoise Meltzer, *Critical Inquiry* 3, no. 4 (Summer 1977): 657–75, orig. pub. in French in 1972. Also see Christian Metz,

*Impersonal Enunciation, or the Place of Film*, trans. Cormac Deane (New York: Columbia University Press, 2016; orig. pub. in French in 1991), 52–59.

32. Païni, "The Wandering Gaze," 69.
33. Mulvey, "Rear-Projection and the Paradoxes of Hollywood Realism," 218.
34. See Turnock, "The Screen on the Set," 161.
35. See Hansen, "The Mass Production of the Senses," 71–72.
36. Turnock, *Plastic Reality*, esp. 9–12.
37. Vivian Sobchack, "*Detour*: Driving in Back Projection, or Forestalled by Film Noir," in *Kiss the Blood off My Hands: On Classic Film Noir*, ed. Robert Miklitsch (Urbana: University of Illinois Press, 2014), 117.
38. Bordwell, "The Classical Hollywood Style, 1917–60," 50, 52–53.
39. See Donald Crafton, *Shadow of a Mouse: Performance, Belief, and World-Making in Animation* (Berkeley: University of California Press, 2013), 144–212; Ray Zone, *Stereoscopic Cinema and the Origins of 3-D Film, 1838–1952* (Lexington: University Press of Kentucky, 2007), 143–49; William Paul, "The Aesthetics of Emergence," *Film History* 5, no. 3 (1993): 333; and Bordwell and Thompson, "Technological Change and Classical Film Style," 136–40.
40. There are also interesting connections between rear projection and animation practice, particularly via the multiplane camera, in the 1930s. Richard Rickitt contends that the multiplane model system developed for *King Kong* was "the stop-motion precursor" to Disney's multiplane camera (Richard Rickitt, *Special Effects: The History and Technique* [New York: Billboard Books, 2007], 184). Indeed, like the miniature set used for *King Kong*, the multiplane camera was presented as compatible with process backgrounds (W. E. Garity and W. C. McFadden, "Multiplane Camera Crane for Animation Photography," *Journal of the Society of Motion Picture Engineers* [hereafter *JSMPE*] 31, no. 2 [August 1938]: 147, and Turnock, *Plastic Reality*, 45).
41. Rick Altman, "Sound Space," in *Sound Theory Sound Practice*, ed. Rick Altman (New York: Routledge, 1992), 46–64; James Lastra, *Sound Technology and the American Cinema: Perception, Representation, Modernity* (New York: Columbia University Press, 2000), esp. 180–215.
42. Lastra, *Sound Technology and the American Cinema*, 201–3.
43. Lastra, *Sound Technology and the American Cinema*, 203, 208–9; Altman, "Sound Space," 53. Rick Altman also argues that rear projection was likely modeled on the practice of playing back recorded music while the primary action and dialogue were being filmed and recorded ("The Evolution of Sound Technology," in *Film Sound: Theory and Practice*, ed. Elisabeth Weis and John Belton [New York: Columbia University Press, 1985], 46–47).
44. John L. Cass, "The Illusion of Sound and Picture," *JSMPE* 14, no. 3 (March 1930): 325.
45. Cass, "The Illusion of Sound and Picture," 324; Altman, "Sound Space," 49–50.
46. See Mary Ann Doane, "Ideology and the Practice of Sound Editing and Mixing," in *Film Sound*, ed. Weis and Belton, 60.
47. Emily Thompson, *The Soundscape of Modernity: Architectural Acoustics and the Culture of Listening in America, 1900–1933* (Cambridge, MA: MIT Press, 2002), 2–4, 7, 277–85.
48. Mulvey, "Rear-Projection and the Paradoxes of Hollywood Realism," 211, 218.
49. See Walter Benjamin's discussion of the interpenetration of "body space" and "image space" in "Surrealism: The Last Snapshot of the European Intelligentsia," in *Selected Writings*, 4 vols., ed. Michael W. Jennings and others, trans. Rodney Livingstone and others (Cambridge, MA: Harvard University Press, 1996–2003), 2:217. Also see

Miriam Bratu Hansen, *Cinema and Experience: Siegfried Kracauer, Walter Benjamin, and Theodor W. Adorno* (Berkeley: University of California Press, 2012), 93, 152–53.

50. Walter Benjamin, "The Work of Art in the Age of Its Technological Reproducibility" (second version), in *Selected Writings*, 3:117, 3:110–13, 3:115–20. I refer to Benjamin's preferred, second version of this text, which he wrote between December 1935 and February 1936. The first version was composed in the fall of 1935; the more well-known third version, which was ultimately published in 1955, was written between the spring of 1936 and the spring of 1939. See the editor's notes in "The Work of Art in the Age of Its Technological Reproducibility" (second version), 3:122, and Walter Benjamin, "The Work of Art in the Age of Its Technological Reproducibility" (third version), in *Selected Writings*, 4:270.
51. Benjamin, "The Work of Art in the Age of Its Technological Reproducibility" (second version), 3:115, 3:117, 3:116.
52. See Bordwell and Thompson, "Technological Change and Classical Film Style," 130–31. For in-depth, nonacademic discussions of effects in Hollywood cinema, see Rickitt, *Special Effects*, and Raymond Fielding, *Special Effects Cinematography*, 4th ed. (Oxford: Focal Press, 1985). Versions of several of the technical articles referenced in subsequent notes are reprinted in George E. Turner, ed., *The ASC Treasury of Visual Effects* (Hollywood, CA: American Society of Cinematographers, 1983).
53. Farciot Edouart, "Paramount Transparency Process Projection Equipment," *JSMPE* 40, no. 6 (June 1943): 369.
54. G. A. Chambers, "Process Photography," *JSMPE* 18, no. 6 (June 1932): 782.
55. Farciot Edouart, "Economic Advantages of Process Photography," *Technical Bulletin*, supplement no. 9 (July 20, 1932): 1, Margaret Herrick Library Digital Collections, Academy of Motion Picture Arts and Sciences, http://digitalcollections.oscars.org/cdm/ref/collection/p15759coll4/id/327.
56. Turnock, *Plastic Reality*, 8.
57. "Biggest Stage on Earth Devoted Entirely to Special Process Work," *AC*, April 1929, 20.
58. Ralph G. Fear, "Projected Background Anematography," *AC*, January 1932, 11–12, 26; Vern Walker, "Special Process Technic [*sic*]," *JSMPE* 18, no. 5 (May 1932): 663–64; Chambers, "Process Photography," 786–87; Farciot Edouart, "The Transparency Projection Process," *AC*, July 1932, 15, 39; H. G. Tasker, "Current Developments in Production Methods in Hollywood," *JSMPE* 24, no. 1 (January 1935): 3–11; G. G. Popovici, "Background Projection for Process Cinematography," *JSMPE* 24, no. 2 (February 1935): 102–9.
59. George E. Turner, "The Evolution of Special Visual Effects," in *The ASC Treasury of Visual Effects*, ed. Turner, 46; Tom Gunning, "Phantasmagoria and the Manufacturing of Wonder: Towards a Cultural Optics of the Cinematic Apparatus," in *The Cinema: A New Technology for the 20th Century*, ed. André Gaudreault, Catherine Russell, and Pierre Véronneau (Lausanne, Switzerland: Éditions Payot Lausanne, 2004), 34. Also see Jennifer Wild, *The Parisian Avant-Garde in the Age of Cinema, 1900–1923* (Oakland: University of California Press, 2015), 50–51.
60. Edouart, "Paramount Transparency Process Projection Equipment," 368–69.
61. Edouart, "Paramount Transparency Process Projection Equipment," 369; Turner, "The Evolution of Special Visual Effects," 45; "Rear Projection Big Advance," *International Photographer* 10, no. 3 (April 1938): 31.
62. Edouart, "The Transparency Projection Process," 15.
63. "Progress in the Motion Picture Industry," *JSMPE* 20, no. 6 (June 1933): 464, 463. Thanks to John Belton for calling my attention to this information.

64. Frank Williams, "Trick Photography," *Transactions of the Society of Motion Picture Engineers* (hereafter *TSMPE*) 12, no. 34 (April 1928): 537–40. Also see John P. Fulton, "How We Made *The Invisible Man*," *AC*, September 1934, 200–201, 214.
65. On the Handschiegl process, see "New Photographic Process Is Launched," *AC*, August 1926, 23. On the Dunning process, see Carroll Dunning, "Composite Photography," *TSMPE* 12, no. 36 (September 1928): 975–79; Carroll Dunning, "Typical Problems in Process Photography," *TSMPE* 13, no. 38 (May 1929): 298–302; and C. H. Dunning, "Dunning Process and Process Backgrounds," *JSMPE* 17, no. 5 (November 1931): 743–48.
66. See, for instance, Walker, "Special Process Technic," and Chambers, "Process Photography." Also see Turnock, *Plastic Reality*, 30.
67. Jonathan Sterne and Tara Rodgers make a similar comparison between the processing of sound and the processing of food in "The Poetics of Signal Processing," *differences* 22, nos. 2–3 (2011): 31–53.
68. Dunning, "Dunning Process and Process Backgrounds," 747.
69. Edouart, "Economic Advantages of Process Photography," 3–5, 10.
70. Walter Blanchard, "Production Economies with Process Photography," *AC*, July 1934, 110–11, 118–19. Although the studio goes unnamed in Blanchard's article, a comparison with the writing of Farciot Edouart of Paramount suggests that his was the studio in question. See Edouart, "Paramount Transparency Process Projection Equipment," 369.
71. Blanchard, "Production Economies with Process Photography," 110.
72. Edouart, "Paramount Transparency Process Projection Equipment," 369–70.
73. Alfred B. Hitchins, "Duplex Optical Printers," *TSMPE* 11, no. 32 (September 1927): 771–74; Herford Tynes Cowling, "For Trick Work," *AC*, March 1928, 7, 22–23; Carl Louis Gregory, "An Optical Printer for Trick Work," *TSMPE* 12, no. 34 (April 1928): 419–26; Lynn Dunn, "Tricks by Optical Printing," *AC*, April 1934, 487, 496; L. Dunn, "Optical Printing and Technic," *JSMPE* 26, no. 1 (January 1936): 54–66; Linwood Dunn, "New Acme-Dunn Optical Printer," *JSMPE* 42, no. 4 (April 1944): 204–10.
74. As of October 1926, Eastman Duplicating Film had been available "for a number of years," but there had not yet been instructions for its use. See J. G. Capstaff and M. W. Seymour, "The Duplication of Motion Picture Negatives," *TSMPE* 10, no. 28 (February 1927): 229, and Barry Salt, *Film Style and Technology: History and Analysis*, 2nd ed. (London: Starword, 1992), 180, 195.
75. Gregory, "An Optical Printer for Trick Work," 420.
76. Fred W. Jackman, "The Special-Effects Cinematographer," *AC*, October 1932, 42; Dunn, "Optical Printing and Technic," 55.
77. On miniatures, see Don Jahraus, "Making Miniatures," *AC*, November 1931, 9–10, 42. On the Schüfftan process, see Katharina Loew, "Magic Mirrors: The Schüfftan Process," in *Special Effects*, ed. North, Rehak, and Duffy, 62–77.
78. Dunning, "Typical Problems in Process Photography," 300–301. Also see "Biggest Stage on Earth Devoted Entirely to Special Process Work," 20.
79. Edwin G. Lindin, "Destroying Pompeii—in Miniature," *AC*, December 1935, 519, 522–23; Fred M. Sersen, "Special Photographic Effects," *JSMPE* 40, no. 6 (June 1943): 376.
80. *The Rains Came* (1939), Foundation Program, September 29, 2007, Title no. 60471, Academy Film Archive, Academy of Motion Picture Arts and Sciences, Beverly Hills, CA. On the Academy Award for Best Special Effects, see the Official Academy Awards Database at http://awardsdatabase.oscars.org.
81. Gordon Jennings, "Special-Effects and Montage for 'Cleopatra,'" *AC*, December 1934, 354.

82. See Walker, "Use of Miniatures in Process Backgrounds," 162, and F. W. Jackman, "Evolution of Special-Effects Cinematography from an Engineering Viewpoint," *JSMPE* 29, no. 3 (September 1937): 299.
83. Goldner and Turner, *The Making of "King Kong,"* 62, 93; Rickitt, *Special Effects*, 184–86.
84. R. Riley, Natalie Kalmus, and W. E. Pohl, "Report for Special Effects Department Committee: Painting Matte Shots," 2–3, *Gone with the Wind* (Special Effects) file, Ronald Haver Collection, Margaret Herrick Library (hereafter MHL), Academy of Motion Picture Arts and Sciences, Beverly Hills, CA.
85. On metaphors for the screen, see Vivian Sobchack, *The Address of the Eye: A Phenomenology of Film Experience* (Princeton, NJ: Princeton University Press, 1992), 14–17, and Friedberg, *The Virtual Window*, 15–18. For an acknowledgment of the actual as well as metaphoric role that mirrors have played in the cinematic apparatus, see Christian Metz, *The Imaginary Signifier: Psychoanalysis and the Cinema*, trans. Celia Britton, Annwyl Williams, Ben Brewster, and Alfred Guzzetti (Bloomington: Indiana University Press, 1982), 51.
86. For discussions of these issues, see Research Council of the Academy of Motion Picture Arts and Sciences, "Recommendations on Process Projection Equipment," *JSMPE* 32, no. 6 (June 1939): 592, and A. F. Edouart, "Work of the Process Projection Equipment Committee of the Research Council," *JSMPE* 33, no. 3 (September 1939): 248–53.
87. Walker, "Special Process Technic," 664; Vernon Walker, "Saunders Cellulose Screen Reduces 'Hot Spot,'" *AC*, October 1932, 11; William Stull, "The Dieterich Process for Composite Photography," *AC*, March 1933, 9; H. D. Hineline, "Composite Photographic Processes," *JSMPE* 20, no. 4 (April 1933): 293–94; Popovici, "Background Projection for Process Cinematography," 102–9; H. Griffin, "New Background Projector for Process Cinematography," *JSMPE* 27, no. 1 (July 1936): 96; A. F. Edouart, "Paramount Triple-Head Transparency Process Projector," *JSMPE* 33, no. 2 (August 1939): 178.
88. Turnock, *Plastic Reality*, 21, 32–33.
89. Friedberg, *The Virtual Window*, esp. 133–38.
90. See Walker, "Saunders Cellulose Screen Reduces 'Hot Spot,'" 11; Hartley Harrison, "Problems of Background Projection," *AC*, January 1934, 353, 386; Popovici, "Background Projection for Process Cinematography," 105; G. G. Popovici, "Recent Developments in Background Projection," *JSMPE* 30, no. 5 (May 1938): 538; and D. B. Joy, W. W. Lozier, and M. R. Null, "Carbons for Transparency Process Projection in Motion Picture Studios," *JSMPE* 33, no. 4 (October 1939): 355.
91. See Edouart, "Economic Advantages of Process Photography," 7.
92. Walker, "Saunders Cellulose Screen Reduces 'Hot Spot,'" 11.
93. Fear, "Projected Background Anematography," 12.
94. See Loew, "Magic Mirrors," 63.
95. Loew, "Magic Mirrors," 73, 77 n. 50.
96. Fielding, *Special Effects Cinematography*, 290–91.
97. Walker, "Saunders Cellulose Screen Reduces 'Hot Spot,'" 11.
98. Walker, "Saunders Cellulose Screen Reduces 'Hot Spot,'" 11.
99. Popovici, "Recent Developments in Background Projection," 539.
100. Walker, "Saunders Cellulose Screen Reduces 'Hot Spot,'" 11; Arthur Campbell, "A Fireproof Process Screen," *AC*, February 1934, 406; Jackman, "Evolution of Special-Effects Cinematography from an Engineering Viewpoint," 296; Popovici, "Recent Developments in Background Projection," 539.
101. Popovici, "Background Projection for Process Cinematography," 107.
102. Popovici, "Recent Developments in Background Projection," 539.

103. Popovici, "Background Projection for Process Cinematography," 105.
104. Douglas Gomery, *Shared Pleasures: A History of Movie Presentation in the United States* (Madison: University of Wisconsin Press, 1992), 72–75; Maggie Valentine, *The Show Starts on the Sidewalk: An Architectural History of the Movie Theatre, Starring S. Charles Lee* (New Haven, CT: Yale University Press, 1996), 91–92.
105. By 1932, matte painting reportedly began replacing glass shots, and by 1937 glass shots were reputedly "seldom used" (Chambers, "Process Photography," 783; J. A. Norling, "Trick and Process Cinematography," *JSMPE* 28, no. 2 [February 1937]: 151). As one cinematographer put it in 1934, matte shots did not interfere with production to the degree that glass shots did, and they allowed "the director and cinematographer much greater freedom in their choice of set-ups" (Arthur J. Campbell, "Making Matte-Shots," *AC*, December 1934, 347).
106. Carl Louis Gregory, "Trick Photography," *TSMPE* 9, no. 25 (September 1926): 104–5.
107. Goldner and Turner, *The Making of "King Kong,"* 109.
108. Turnock notes a similar effect with the rise in color rear projection in the 1950s, arguing that the contrast between foreground and background planes disrupted "the fully convincing illusion of a whole" (*Plastic Reality*, 39). The point I am making here is meant to be broader, encompassing films that read as convincing illusions as well as films that don't.
109. Goldner and Turner, *The Making of "King Kong,"* 109–10.
110. See Sigfried Giedion, *Space, Time, and Architecture: The Growth of a New Tradition*, 5th rev. ed. (Cambridge, MA: Harvard University Press, 2008), 493–95; Anthony Vidler, *The Architectural Uncanny: Essays in the Modern Unhomely* (Cambridge, MA: MIT Press, 1992), 217–25; Friedberg, *The Virtual Window*, 117–23; Eve Blau, "Transparency and the Irreconcilable Contradictions of Modernity," *Praxis* 9 (Fall 2007): 50–59; and Wild, *The Parisian Avant-Garde in the Age of Cinema*, 23–61.
111. Judith Brown, *Glamour in Six Dimensions: Modernism and the Radiance of Form* (Ithaca, NY: Cornell University Press, 2009), 160, 162.
112. Ella Shohat and Robert Stam, *Unthinking Eurocentrism: Multiculturalism and the Media*, 2nd ed. (New York: Routledge, 2014), 104. On modern travel and the construction of geopolitical space, also see Stephen Kern, *The Culture of Time and Space, 1880–1918* (Cambridge, MA: Harvard University Press, 2003), 211–40.
113. On the way in which rear projection contributes to a colonial worldview in *Tarzan the Ape Man* (W. S. Van Dyke, 1932), also see Jacob Smith, *The Thrill Makers: Celebrity, Masculinity, and Stunt Performance* (Berkeley: University of California Press, 2012), 113.
114. See Balio, *Grand Design*, 193–94.
115. Cecil DeMille, "A Director Looks at 'Process-Shots,'" *AC*, November 1936, 458.
116. "*Union Pacific* Location Notes: Discussion: Mr. deMille [*sic*], Mr. Rosson, Mr. Pine," September 4, 1938, *Union Pacific* folder 3 (Production 1938), Paramount Production Records, MHL.
117. "Request for: (Retake) Added Money," December 21, 1938, *Union Pacific* folder 3 (Production 1938), Paramount Production Records.
118. See Beverly R. Singer, *Wiping the War Paint off the Lens: Native American Film and Video* (Minneapolis: University of Minnesota Press, 2001), 18–19, and Angela Aleiss, *Making the White Man's Indian: Native Americans and Hollywood Movies* (Westport, CT: Praeger, 2005), 40, 54–57.
119. Michelle H. Raheja, *Reservation Reelism: Redfacing, Visual Sovereignty, and Representations of Native Americans in Film* (Lincoln: University of Nebraska Press, 2010), 43, 44.

120. Balio, *Grand Design*, 194–95.
121. Raheja, *Reservation Reelism*, 39, 44.
122. See Raheja, *Reservation Reelism*, 1–45; Singer, *Wiping the War Paint Off the Lens*, 19–20; and Aleiss, *Making the White Man's Indian*, 54–57.
123. Danks, "Being in Two Places at the Same Time," 72.
124. Edouart, "The Transparency Projection Process," 15; Edouart, "Paramount Triple-Head Transparency Process Projector," 180.
125. Edouart, "Paramount Triple-Head Transparency Process Projector," 180.
126. Edouart, "Economic Advantages of Process Photography," 8–10. On average sizes for exhibition screens, see chapter 2, note 44.
127. Walker, "Saunders Cellulose Screen Reduces 'Hot Spot,'" 11.
128. Dunn, "Creating Film Magic for the Original 'King Kong,'" 96; King Kong 1932–1933, Recollections by Harold E. Wellman, A.S.C., folder 1015 (*King Kong*—Miscellaneous), Linwood G. Dunn Papers, MHL.
129. Walker, "Saunders Cellulose Screen Reduces 'Hot Spot,'" 11.
130. Earl Theisen, "Rear Projection," *Movie Makers*, September 1936, 404.
131. Edouart, "Paramount Triple-Head Transparency Process Projector," 180.
132. Edouart, "Paramount Triple-Head Transparency Process Projector," 180.
133. Walker, "Saunders Cellulose Screen Reduces 'Hot Spot,'" 11.
134. William Stull, "Process Shots Aided by Triple Projector," *AC*, August 1939, 363.
135. Edouart, "Paramount Triple-Head Transparency Process Projector," 171–84; Byron Haskin, "Development and Practical Application of the Triple-Head Background Projector," *JSMPE* 34, no. 3 (March 1940): 252–58; Stull, "Process Shots Aided by Triple Projector," 363–66, 376.
136. Edouart, "Paramount Triple-Head Transparency Process Projector," 172; Edouart, "Work of the Process Projection Equipment Committee of the Research Council," 252.
137. Research Council of the Academy of Motion Picture Arts and Sciences, "Recommendations on Process Projection Equipment," 608.
138. On the introduction and use of three-strip Technicolor in the 1930s, see Scott Higgins, *Harnessing the Technicolor Rainbow: Color Design in the 1930s* (Austin: University of Texas Press, 2007).
139. Stull, "Process Shots Aided by Triple Projector," 364.
140. Stull, "Process Shots Aided by Triple Projector," 364; Haskin, "Development and Practical Application of the Triple-Head Background Projector," 252–53.
141. Stull, "Process Shots Aided by Triple Projector," 376.
142. Edouart, "Paramount Transparency Process Projection Equipment," 371.
143. Winton Hoch, "Technicolor Cinematography," *JSMPE* 39, no. 2 (August 1942): 102.
144. Hoch, "Technicolor Cinematography," 102; "Rough Draft of Notes Taken at 2:30pm Meeting on 'Reap the Wild Wind' Today," October 10, 1940, *Reap the Wild Wind* folder 6 (Production), Paramount Pictures Production Records, MHL; Frank Caffey to Frank Gilbert, May 24, 1941, *Reap the Wild Wind* folder 6 (Production), Paramount Pictures Production Records. Although Hoch claimed that *Reap the Wild Wind* employed 28-foot-wide background screens, I believe this might have been a slight exaggeration because Farciot Edouart reported in 1943 that Paramount had worked successfully only "on a 24-foot screen in Technicolor" ("Paramount Transparency Process Projection Equipment," 372).
145. Edouart, "Paramount Transparency Process Projection Equipment," 372–73; Farciot Edouart, "The Evolution of Transparency Process Photography," *AC*, October 1943, 382.
146. Sersen, "Special Photographic Effects," 379.

147. Ralph Hammeras, quoted in Edouart, "Economic Advantages of Process Photography," 10.
148. Edouart, "Economic Advantages of Process Photography," 10, 9.
149. Edouart, "Economic Advantages of Process Photography," 8, 10, 9.
150. Fred Pelton, quoted in Edouart, "Economic Advantages of Process Photography," 9.
151. Edouart, "Economic Advantages of Process Photography," 10.
152. Mr. F. Gabourie to Messrs. Pelton, Cohn, Gibbons, Besevi, August 25, 1934, folder 66 (Set Construction—General), MGM Art Department Records, MHL.
153. See "Notes on Projection Process and Paint Frame," November 18, 1937, folder 18 (Mannix, E. J.), MGM Art Department Records. Also see Jack Cosgrove to Mr. Ginsberg, October 13, 1938, *Gone with the Wind* (Special Effects) file, Ronald Haver Collection, MHL.
154. Edouart, "Paramount Triple-Head Transparency Process Projector," 181, 183.
155. R. W. Henderson, "Developments in Time-Savings Process Projection Equipment," *JSMPE* 39, no. 4 (October 1942): 251; Edouart, "Paramount Transparency Process Projection Equipment," 372–73.
156. "Explanation of Transparency Process Prepared by Raymond A. Klune for Mr. William Paley, at Request of Mr. Selznick," July 3, 1941, *Gone with the Wind* (Special Effects) file, Ronald Haver Collection. On optical printing in *Citizen Kane*, see David Bordwell, "Film Style and Technology, 1930–60," in Bordwell, Staiger, and Thompson, *The Classical Hollywood Cinema*, 349.
157. Edouart, "Paramount Triple-Head Transparency Process Projector," 180–81; Haskin, "Development and Practical Application of the Triple-Head Background Projector," 258.
158. Farciot Edouart to unknown addressee, memo, April 19, 1939, *Dr. Cyclops* folder 2, Paramount Pictures Production Records; Frank Caffey to T. K. Glennan, May 26, 1939, *Dr. Cyclops* folder 2, Paramount Pictures Production Records; Budget for Transparencies, *Dr. Cyclops* folder 3, Paramount Pictures Production Records; A. J. Carpenter, "Reasons for Increases in Estimate on 'Doctor Cyclops,'" June 3, 1939, *Dr. Cyclops* folder 2, Paramount Pictures Production Records; 1st White Shooting Schedule, June 19, 1939, *Dr. Cyclops* folder 3, Paramount Pictures Production Records.
159. DeMille, "A Director Looks at 'Process-Shots,'" 458–59.
160. Sobchack, "*Detour*," 117.
161. Westerns continued to be made on location even after most other films were using rear projection. A report on one studio's use of process photography in 1933, for instance, indicated that fifty-one of the fifty-eight features made that year entailed process shots; of the remaining seven productions, six were Westerns "made almost exclusively on location, with very few studio-made scenes, and little or no need for any type of composite photography" (Blanchard, "Production Economies with Process Photography," 110). By the end of the 1930s, however, as I have mentioned, Westerns relied heavily on rear projection.
162. C. Wilson, "Production Problems of the Writer as Related to the Technician," *JSMPE* 26, no. 6 (June 1936): 672–73; A. Arnold Gillespie, *The Wizard of MGM: Memoirs of A. Arnold Gillespie, Art Director/Head of Special Effects from 1924–1965*, ed. Philip J. Riley and Robert A. Welch (Duncan, OK: BearManor Media, 2011), 290.
163. Jackman, "Evolution of Special-Effects Cinematography from an Engineering Viewpoint," 298.
164. Edouart, "Work of the Process Projection Equipment Committee of the Research Council," 252; 1st White Shooting Schedule, May 26, 1941, *Reap the Wild Wind* folder

11 (Schedules), Paramount Pictures Production Records; Fred W. Jackman, "'Process-Shot' Economies Made 'Captain Blood' Possible," *AC*, February 1936, 48–49, 61–62; Frank Mattison to T. C. Wright, March 30, 1940, folder 1486 (*The Sea Hawk* Curtiz), F002581, Warner Bros. Archives (WBA), School of Cinematic Arts, University of Southern California, Los Angeles; T. C. Wright to Mr. Wallis, March 27, 1940, folder 2230 (*The Sea Hawk* Story—Memos and Correspondence), F027652, WBA. *The Sea Hawk*, which was considered a "follow-up" to *Captain Blood*, also employed water backings animated through Warner Bros.' new Water Ripple and Wave Illusion Machine. See Anton Grot and Leo Kuter, "Freedom of the Seas—on a Sound Stage," *AC*, April 1941, 165, 186, and Hal Wallis to Harry Joe Brown, December 14, 1935, folder 2230 (*The Sea Hawk* Story—Memos and Correspondence), F000763, WBA.

165. Haskin, "Development and Practical Application of the Triple-Head Background Projector," 252. Also see Byron Haskin to Hal Wallis, November 10, 1937, *Gold Is Where You Find It* folder 1928, F002956, WBA; Hal Wallis to Haskins [*sic*], November 11, 1937, *Gold Is Where You Find It* folder 1928, F002956, WBA; Byron Haskin to Hal Wallis, November 12, 1937, *Gold Is Where You Find It* folder 1928, F002956, WBA; Frank Mattison to T. C. Wright, November 26, 1937, *Gold Is Where You Find It* folder 1494, F027619, WBA.
166. *Gone with the Wind* Set List, January 23, 1939, *Gone with the Wind* folder 45 (Misc.), Vertical File Collection, MHL; *Rebecca* Break-Down Script, August 24, 1939, pp. A (Ext. Manderley—Miniatures) and B (Process Plates and Stock Shots), *Rebecca* folder 628 (Prod.), Alfred Hitchcock Collection, MHL; numbered list of effects shots for *Aloma of the South Seas*, March 11, 1941, *Aloma of the South Seas* folder 89 (Production), Paramount Pictures Production Records; Shooting Schedule: 2nd White—Miniature Shooting Schedule, *The Forest Rangers* folder 6, Paramount Pictures Production Records.
167. "Notes on Typhoon," July 24, 1939, *Typhoon* folder 3 (Production), Paramount Pictures Production Records. Also see Request for Added Money, October 18, 1939, *Typhoon* folder 3 (Production), Paramount Pictures Production Records; "Additional transparency and straight shots required for TYPHOON as outlined by Louis King, Oct. 16th," *Typhoon* folder 3 (Production), Paramount Pictures Production Records; and Roy Hunter to Bert Heath, September 19, 1939, *Typhoon* folder 3 (Production), Paramount Production Records.
168. Gillespie, *The Wizard of MGM*, 293–94, 80, 76; *Test Pilot* Dialogue Cutting Continuity (Trailer), March 30, 1938, T-596: *Test Pilot* (1938), Turner/MGM Scripts, MHL.
169. John Durham Peters, *The Marvelous Clouds: Toward a Philosophy of Elemental Media* (Chicago: University of Chicago Press, 2015), 2, 101–2, 111–12.
170. Popovici, "Background Projection for Process Cinematography," 102–4; Griffin, "New Background Projector for Process Cinematography," 96; G. H. Worrall, "New Background Projector for Process Cinematography," *JSMPE* 32, no. 4 (April 1939): 442–44; Research Council of the Academy of Motion Picture Arts and Sciences, "Recommendations on Process Projection Equipment," 593; F. Gabourie to Mr. Gibbons, April 10, 1939, folder 1 (Camera Dept.), MGM Art Department Records; J. A. Gaylord to Mr. F. Gabourie, April 11, 1939, folder 1 (Camera Dept.), MGM Art Department Records.
171. Film historians such as David Bordwell, Kristin Thompson, and Donald Crafton have argued that camera movement was not as constrained by the introduction of synchronized sound as legend has it. See Bordwell and Thompson, "Technological Change and Classical Film Style," 116–17, 126, and Donald Crafton, *The Talkies: American Cinema's Transition to Sound, 1926–1931* (Berkeley: University of California Press, 1997), 230–32. Also see Salt, *Film Style and Technology*, 185.

172. Popovici, "Background Projection for Process Cinematography," 105–6. On Eastern Service Studios and its relationship to Paramount, see "Motion Pictures," *Brooklyn Daily Eagle*, May 30, 1936, http://bklyn.newspapers.com/image/52660681.
173. Popovici, "Background Projection for Process Cinematography," 107–8.
174. Tasker, "Current Developments in Production Methods in Hollywood," 5–6.
175. Cedric Gibbons to E. J. Mannix, November 17, 1937, folder 18 (Mannix, E. J.), MGM Art Department Records; A. Arnold Gillespie, "Big Ones out of Little Ones," manuscript, 10 vols., n.d., 7:7–8, 10:32, folders 1–3, A. Arnold Gillespie Manuscript Collection, MHL.
176. Tasker, "Current Developments in Production Methods in Hollywood," 6.
177. Edouart, "Economic Advantages of Process Photography," 6.
178. Gillespie, *The Wizard of MGM*, 290.
179. Fear, "Projected Background Anematography," 12.
180. Edouart, "Economic Advantages of Process Photography," 8.
181. DeMille, "A Director Looks at 'Process-Shots,'" 459.
182. Frank B. Good, "A Super-Portable Background-Projector," *AC*, August 1933, 148.
183. "Improved Mount for Process Projector," *AC*, June 1934, 71.
184. Gillespie, *The Wizard of MGM*, 294–97.
185. Jackman, "'Process-Shot' Economies Made 'Captain Blood' Possible," 49; Grot and Kuter, "Freedom of the Seas—on a Sound Stage," 165.
186. Hal Wallis to Michael Curtiz, September 30, 1935, folder 1788 (*Captain Blood* Story—Memos), F002891, WBA. Wallis continued to urge Curtiz to make sea backings more visible during production of *The Sea Hawk* in 1940. See Wallis to Curtiz, February 15, 1940, folder 2230 (*The Sea Hawk*—Story—Memos and Correspondence), F000761, WBA.
187. Friedberg, *The Virtual Window*, 160–62; Verhoeff, *Mobile Screens*, esp. 44–50.
188. On approaches to lighting women and the difficulty in maintaining precise control over lighting when shooting exteriors in this period, see Patrick Keating, *Hollywood Lighting: From the Silent Era to Film Noir* (New York: Columbia University Press, 2010), 127–33. On Hollywood lighting and the construction of white femininity, see Richard Dyer, *White* (New York: Routledge, 1997), 122–42. On Fay Wray's role in *King Kong* (and the argument that Ann Darrow's relation to race and gender is ambiguous), see Rhona J. Berenstein, *Attack of the Leading Ladies: Gender, Sexuality, and Spectatorship in Classic Horror Cinema* (New York: Columbia University Press, 1996), 185–96.

## 2. Theatrical Screens, 1926–1931: Transforming the Screen

1. Clarence W. D. Slifer, "Creating Visual Effects for *G.W.T.W.*," *American Cinematographer* (hereafter *AC*), August 1982, 788–91, 835–48; Ronald Haver, *David O. Selznick's Hollywood* (New York: Knopf, 1980), 252–58; Alan David Vertrees, *Selznick's Vision: "Gone with the Wind" and Hollywood Filmmaking* (Austin: University of Texas Press, 1997), 69–115; *Gone with the Wind* Set List, January 23, 1939, *Gone with the Wind* folder 45 (Misc.), Vertical File Collection, Margaret Herrick Library (hereafter MHL), Academy of Motion Picture Arts and Sciences, Beverly Hills, CA; *Gone with the Wind* Schedule of Cosgrove Shots, May 27, 1939, *Gone with the Wind* (Special Effects) file, Ronald Haver Collection, MHL.
2. See Tino Balio, *Grand Design: Hollywood as a Modern Business Enterprise, 1930–1939* (Berkeley: University of California Press, 1993), 1.

3. See Thomas Schatz, *The Genius of the System: Hollywood Filmmaking in the Studio Era* (New York: Pantheon, 1988), 176–81, and Balio, *Grand Design*, 207–9.
4. On the film's use of Technicolor, see Scott Higgins, *Harnessing the Technicolor Rainbow: Color Design in the 1930s* (Austin: University of Texas Press, 2007), 172–207.
5. John Belton, *Widescreen Cinema* (Cambridge, MA: Harvard University Press, 1992), 100.
6. Vertrees, *Selznick's Vision*, 73; Haver, *David O. Selznick's Hollywood*, 292.
7. David O. Selznick to Mr. Klune, May 4, 1939, *Gone with the Wind* folder 45 (Misc.), Vertical File Collection; David O. Selznick to Mr. Klune, May 8, 1939, *Gone with the Wind* folder 45 (Misc.), Vertical File Collection.
8. R. A. Klune to Mr. Selznick, May 12, 1939, *Gone with the Wind* (Special Effects) file, Ronald Haver Collection.
9. Slifer, "Creating Visual Effects for *G.W.T.W.*," 838; Belton, *Widescreen Cinema*, 100.
10. David O. Selznick to Mr. Kenneth Macgowan, October 14, 1955, in *Memo from David O. Selznick*, ed. Rudy Behlmer (New York: Random House, 2000), 454.
11. Slifer, "Creating Visual Effects for *G.W.T.W.*," 838.
12. Miriam J. Petty, *Stealing the Show: African American Performers and Audiences in 1930s Hollywood* (Oakland: University of California Press, 2016), 44.
13. Walter White to David O. Selznick, June 28, 1938, in Steve Wilson, *The Making of "Gone with the Wind"* (Austin: University of Texas Press, 2014), 58.
14. David O. Selznick, quoted in Wilson, *The Making of "Gone with the Wind*," 58; Wilson, *The Making of "Gone with the Wind*," 116.
15. Wilson, *The Making of "Gone with the Wind*," 59.
16. See Thomas Cripps, *Slow Fade to Black: The Negro in American Film, 1900–1942* (1977; reprint, New York: Oxford University Press, 1993), 363–64.
17. Melvin B. Tolson, "*Gone with the Wind* Is More Dangerous Than *Birth of a Nation*," *Washington Tribune*, March 23, 1940, reprinted in *Caviar and Cabbage: Selected Columns by Melvin B. Tolson from the "Washington Tribune," 1937–1944*, ed. Robert M. Farnsworth (Columbia: University of Missouri Press, 1982), 213–17. Also see Anna Everett, *Returning the Gaze: A Genealogy of Black Film Criticism, 1909–1949* (Durham, NC: Duke University Press, 2001), 290–99.
18. Selznick, quoted in Belton, *Widescreen Cinema*, 100; "Concerning the Presentation of *Gone with the Wind*," *Gone with the Wind* folder 45 (Misc.), Vertical File Collection.
19. I am borrowing this periodization from Donald Crafton, *The Talkies: American Cinema's Transition to Sound, 1926–1931* (Berkeley: University of California Press, 1997).
20. Belton, *Widescreen Cinema*, esp. 34–51; William Paul, *When Movies Were Theater: Architecture, Exhibition, and the Evolution of American Film* (New York: Columbia University Press, 2016), esp. 230–74.
21. H. F. Hopkins, "Considerations in the Design and Testing of Motion Picture Screens for Sound Picture Work," *Journal of the Society of Motion Picture Engineers* (hereafter *JSMPE*) 15, no. 3 (September 1930): 320–21; E. I. Sponable, "Historical Development of Sound Films," part 2, *JSMPE* 48, no. 4 (April 1947): 300; Rick Altman, "Sound Space," in *Sound Theory Sound Practice*, ed. Rick Altman (New York: Routledge, 1992), 47–49; Crafton, *The Talkies*, 235.
22. "Report on Projection Screens," *JSMPE* 20, no. 6 (June 1933): 515.
23. Paul, *When Movies Were Theater*, 235.
24. E. I. Sponable, "Historical Development of Sound Films," part 3, *JSMPE* 48, no. 5 (May 1947): 408–9.
25. See H. Rubin, "Some Problems in the Projection of Sound Movies," *Transactions of the Society of Motion Picture Engineers* (hereafter *TSMPE*) 12, no. 35 (September 1928):

867, and A. L. Raven, "Permeable Projection Screens for Sound Pictures," *TSMPE* 13, no. 38 (May 1929): 466.

26. F. M. Falge, "Motion Picture Screens—Their Selection and Use for Best Picture Presentation," *JSMPE* 17, no. 3 (September 1931): 349; Francis M. Falge, "How to Select a Proper Screen," *Motion Picture Projectionist*, October 1932, 9.
27. Paul, *When Movies Were Theater*, 195, 202–4, 237, 239.
28. For a similar formulation vis-à-vis the transparent screens employed for rear projection in France in earlier decades, see Jennifer Wild, *The Parisian Avant-Garde in the Age of Cinema, 1900–1923* (Oakland: University of California Press, 2015), esp. 33, 44.
29. On Transvox screens, see "Layout for W. E. Sound Equipment," *Motion Picture Projectionist*, November 1928, 16, and "New Motion Picture Apparatus," *TSMPE* 12, no. 36 (1928): 1168.
30. Hopkins, "Considerations in the Design and Testing of Motion Picture Screens for Sound Picture Work," 326.
31. Rubin, "Some Problems in the Projection of Sound Movies," 867.
32. Hopkins, "Considerations in the Design and Testing of Motion Picture Screens," 323–26.
33. "Report of the Screens Committee," *Motion Picture Projectionist*, November 1931, 35. The committee reported that the perforated screens that were currently in use, by contrast, had ratios of open to solid space of about 8 percent (34).
34. F. H. Richardson allowed an exception to this recommendation when "a considerable percentage of the audience views a picture from an angle greater than thirty-five degrees in relation to the screen" (*F. H. Richardson's Bluebook of Projection*, 6th ed. [New York: Quigley, 1935], 157). Also see F. H. Richardson, "The Case for Unperforated Screens," *Motion Picture Herald*, Better Theatres sec., September 24, 1932, 23–24; F. H. Richardson, "Perforated Screens and Their Faults," *JSMPE* 30, no. 3 (March 1938): 339–41; and Paul, *When Movies Were Theater*, 235–36.
35. Raven, "Permeable Projection Screens for Sound Pictures," 467.
36. On windows' performance of these functions, see Anne Friedberg, *The Virtual Window: From Alberti to Microsoft* (Cambridge, MA: MIT Press, 2006), esp. 111–13, 117.
37. On the term *sound projector*, taken to encompass both the horn and the receiver, see F. H. Richardson, "The Bluebook School," *Motion Picture Herald*, Better Theatres sec., April 25, 1931, 73.
38. James Lastra, *Sound Technology and the American Cinema: Perception, Representation, Modernity* (New York: Columbia University Press, 2000), 201–3.
39. Belton, *Widescreen Cinema*, 36–50.
40. The *New York Herald Tribune* explained: "A camera with two lenses, both of which record impressions on film through a single aperture[,] and new methods of development, projection and sound recording are features of the process. The double exposure camera places its identical impressions on the film in 'staggered' formation. This feature, Mr. Spoor said, gives the illusion of depth on the screen" ("Natural Vision Process for Films Is Perfected," *New York Herald Tribune*, August 5, 1930, 14). Also see, for instance, "Natural Vision Picture," *New York Times*, August 21, 1923, 6, and "New Type Movie Camera Promises Great Things," *Los Angeles Times*, January 9, 1927, B6.
41. A. S. Howell and J. A. Dubray, "Some Practical Aspects and Recommendations on Wide Film Standards," *JSMPE* 14, no. 1 (January 1930): 60, emphasis in original.
42. B. Schlanger, "Method of Enlarging the Visual Field of the Motion Picture," *JSMPE* 30, no. 5 (May 1938): 503.

43. Although industrial discourse at the time identified this screening at the Rivoli Theatre in December 1926 as the first use of Magnascope, the Magnascope lens was employed as early as February 1925 (H. Rubin, "The Magnascope," *TSMPE* 12, no. 34 [April 1928]: 403; Belton, *Widescreen Cinema*, 36, 244 n. 10).
44. Although regular (non-Magnascope) screens could reach 20–24 feet wide, the standard screen in this period was about 16–18 feet wide. As of 1932, the average four-thousand-seat picture palace still employed an 18-foot-wide screen. For its East Coast premiere at the Rivoli and its West Coast premiere at the Egyptian, *Old Ironsides* began with an 18-foot-wide screen image that expanded to 40 feet wide. On screen size, see Belton, *Widescreen Cinema*, 36; Ben M. Hall, *The Best Remaining Seats: The Story of the Golden Age of the Movie Palace* (New York: Bramhall House, 1961), 200; "Motion-Picture Invention Promises Startling Effects on Silver Sheet," *Los Angeles Times*, February 23, 1930, B9; Mayme Ober Peak, "Wide Film Makes Its Debut Auspiciously at Movie Capital," *Boston Daily Globe*, March 13, 1930, 29; F. H. Richardson, *Richardson's Handbook of Projection: The Blue Book of Projection*, 3 vols., 5th ed. (New York: Chalmers, 1927–1930), 1:243–45; "The World's Largest Screen," *Motion Picture Projectionist*, November 1932, 4; and Falge, "Motion Picture Screens," 352–54. On Magnascope's premieres at the Rivoli and the Egyptian, see Mordaunt Hall, "Pictures from Three Countries," *New York Times*, December 12, 1926, X9, and "'Magnascope' Is Amazing," *Los Angeles Times*, March 5, 1927, A7.
45. For discussions of the use of Magnascope in exhibiting these films, see Edwin Schallert, "'Old Ironsides' Stirring Film," *Los Angeles Times*, January 31, 1927, A7; Richard Watts Jr., "'Chang' Brings Life of Jungle Right Up Before Your Eyes," *New York Herald Tribune*, April 30, 1927, 12; and "'Wings' Film of Vivid Realism," *Boston Daily Globe*, December 20, 1927, 10.
46. Rubin, "The Magnascope," 404.
47. Hall, "Pictures from Three Countries," X9.
48. Paul, *When Movies Were Theater*, 231–33; "Technique of Wide Screen," *Times* (London), March 26, 1930, "Cinema: Screens" folder, Clippings Files, Billy Rose Theatre Division, New York Public Library for the Performing Arts, Lincoln Center.
49. "'Magnascope' Is Amazing," A7.
50. Quinn Martin, "Old Ironsides," no source or date, Scrapbook 5, p. 68, Audrey Chamberlin Scrapbooks, MHL.
51. "Movie Facts and Fancies," *Boston Daily Globe*, July 24, 1927, B14.
52. "'Wings' Film of Vivid Realism," 10.
53. Review of *Old Ironsides*, no source or date, Scrapbook 5, p. 67, Audrey Chamberlin Scrapbooks. On the shooting of parts of *Old Ironsides* at sea (the production also employed a set placed on rockers in the studio), see Production Detail Sheets 3–12, folder 147.f-2 (*Old Ironsides*—Production 1926, 1959), Paramount Pictures Production Records, MHL.
54. Richard Watts Jr., "Chang," no source or date, Scrapbook 19, p. 130, Audrey Chamberlin Scrapbooks. For a hyperbolic account of the dangers faced by the crew when shooting *Chang* on location, also see "Tremendous Forces of Nature Provide Menace in 'Chang,'" *Los Angeles Times*, June 26, 1927, C13, C14.
55. Seven-column Newspaper Strip Advertisement, 7AX, *WINGS*—Pressbook, 1928–1929, Paramount Pictures Press Sheets, MHL. On the production of *Wings*, also see Shooting Schedules, *Wings* (Prod. 618), folder 237.f-1 (*Wings*—General 1927), Paramount Pictures Production Records; Harry Perry, "Aerial Cinematography," *TSMPE* 12, no. 33 (April 1928): 162–66; and Jacob Smith, *The Thrill Makers: Celebrity, Masculinity, and Stunt Performance* (Berkeley: University of California Press, 2012), 166–68.

56. See Smith, *The Thrill Makers*, 169, 178–80. On the use of rear projection in *Test Pilot*, including the employment of miniatures in the backgrounds, see A. Arnold Gillespie, *The Wizard of MGM: Memoirs of A. Arnold Gillespie, Art Director/Head of Special Effects from 1924–1965*, ed. Philip J. Riley and Robert A. Welch (Duncan, OK: BearManor Media, 2011), 294.
57. Dominique Païni, "The Wandering Gaze: Hitchcock's Use of Transparencies," in *Hitchcock and Art: Fatal Coincidences*, ed. Dominique Païni and Guy Cogeval (Montreal: Montreal Museum of Fine Arts, 2000), 56. See also Laura Mulvey, "Rear-Projection and the Paradoxes of Hollywood Realism," in *Theorizing World Cinema*, ed. Lúcia Nagib, Chris Perriam, and Rajinder Kumar Dudrah (London: I. B. Tauris, 2012), 212.
58. "Alaska's Salmon War" played at the Washington, DC, Trans-Lux theater the week of December 10, 1937. *Spawn of the North* was released in August 1938. See Washington, DC, Trans-Lux program beginning Friday, December 10, [1937], Theatre Historical Society of America Archives, Pittsburg, PA. On the exhibition arrangement at the Trans-Lux in Washington, DC, see Ray Gingell, "Trans-Lux," *Marquee* 7, no. 3 (1975): 20–21.
59. On the use of 36-foot screens for *Spawn of the North*, see Farciot Edouart, "The Work of the Process Projection Equipment Committee of the Research Council," *JSMPE* 33, no. 3 (September 1939): 252.
60. Falge, "Motion Picture Screens," 355; "Large Versus Small Picture," *Motion Picture Herald*, Better Theatres sec., May 7, 1932, 32–33; Ben Schlanger, "Cinemas," *Architectural Record*, June 1938, 113.
61. Academy of Motion Picture Arts and Sciences, *Analysis of the Wide Film Problem*, Technical Bureau Report no. 17, September 16, 1930, box 10 (General Files Fr–Gr), "Grandeur 1928–1930" folder, Earl I. Sponable Papers (hereafter EISP), Rare Book and Manuscript Library, Columbia University Libraries, New York.
62. Academy of Motion Picture Arts and Sciences, *Analysis of the Wide Film Problem*.
63. "Introduction of Wide Film Helped to Release Photography from Its Old Limitations," *New York Herald Tribune*, April 20, 1930, C2.
64. W. B. Rayton, "The Optical Problems of Wide Film Motion Pictures," *JSMPE* 14, no. 1 (January 1930): 50.
65. Academy of Motion Picture Arts and Sciences, *Analysis of the Wide Film Problem*.
66. On 50mm film, see "Symposium on Large Screen Pictures," *Motion Picture Projectionist*, February 1931, 20, and John Belton, "Fox and 50mm Film," in *Widescreen Worldwide*, ed. John Belton, Sheldon Hall, and Steve Neale (New Barnet, Herts, UK: John Libbey, 2010), 14–15.
67. Loyd A. Jones, "Rectangle Proportions in Pictorial Composition," *JSMPE* 14, no. 1 (January 1930): 32–49; "Symposium on Large Screen Pictures," 20; Ben Schlanger, "The Screen: A Problem in Exhibition," *Motion Picture Herald*, Better Theatres sec., October 24, 1931, 66; Belton, "Fox and 50mm Film," 15–17.
68. Richard Watts Jr., "3d Dimension Effect in Films on Wide Screen," *New York Herald Tribune*, September 18, 1929, 23. Also see L. M. Jr., rough draft of text for Grandeur leaflet, n.d., box 10 (General Files Fr–Gr), "Grandeur 1928–1930" folder, EISP; "'Magnascope' Is Amazing," A7. For the complaint that the Grandeur screen did not fill the proscenium at Grauman's Chinese Theatre in Hollywood, see J. A. Dubray, Acting Manager of the Bell and Howell Company, to Professor A. C. Hardy, October 7, 1930, box 24 (General Files S.M.P.E.: Fe–Wi), "S.M.P.E.—Wide-Film Standardization Subcommittee 1930–1931" folder, EISP. For a discussion of the decision to install a Grandeur screen that did not fully span the proscenium at the Roxy, also see "Introduction of Wide Film Helped to Release Photography from Its Old Limitations," C2.

69. List of Exhibitions, 1929–1930, August 3, 1955, box 10 (General Files Fr–Gr), "Grandeur (History) 1960" folder, EISP. Also see "World's Largest Screen Shown at Roxy with Grandeur Film," *New York Herald Tribune*, February 16, 1930, G3, and "The Roxy's Grandeur Picture," *New York Times*, February 16, 1930, 123.
70. Souvenir program for *Happy Days* at Fox Carthay Circle Theatre, folder 464 (California, Los Angeles: Carthay Circle Theatre—Programs), Tom B'hend and Preston Kaufmann Collection, MHL. Also see Edwin Schallert, "Grandeur Film Arrival Event," *Los Angeles Times*, March 3, 1930, A7.
71. Advertisement for *Happy Days* at the Roxy Theatre, *New York Times*, February 16, 1930, X7. Also see advertisement for *Happy Days* at the Roxy Theatre, *New York Times*, February 21, 1930, folder 2829 (New York, Manhattan [Roxy Theatre]—Advertising 1927–1930), Tom B'hend and Preston Kaufmann Collection.
72. "Wide Screen Contest On," *Variety*, October 22, 1930, 42.
73. Program for *Fox Movietone Follies of 1929* at the Gaiety Theatre, box 10 (General Files Fr–Gr), "Grandeur 1928–1930" folder, EISP.
74. Belton, *Widescreen Cinema*, 38; Crafton, *The Talkies*, 94. This was well before the date Douglas Gomery identifies as the turning point in the coming of sound, May 11, 1928, when several studios signed contracts with Electrical Research Products Incorporated (ERPI), and it was months before the October premiere of *The Jazz Singer* (Alan Crosland, 1927) (*The Coming of Sound: A History* [New York: Routledge, 2005], 1). For a timeline of technological developments at Fox in 1926–1929, see Sponable, "Historical Development of Sound Films," part 3, 407–14.
75. E. I. Sponable to W. C. Michel, June 24, 1932, box 10 (General Files Fr–Gr), "Grandeur 1932" folder, EISP; Belton, "Fox and 50mm Film," 12.
76. Courtland Smith to William Fox, October 8, 1927, emphasis removed, box 10 (General Files Fr–Gr), "Grandeur 1932" folder, EISP.
77. Earl Sponable, memo on Grandeur conference, August 3, 1928, box 10 (General Files Fr–Gr), "Grandeur 1932" folder, EISP. This is consistent with Douglas Gomery's claim that "in May 1928 William Fox declared that 100 percent of the upcoming production schedule would be 'Movietoned'" ("The Coming of Sound: Technological Change in the American Film Industry," in *Film Sound: Theory and Practice*, ed. Elisabeth Weis and John Belton [New York: Columbia University Press, 1985], 18), although it should be noted that Fox continued to make dual versions (silent and sound) of its films in this period. See Crafton, *The Talkies*, 170.
78. Belton, "Fox and 50mm Film," 11.
79. See Erik Barnouw, *Tube of Plenty: The Evolution of American Television*, 2nd rev. ed. (New York: Oxford University Press, 1990), 61, and Joseph H. Udelson, *The Great Television Race: A History of the American Television Industry, 1925–1941* (Tuscaloosa: University of Alabama Press, 1982), 31.
80. C. Francis Jenkins, "Radio Movies and the Theater," *TSMPE* 11, no. 29 (July 1927): 47, 45.
81. Belton, "Fox and 50mm Film," 11.
82. Arthur Edeson, A.S.C., "Wide Film Cinematography," *AC*, September 1930, 8.
83. Program for *Fox Movietone Follies of 1929* at Gaiety Theatre.
84. L. M. Jr., rough draft of text for Grandeur leaflet
85. Mordaunt Hall, "Grandeur Films Thrill Audience," *New York Times*, September 18, 1929, 34.
86. Peak, "Wide Film Makes Its Debut Auspiciously at Movie Capital," 29; Schallert, "Grandeur Film Arrival Event," A7. On *Happy Days*'s status as the first film made specifically for Grandeur, see "The Roxy's Grandeur Picture," 123.

87. For discussions of this film's visual and aural style, see in particular Paul, *When Movies Were Theater*, 255–69, and Luke Stadel, "Natural Sound in the Early Talkie Western," *Music, Sound, and the Moving Image* 5, no. 2 (Autumn 2011): 113–36.
88. See Edwin Schallert, "Scenic Backgrounds High Light Vidor's 'Billy the Kid,'" *Los Angeles Times*, October 19, 1930, H3; Rob Reel, "Natural Vision Puts Chicago on Film Map Again," manuscript, no source, n.d., folder 11 (Scrap Book Clippings About Mr. Spoor), George Spoor and Essanay Film Manufacturing Company Manuscript Collection, 1918–1972, Chicago History Museum; and Tom Pettey, "New Photography Produces Lifelike Films for Screen," *Chicago Daily Tribune*, December 21, 1930, E1.
89. See Reel, "Natural Vision Puts Chicago on Film Map Again."
90. Mordaunt Hall, "The Screen," *New York Times*, March 1, 1930, 25.
91. H. F. Jermain to E. I. Sponable, October 7, 1929, box 10 (General Files Fr–Gr), "Grandeur 1928–1930" folder, EISP; Watts, "3d Dimension Effect in Films on Wide Screen," 23; Hall, "Grandeur Films Thrill Audience," 34.
92. Review of *Happy Days*, *Variety*, February 19, 1930, 21.
93. L. M. Jr., rough draft of text for Grandeur leaflet; Belton, *Widescreen Cinema*, 50. This publicity strategy was prescient of Fox's exploitation of CinemaScope in 1953. See Ariel Rogers, *Cinematic Appeals: The Experience of New Movie Technologies* (New York: Columbia University Press, 2013), 39–41.
94. See Paul, *When Movies Were Theater*, 256, 265.
95. Review of *Happy Days*, *Variety*, 21. See also Belton, *Widescreen Cinema*, 50, and Crafton, *The Talkies*, 315.
96. Hall, "Grandeur Films Thrill Audience," 34.
97. Watts, "3d Dimension Effect in Films on Wide Screen," 23.
98. "Wide Screen, Sponsored by Fox, to Be Demonstrated Tuesday," *New York Herald Tribune*, September 15, 1929, G3.
99. Siegfried Kracauer, "The Mass Ornament," in *The Mass Ornament: Weimar Essays*, ed. and trans. Thomas Y. Levin (Cambridge, MA: Harvard University Press, 1995), 75–86.
100. Ralph Fear, quoted in "Methods of Securing a Large Screen Picture (Open Discussion at the October, 1930 Meeting at New York, NY)," *JSMPE* 16, no. 1 (January 1931): 82.
101. Edeson, "Wide Film Cinematography," 21.
102. Philip K. Scheuer, "New Hit Made by Wide Film," *Los Angeles Times*, October 5, 1930, B13.
103. Review of *The Big Trail*, no source, n.d., Scrapbook 17, p. 136, Audrey Chamberlin Scrapbooks.
104. Paul, *When Movies Were Theater*, 265.
105. On the similar presentation of a minstrel troupe in Al Jolson's (nonwidescreen) film *Mammy* (Michael Curtiz, 1930), see Michael Rogin, *Blackface, White Noise: Jewish Immigrants in the Hollywood Melting Pot* (Berkeley: University of California Press, 1996), 119–20.
106. See Rogin, *Blackface, White Noise*; Eric Lott, *Love and Theft: Blackface Minstrelsy and the American Working Class* (1993; reprint, New York: Oxford University Press, 2013); and Arthur Knight, *Disintegrating the Musical: Black Performance and American Musical Film* (Durham, NC: Duke University Press, 2002).
107. Petty, *Stealing the Show*, 9. See also Jacqueline Najuma Stewart, *Migrating to the Movies: Cinema and Black Urban Modernity* (Berkeley: University of California Press, 2005), 57, and Cripps, *Slow Fade to Black*, 118.
108. Knight, *Disintegrating the Musical*, 33–34, 49–91; Petty, *Stealing the Show*.

109. For formulations of blackface as distorted mirror, disguise, barrier, and mode of display or presencing, see Lott, *Love and Theft*, 8, 37, 50; Rogin, *Blackface, White Noise*, 37, 79, 99, 112; and Knight, *Disintegrating the Musical*, 30–31, 47.
110. Siegfried Kracauer, "Cult of Distraction," in *The Mass Ornament*, 325, emphasis removed; Kracauer, "The Mass Ornament," 76.
111. On segregation in American movie theaters, see Douglas Gomery, *Shared Pleasures: A History of Movie Presentation in the United States* (Madison: University of Wisconsin Press, 1992), 155–63, and Gregory A. Waller, *Main Street Amusements: Movies and Commercial Entertainment in a Southern City, 1896–1930* (Washington, DC: Smithsonian Institution Press, 1995).
112. "Screen Masks," *Motion Picture Herald*, Better Theatres sec., April 11, 1931, 113.
113. "World's Largest Screen Shown at Roxy with Grandeur Film," G3.
114. R. H. McCullough, "*Grandeur* Wide Film System," *Motion Picture Projectionist*, April 1930, 28. On a similar use of curtains, see Irvin L. Scott, "A Motion Picture Theater for a Suburban Town in New York," *Architectural Record*, August 1931, 114.
115. Francis M. Falge, "Selection and Maintenance of Screens," *Motion Picture Herald*, Better Theatres sec., August 1, 1931, 76; Francis M. Falge, "Selection and Use of Screens," *Motion Picture Projectionist*, October 1931, 22.
116. Belton, *Widescreen Cinema*, 49.
117. Advertisement for Vallen Automatic Screen Modifier, *Motion Picture Herald*, Better Theatres sec., April 11, 1931, 85.
118. "Screen Masks," 113.
119. Hall, *The Best Remaining Seats*, 200. Also see, for instance, "Throng Attends Uptown Theatre on Opening Night to Enjoy Latest Devices of Talkies and Movies," *Boston Daily Globe*, August 3, 1929, 2.
120. Reporting on the RKO Roxy's screen, *Motion Picture Projectionist* claimed that screen modifiers had previously "been limited to two dimensions, one for the ordinary picture, and the second for the so-called wide-film magnifying size" ("The World's Largest Screen," 4), but the Los Angeles Theatre featured a three-position screen modifier when it opened in 1931 (William Crouch, "Ultra-modern Facility in Period Design," *Motion Picture Herald*, Better Theatres sec., February 14, 1931, 29). On the screen modifier at RKO Roxy, see Arthur E. Clark, "Mechanizing the Modern Stage," *Motion Picture Herald*, Better Theatres sec., August 27, 1932.
121. F. H. Richardson, "Projection in Radio City," *Motion Picture Herald*, Better Theatres sec., December 17, 1932, 18.
122. Mordaunt Hall, "New RKO Roxy Opens with an Unusually Interesting Stage Show and an Excellent Picture," *New York Times*, December 30, 1932, 14.
123. See, for instance, *Motion Picture Herald*, Better Theatres sec., February 9, 1935, 33, and November 11, 1939, 34.
124. Sergei Eisenstein, "The Dynamic Square," in *Film Essays and a Lecture*, ed. Jay Leyda (New York: Praeger, 1970), 52.
125. Belton, *Widescreen Cinema*, 40–41.
126. Will Whitmore, "Combining Talking Pictures and the Stage," *Motion Picture Herald*, Better Theatres sec., April 11, 1931, 61. See also Richard Koszarski, *Hollywood on the Hudson: Film and Television in New York from Griffith to Sarnoff* (New Brunswick, NJ: Rutgers University Press, 2008), 258.
127. Fred Westerberg, "Symposium on Wide Film Proportions," *Motion Picture Projectionist*, November 1930, 21.
128. Westerberg, "Symposium on Wide Film Proportions," 21.

129. "Report of the Screens Committee," *Motion Picture Projectionist*, December 1931, 25. Also see "Report of the Screens Committee," *Motion Picture Projectionist*, November 1931, 33–36.
130. Frederic Arden Pawley, "Design of Motion Picture Theaters," *Architectural Record*, June 1932, 432.
131. "Wide Film Goes Back on the Shelf as Lens Makers Quit," *Motion Picture Herald*, Better Theatres sec., May 9, 1931, 23; "Progress in the Motion Picture Industry," *JSMPE* 17, no. 1 (July 1931): 71; Belton, *Widescreen Cinema*, 52–68; Belton, "Fox and 50mm Film," 14–18.
132. Ross Melnick, *American Showman: Samuel "Roxy" Rothafel and the Birth of the Entertainment Industry* (New York: Columbia University Press, 2012), 277, 268. See also advertisement for the opening of the Roxy Theatre, no source or date, folder 79.f-2829 (New York, Manhattan [Roxy Theatre]—Advertising 1927–1930), Tom B'hend and Preston Kaufmann Collection; Hall, *The Best Remaining Seats*, 5–10; and "A Trip Through the New Roxy Theatre," *New York Times*, February 27, 1927, X7.
133. Crafton, *The Talkies*, 93.
134. "Announce New Features for Roxy Theatre," *Motion Picture News*, February 18, 1927, 564.
135. Advertisement for the opening of the Roxy Theatre, Tom B'hend and Preston Kaufmann Collection; "New Roxy Theatre Will Open Tonight," *New York Times*, March 11, 1927, 24.
136. Mordaunt Hall, "New Roxy Theatre Has Gala Opening," *New York Times*, March 12, 1927, 12.
137. "New Type Movie Camera Promises Great Things," B6.
138. "Announce New Features for Roxy Theatre," 564.
139. "Arthur Smith with Roxy," *Motion Picture News*, February 18, 1927, 606. This mention of a 44-foot-wide screen occurred in the buildup to the theater's opening. Because press coverage of the Roxy's opening does not, as far as I have seen, mention the size of the screen, I do not believe the screen was as record breaking as predicted. When Fox installed its 42-foot-wide Grandeur screen at the Roxy in 1930, moreover, that screen was reported to be the "largest screen ever constructed"—and to have replaced this theater's standard 24-foot-wide screen. See Mordaunt Hall, "Grandeur's First Real Talker Shown," *New York Times*, February 14, 1930, 26, and "World's Largest Screen Shown at Roxy with Grandeur Film," G3.
140. "Roxy Theatre Opening Set for March 11," *Motion Picture News*, March 11, 1927, 865.
141. "'Magnascope' Is Amazing," A7.
142. At the time of the Roxy's opening, it was claimed that the first film in Natural Vision "now is being produced"; however, shorts in Natural Vision had already been completed by this point. See "Roxy Theatre Opening Set for March 11," 865; Maurice Kann, "A House Built on Merit," *Film Daily*, special section on Roxy, March 13, 1927, 26; Belton, *Widescreen Cinema*, 49.
143. Melnick, *American Showman*, 279; Crafton, *The Talkies*, 93.
144. "The Roxy's Grandeur Picture," 123.
145. See "C'Scope Gets Spotlight as 'Robe' Preem Cues 20th's 'Biggest Gamble,'" *Variety*, September 16, 1953, 1.
146. Melnick, *American Showman*, 51, 60–64.
147. "Roxy Theatre Opening Set for March 11," 865. Although this was the first theatrical use of a Trans-Lux screen, it was not the first time rear projection was used

theatrically. Rear projection had been employed regularly in Europe for years by this time, and it had been used for a presentation at Madison Square Garden in 1925. In 1927, F. H. Richardson listed translucent screens as one of "seven classes of screens"; however, he claimed that rear projection was "but very little used" in exhibition, and, when employed, it was "as more or less of a makeshift." See William Mayer, "Trans Lux Rear Stage Projection," *Motion Picture Projectionist*, October 1931, 12; "Report of the Projection Screens Committee," *JSMPE* 18, no. 2 (February 1932): 248; Wild, *The Parisian Avant-Garde in the Age of Cinema, 1900–1923*, 23–61; and Richardson, *Richardson's Handbook of Projection*, 1:226, 1:233.

148. Review of Roxy Theatre opening, *Variety*, March 16, 1927, 22; Hall, *The Best Remaining Seats*, 10.
149. "Bledsoe, Jules," in *Encyclopedia of the Harlem Renaissance*, vol. 1, ed. Cary D. Wintz and Paul Finkelman (New York: Routledge, 2004), 155; Knight, *Disintegrating the Musical*, 59. Whether the Roxy Theatre should be considered "Broadway" is perhaps debatable. Bledsoe, however, would also originate the role of Joe in *Show Boat*, which opened on Broadway in December 1927, and tour Europe in the opera version of *The Emperor Jones*. On Bledsoe, also see Katie N. Johnson, "Brutus Jones's Remains: The Case of Jules Bledsoe," *Eugene O'Neill Review* 36, no. 1 (2015): 1–28.
150. Melnick, *American Showman*, 273; George E. Turner, "The Evolution of Special Visual Effects," in *The ASC Treasury of Visual Effects*, ed. George E. Turner (Hollywood, CA: American Society of Cinematographers, 1983), 46. On the opening of *Metropolis*, see Harriette Underhill, "'Metropolis' a Cinema 'R. U. R.' Weird and Fantastic but Interesting," *New York Times*, March 7, 1927, 12. On the depiction of television in *Metropolis*, also see Richard Koszarski, "Coming Next Week: Images of Television in Pre-war Motion Pictures," *Film History* 10, no. 2 (1998): 131.
151. Hall, *The Best Remaining Seats*, 10.
152. Jack Alicoate, "The Romance of the Roxy," *Film Daily*, special section on Roxy, March 13, 1927, 12. Also see Paul, *When Movies Were Theater*, 249.
153. See Melnick, *American Showman*, 48.
154. Samuel L. Rothafel, "What the Public Wants in the Picture Theater" (1925), in *Moviegoing in America: A Sourcebook in the History of Film Exhibition*, ed. Gregory A. Waller (Malden, MA: Blackwell, 2002), 101, 102.
155. Melnick, *American Showman*, 209, 210.
156. Melnick, *American Showman*, 272; Michele Hilmes, *Hollywood and Broadcasting: From Radio to Cable* (Urbana: University of Illinois Press, 1990), 36.
157. "Far-off Speakers Seen as Well as Heard Here in Test of Television," *New York Times*, April 8, 1927, 1, 20.

## 3. Theatrical Screens, 1931–1940: Integrating the Screen

1. See, in particular, Maggie Valentine, *The Show Starts on the Sidewalk: An Architectural History of the Movie Theatre, Starring S. Charles Lee* (New Haven, CT: Yale University Press, 1996), 90–127; Amir H. Ameri, "The Architecture of the Illusive Distance," *Screen* 54, no. 4 (Winter 2013): 439–62; William Paul, *When Movies Were Theater: Architecture, Exhibition, and the Evolution of American Film* (New York: Columbia University Press, 2016), 230–74; and Jocelyn Szczepaniak-Gillece, *The Optical Vacuum: Spectatorship and Modernized American Theater Architecture* (New York: Oxford University Press, 2018), 19–100.

2. Benjamin Schlanger, "Reversing the Form and Inclination of the Motion Picture Theater Floor for Improving Vision," *Journal of the Society of Motion Picture Engineers* (hereafter *JSMPE*) 17, no. 2 (August 1931): 165, 167.
3. See Donald Albrecht, *Designing Dreams: Modern Architecture in the Movies* (New York: Harper and Row, 1986).
4. On exhibition practices during this period, see Douglas Gomery, *Shared Pleasures: A History of Movie Presentation in the United States* (Madison: University of Wisconsin Press, 1992), 57–82. On Hollywood's industrial, representational, and technological practices more broadly, see Tino Balio, *Grand Design: Hollywood as a Modern Business Enterprise, 1930–1939* (Berkeley: University of California Press, 1993).
5. See, for instance, Anne Friedberg, *The Virtual Window: From Alberti to Microsoft* (Cambridge, MA: MIT Press, 2006), 170. The notion of distraction was most famously theorized by Siegfried Kracauer in the 1920s in relation to Berlin's picture palaces. See Siegfried Kracauer, "Cult of Distraction: On Berlin's Picture Palaces," in *The Mass Ornament: Weimar Essays*, ed. and trans. Thomas Y. Levin (Cambridge, MA: Harvard University Press, 1995), 323–28.
6. Emily Thompson, *The Soundscape of Modernity: Architectural Acoustics and the Culture of Listening in America, 1900–1933* (Cambridge, MA: MIT Press, 2002), 7.
7. G. T. Stanton, quoted in Thompson, *The Soundscape of Modernity*, 262.
8. George E. Turner, "The Evolution of Special Visual Effects," in *The ASC Treasury of Visual Effects*, ed. George E. Turner (Hollywood, CA: American Society of Cinematographers, 1983), 45. Also see Ralph G. Fear, "Projected Background Anematography," *American Cinematographer* (hereafter *AC*), January 1932, 11.
9. See advertisement for Vallen Automatic Screen Modifier, *Motion Picture Herald*, Better Theatres sec., April 11, 1931, 85; advertisement for Vallen Automatic Curtain Control, *Motion Picture Herald*, Better Theatres sec., July 4, 1931, 48; and "Screen Masks," *Motion Picture Herald*, Better Theatres sec., April 11, 1931, 113.
10. Stephen Kern, *The Culture of Time and Space, 1880–1918* (Cambridge, MA: Harvard University Press, 2003), esp. 65–88. Also see Donald Crafton, *The Talkies: American Cinema's Transition to Sound, 1926–1931* (Berkeley: University of California Press, 1997), esp. 19–61.
11. Michel Chion, *Film: A Sound Art*, trans. Claudia Gorbman (New York: Columbia University Press, 2009), 37, 23.
12. Mary Ann Doane, *The Emergence of Cinematic Time: Modernity, Contingency, the Archive* (Cambridge, MA: Harvard University Press, 2002), 30, 222.
13. For such a historicization of the cinematic apparatus, focusing specifically on the design of theater chairs, see Jocelyn Szczepaniak-Gillece, "Revisiting the Apparatus: The Theatre Chair and Cinematic Spectatorship," *Screen* 57, no. 3 (Autumn 2016): 253–76. On apparatus theory's relation to historical approaches to theater design, see also Szczepaniak-Gillece, *The Optical Vacuum*, 14–15.
14. On the elimination of silent-film theaters between 1930 and 1935, see Douglas Gomery, *The Coming of Sound: A History* (New York: Routledge, 2005), 92.
15. Gomery, *Shared Pleasures*, 72–75; Valentine, *The Show Starts on the Sidewalk*, 88–95; Paul, *When Movies Were Theater*, 93–94, 177–79. On the new theaters' relationship to the stock market crash in 1929, see Larry May, *The Big Tomorrow: Hollywood and the Politics of the American Way* (Chicago: University of Chicago Press, 2000), 101–35. On the movie theaters' relationship to architectural modernism, see Szczepaniak-Gillece, *The Optical Vacuum*, esp. 27–29. On the "little cinema" movement, see Peter Decherney, "Cult of Attention: An Introduction to Seymour Stern and Harry

Alan Potamkin (Contra Kracauer) on the Ideal Movie Theater," *Spectator* 18, no 2 (Spring–Summer 1998): 18–25, and Barbara Wilinsky, *Sure Seaters: The Emergence of Art House Cinema* (Minneapolis: University of Minnesota Press, 2001), 46–55.

16. Szczepaniak-Gillece, *The Optical Vacuum*, 6, 21. See also Ben Schlanger, "Motion Picture Theatres of Tomorrow," *Motion Picture Herald*, Better Theatres sec., February 14, 1931, 12–13, 56–57; George Schutz, "Reversing the Slope of the Main Floor: An Examination of the Schlanger Plan," *Motion Picture Herald*, Better Theatres sec., July 4, 1931, 13; Irvin L. Scott, "A Motion Picture Theater for a Suburban Town in New York," *Architectural Record*, August 1931, 111–15; "Trans-Lux Theater: An Innovation in Film Projection," *Architectural Record*, August 1931, 118–20; "The Unique Studio Theatre: The Turnstile Idea in a New Interpretation," *Motion Picture Herald*, Better Theatres sec., August 1, 1931, 17–18; Irving H. Bowman, "Modern Theatre Construction: An Architect Envisions Today's Advantages," *Motion Picture Herald*, Better Theatres sec., August 1, 1931, 19, 82; Ben Schlanger, "Architecture and the Engineer: What Their Coordination Means in Theatre Planning," *Motion Picture Herald*, Better Theatres sec., September 26, 1931, 20–21; George Schutz, "The Motion Picture in Rockefeller City," *Motion Picture Herald*, Better Theatres sec., February 13, 1932, 13–17; George Schutz, "The Reversed Floor Slope in Practice," *Motion Picture Herald*, Better Theatres sec., April 9, 1932, 29–32, 116; and Frederic Arden Pawley, "Design of Motion Picture Theaters," *Architectural Record*, June 1932, 429–38, 38–40. On slightly later implementations of these ideas, also see "Will Rogers Theater, Chicago, ILL.," *Architectural Forum*, October 1936, 374; "Oak Park, Illinois," *Architectural Forum*, October 1936, 374A; "Pic Theater," *Architectural Forum*, October 1936, 374B; and "The Esquire Theater, Chicago, IL," *Architectural Forum*, April 1938, 271–80.
17. Valentine, *The Show Starts on the Sidewalk*, 95.
18. Jocelyn Szczepaniak-Gillece's book *The Optical Vacuum*, which was published as I was finalizing my manuscript, offers an exception to this emphasis, situating the redesigned theater in relation to broader trends in modernism as well as to discourses on immersion.
19. Paul, *When Movies Were Theater*, 237–39.
20. Robert Spadoni, *Uncanny Bodies: The Coming of Sound Film and the Origins of the Horror Genre* (Berkeley: University of California Press, 2007), 17, 91–92.
21. Ameri, "The Architecture of the Illusive Distance," esp. 452–62.
22. L. M. Dieterich, Ph.D., "Screen Characteristics and Natural Vision," *Motion Picture Projectionist*, January 1931, 18.
23. Schlanger, "Architecture and the Engineer," 20, 21.
24. See, in particular, Ameri, "The Architecture of the Illusive Distance," 454–59; Paul, *When Movies Were Theater*, 239–54; and Szczepaniak-Gillece, *The Optical Vacuum*.
25. Le Corbusier, "Twentieth Century Building and Twentieth Century Living" (1930), in *Raumplan Versus Plan Libre*, ed. Max Risselada (Delft: Delft University Press, 1987), 146; László Moholy-Nagy, "Light-Architecture," *Industrial Arts* 1, no. 1 (Spring 1936): 15–17; Beatriz Colomina, *Privacy and Publicity: Modern Architecture as Mass Media* (Cambridge, MA: MIT Press, 1994), 6–7, 283–335; Noam M. Elcott, "Rooms of Our Time: László Moholy-Nagy and the Stillbirth of Multi-media Museums," in *Screen/Space: The Projected Image in Contemporary Art*, ed. Tamara Trodd (Manchester: Manchester University Press, 2011), 25–52.
26. Paul, *When Movies Were Theater*, 241.
27. "Report on Projection Screens," *JSMPE* 20, no. 6 (June 1933): 511; "Big Screen in Music Hall," *New York Times*, January 8, 1933, 26.

28. See, for instance, Schutz, "The Motion Picture in Rockefeller City," 15; Ben Schlanger, "Two Late Theatre Forms: A Criticism," *Motion Picture Herald*, Better Theatres sec., February 11, 1933, 8–9, 28–29; Clifton Tuttle, "Distortion in the Projection and Viewing of Motion Pictures," *Motion Picture Herald*, Better Theatres sec., September 23, 1933, 17–19; and Benjamin Schlanger, quoted in S. K. Wolf, "Analysis of Theater and Screen Illumination Data," *JSMPE* 26, no. 5 (May 1936): 544.
29. See Francis M. Falge, "Motion Picture Screens—Their Selection and Use for Best Picture Presentation," *JSMPE* 17, no. 3 (September 1931): 352–55; "Large Versus Small Picture," *Motion Picture Herald*, Better Theatres sec., May 7, 1932, 32–33; Pawley, "Design of Motion Picture Theaters," 432–33; F. H. Richardson, *F. H. Richardson's Bluebook of Projection*, 6th ed. (New York: Quigley, 1935), 160–61; Ben Schlanger, "Motion Picture Theaters," *Architectural Record*, February 1937, 19; and Ben Schlanger, "Cinemas," *Architectural Record*, June 1938, 113.
30. R. H. McCullough, "Care and Setting of Sound Screens," *Motion Picture Projectionist*, June 1930, 52.
31. Falge, "Motion Picture Screens," 355; Pawley, "Design of Motion Picture Theaters," 431.
32. F. H. Richardson, "F. H. Richardson's Comment and Answers to Inquiries," *Motion Picture Herald*, Better Theatres sec., November 21, 1931, 52.
33. Pawley, "Design of Motion Picture Theaters," 430.
34. "The Accomplishments of Ben Schlanger, Architect," in *Theatre Catalog 1953–1954*, 11th ed. (Philadelphia: Jay Emanuel, 1953), xi, Theatre Historical Society of America Archives, Pittsburgh, PA.
35. Schlanger, "Motion Picture Theatres of Tomorrow," 12, 13.
36. Schlanger, "Motion Picture Theatres of Tomorrow," 56. The abandonment of wide-gauge filmmaking took place between the late fall of 1930 and the spring of 1931. In February 1931, the *Journal of the Society of Motion Picture Engineers*, for instance, still ran an article, summarizing an open discussion at the December 1930 meeting, that treated wide-gauge filmmaking and widescreen exhibition as active and ongoing concerns. See "Methods of Securing a Large Screen Picture," *JSMPE* 16, no. 2 (February 1931): esp. 180–82; see also "Progress in the Motion Picture Industry," *JSMPE* 17, no. 1 (July 1931): 71. For Schlanger's later suggestion of a 22-foot limit on screen width, see Schlanger, "Cinemas," 113.
37. Schlanger, "Motion Picture Theatres of Tomorrow," 56.
38. For the argument within the discourse on widescreen that existing theaters limited screen height, see "Symposium on Large Screen Pictures," *Motion Picture Projectionist*, February 1931, 20.
39. Schlanger, "Motion Picture Theatres of Tomorrow," 56.
40. Schlanger, "Reversing the Form and Inclination of the Motion Picture Theater Floor for Improving Vision," 161–71. Also see Schutz, "Reversing the Slope of the Main Floor," 12–13; Ben Schlanger, "Planning Today's Simplified Cinema," *Motion Picture Herald*, Better Theatres sec., November 21, 1931, 18–21, 138; B. Schlanger, "Utilization of Desirable Seating Areas in Relation to Screen Shapes and Sizes and Theater Floor Inclinations," *JSMPE* 18, no. 2 (February 1932): 189–98; and Ben Schlanger, "The Floor and the Screen," *Motion Picture Projectionist*, December 1932, 15–19.
41. Schlanger, "Reversing the Form and Inclination of the Motion Picture Theater Floor for Improving Vision," 166; Schlanger, "Planning Today's Simplified Cinema," 21.
42. Schlanger, "Reversing the Form and Inclination of the Motion Picture Theater Floor for Improving Vision," 163–64.

43. Ben Schlanger, "Auditorium Floor Slopes for Motion Picture Theatres Today," *Motion Picture Herald*, Better Theatres sec., September 20, 1947, 17.
44. Ben Schlanger, "New Theaters for the Cinema," *Architectural Forum*, September 1932, 253–60; "Changing a Bank Into a Theatre," *Motion Picture Herald*, Better Theatres sec. (July 28, 1934), 6; "Forum of Events," *Architectural Forum*, February 1936, 13, 43.
45. Ben Schlanger, "Production Methods and the Theatre," *Motion Picture Herald*, Better Theatres sec., April 8, 1933, 8, 9, 9–10, 10, 61. Also see Ben Schlanger, "Use of the Full Screen Area Today," *Motion Picture Herald*, Better Theatres sec., June 3, 1933, 11–13, and Ben Schlanger, "Changing Factors in Theatre Vision," *Motion Picture Herald*, Better Theatres sec., December 15, 1934, 6–7, 31, 34.
46. "Screen Brightness Report (Comprehensive Report of Studies on Screen Brightness)," *JSMPE* 27, no. 2 (August 1936): 131–32, 138. A foot-lambert, defined as "the brightness of a perfect diffuser emitting or reflecting one lumen per square foot," is the product of the illumination of a surface (measured in foot-candles) and the reflection factor of that surface (E. M. Lowry, "Screen Brightness and the Visual Functions," *JSMPE* 26, no. 5 [May 1936]: 492).
47. "Screen Brightness Report (Comprehensive Report of Studies on Screen Brightness)," 135; B. Schlanger, "Motion Picture Auditorium Lighting," *JSMPE* 34, no. 3 (March 1940): 262.
48. Wolf, "Analysis of Theater and Screen Illumination Data," 536, 539; "Screen Brightness Report (Comprehensive Report of Studies on Screen Brightness)," 127–28, 130, 135. Also see B. O'Brien and C. M. Tuttle, "An Experimental Investigation of Projection Screen Brightness," *JSMPE* 26, no. 5 (May 1936): 505–17.
49. Wolf, "Analysis of Theater and Screen Illumination Data," 535.
50. Francis M. Falge, "Selection and Use of Screens," *Motion Picture Projectionist*, October 1931, 20; W. F. Little and A. T. Williams, "Résumé of Methods of Determining Screen Brightness and Reflectance," *JSMPE* 26, no. 5 (May 1936): 571.
51. On the development of a meter to measure screen brightness, see "Projection Practice Report (Theater Survey and Screen Illumination)," *JSMPE* 30, no. 6 (June 1938): 648; "Projection Practice Report (Theater Survey, Screen Illumination, Projection Room Plans, and Underwriters Regulations)," *JSMPE* 31, no. 5 (November 1938): 483; and F. M. Falge and W. D. Riddle, "Lighting of Motion Picture Theater Auditoriums," *JSMPE* 32, no. 2 (February 1939): 202.
52. Little and Williams, "Résumé of Methods of Determining Screen Brightness and Reflectance," 571.
53. O'Brien and Tuttle, "Experimental Investigation of Projection Screen Brightness," 506. Also see "Theater Lighting," *JSMPE* 14, no. 4 (April 1930): 441–43; M. Luckiesh and F. K. Moss, "The Motion Picture Screen as a Lighting Problem," *JSMPE* 26, no. 5 (May 1936): 583–85; and Schlanger, "Motion Picture Auditorium Lighting," 262.
54. "Report of the Screens Committee," *Motion Picture Projectionist*, November 1931, 34. Also see Richardson, "F. H. Richardson's Comment and Answers to Inquiries," November 21, 1931, 51, and B. Schlanger, "Projectionist's Interest in Auditorium Viewing Conditions," *JSMPE* 34, no. 6 (June 1940): 586–87.
55. F. H. Richardson, "F. H. Richardson's Comment and Answers to Inquiries," *Motion Picture Herald*, Better Theatres sec., September 24, 1932, 22; F. H. Richardson, *Richardson's Handbook of Projection: The Blue Book of Projection*, 3 vols., 5th ed. (New York: Chalmers, 1927–1930), 1:241.
56. Pawley, "Design of Motion Picture Theaters," 433; O'Brien and Tuttle, "Experimental Investigation of Projection Screen Brightness," 506.

57. Pawley, "Design of Motion Picture Theaters," 433; "Report of the Screens Committee," *Motion Picture Projectionist*, November 1931, 34.
58. Richardson, "F. H. Richardson's Comment and Answers to Inquiries," September 24, 1932, 22–23.
59. Eugene Clute, "New Schemes in Modern Remodeling," *Motion Picture Herald*, Better Theatres sec., November 17, 1934, 18.
60. Dr. R. L. Whitley, quoted in Henry James Forman, *Our Movie Made Children* (New York: Macmillan, 1933), 259. On the call for lighted theaters in the nickelodeon era, see Eileen Bowser, *The Transformation of Cinema, 1907–1915* (Berkeley: University of California Press, 1990), 39, and Tom Gunning, *D. W. Griffith and the Origins of American Narrative Film: The Early Years at Biograph* (Urbana: University of Illinois Press, 1991), 147. On regulation efforts in the 1930s, see Richard Maltby, "The Production Code and the Hayes Office," in Balio, *Grand Design*, 37–72. On the Payne Fund Studies, see Robert Sklar, *Movie-Made America: A Cultural History of American Movies*, rev. ed. (New York: Vintage, 1994), 134–40, and Garth S. Jowett, Ian C. Jarvie, and Kathryn H. Fuller, *Children and the Movies: Media Influence and the Payne Fund Controversy* (New York: Cambridge University Press, 1996).
61. On the enhancement to visibility, see "Theater Lighting," 442. On the benefit for concentration, see the quote from President Crabtree in "Report of the Theater Lighting Committee," *JSMPE* 16, no. 2 (February 1931): 242.
62. "Theater Lighting," 442; Wolf, "Analysis of Theater and Screen Illumination Data," 533–34; Luckiesh and Moss, "The Motion Picture Screen as a Lighting Problem," 583–87; Falge and Riddle, "Lighting of Motion Picture Theater Auditoriums," 201–2.
63. Luckiesh and Moss, "The Motion Picture Screen as a Lighting Problem," 588.
64. See, for instance, "Report of the Projection Screens Committee," *JSMPE* 18, no. 2 (February 1932): 245–46; "Theater Lighting," 441–42; and Falge and Riddle, "Lighting of Motion Picture Theater Auditoriums," 201–2.
65. "Report of the Theater Lighting Committee," 240.
66. Wolf, "Analysis of Theater and Screen Illumination Data," 534–35.
67. "Report on Projection Screens," 513–14.
68. O'Brien and Tuttle, "Experimental Investigation of Projection Screen Brightness," 506.
69. O'Brien and Tuttle, "Experimental Investigation of Projection Screen Brightness," 515.
70. Luckiesh and Moss, "The Motion Picture Screen as a Lighting Problem," 588–89.
71. Schlanger, "The Floor and the Screen," 19.
72. B. Schlanger, "On the Relation Between the Shape of the Projected Picture, the Areas of Vision, and Cinematographic Technic," *JSMPE* 24, no. 5 (May 1935): 403, 404, 408, 404, 406. Although this paper was published in 1935, it was first presented at the SMPE meeting in Atlantic City in the spring of 1934.
73. Schlanger, "On the Relation Between the Shape of the Projected Picture, the Areas of Vision, and Cinematographic Technic," 409.
74. Ben Schlanger, "The Screen: A Problem in Exhibition," *Motion Picture Herald*, Better Theatres sec., October 24, 1931, 66.
75. Schlanger, "Motion Picture Theatres of Tomorrow," 13, 56.
76. The demonstration of the "screen synchrofield" took place at the SMPE meeting in New York City in the fall of 1937. See B. Schlanger, "A Method of Enlarging the Visual Field of the Motion Picture," *JSMPE* 30, no. 5 (May 1938): 503.
77. Schlanger, "A Method of Enlarging the Visual Field of the Motion Picture," 506.
78. R. B. Hunter, "Combining Dimmer Units for Increased Flexibility," *Motion Picture Herald*, Better Theatres sec., August 1, 1931, 14.

79. Hunter, "Combining Dimmer Units for Increased Flexibility," 82. Also see "The Clavilux," *Motion Picture News*, December 1, 1928, 1668, 1670; Douglas Fox, "Symphonies of Color," *Exhibitors Herald-World*, Better Theatres sec., April 13, 1929, 20–21; "Notes on Equipment of the Los Angeles Theatre," *Motion Picture Herald*, Better Theatres sec., February 14, 1931, 30, 32; B. S. Burke, "Play Color Symphonies on 'Color Console,'" *Motion Picture Projectionist*, April 1931, 23–24; Gomery, *Shared Pleasures*, 48; and Andrew Robert Johnston, "The Color of Prometheus: Thomas Wilfred's Lumia and the Projection of Transcendence," in *Color and the Moving Image: History, Theory, Aesthetics, Archive*, ed. Simon Brown, Sarah Street, and Liz Watkins (New York: Routledge, 2013), 67–78.
80. Advertisement for Frank Adam Electric Company, *Motion Picture Herald*, Better Theatres sec., July 4, 1931, 55.
81. Ross Melnick, *American Showman: Samuel "Roxy" Rothafel and the Birth of the Entertainment Industry* (New York: Columbia University Press, 2012), 14. In developing one of his Clavilux systems, Thomas Wilfred drew on a patent from D. W. Griffith, who had employed the play of colored lights onscreen for the premiere of *Broken Blossoms* (1919) (Johnston, "The Color of Prometheus," 74; Paul, *When Movies Were Theater*, 199). On Roxy's use of light in conjunction with film projection, also see "New Recognition for Roxy as Authority on Stage Design," *Motion Picture News*, October 5, 1929, 1238, and Fairfax Downey, "The Master of Light," *Popular Mechanics*, November 1928, 779, 781.
82. Samuel L. Rothafel, "What the Public Wants in the Picture Theater" (1925), in *Moviegoing in America: A Sourcebook in the History of Film Exhibition*, ed. Gregory A. Waller (Malden, MA: Blackwell, 2002), 101.
83. Schlanger, "A Method of Enlarging the Visual Field of the Motion Picture," 506.
84. Schlanger, "Motion Picture Auditorium Lighting," 262–63.
85. "The Accomplishments of Ben Schlanger, Architect," xxviii.
86. William Paul, "The Aesthetics of Emergence," *Film History* 5, no. 3 (1993): 335. On the RCA Synchro Screen and its relation to both Schlanger's career and widescreen cinema, see also Szczepaniak-Gillece, *The Optical Vacuum*, esp. 85–121.
87. "The Accomplishments of Ben Schlanger, Architect," xxx; Morris Ketchum Jr., "A New Theatre in a Regional Shopping Center," *Theatre Catalog 1952*, 10th ed. (Philadelphia: Jay Emanuel, 1952), 24–31, Theatre Historical Society of America Archives; Marjory Adams, "Theatre at Shoppers' World Is Last Word in Film Houses," *Boston Daily Globe*, October 29, 1951, 9.
88. Benjamin Schlanger, William A. Hoffberg, and Charles R. Underhill Jr., "The Synchroscreen as a Stage Setting for Motion Picture Presentation," *Journal of the Society of Motion Picture and Television Engineers* 58, no. 6 (June 1952): 528; "Offer Improved Viewing Conditions," *Boxoffice*, April 5, 1952, 47.
89. Schlanger, "Motion Picture Auditorium Lighting," 261.
90. Schlanger, "Cinemas," 112.
91. Schlanger, "Motion Picture Auditorium Lighting," 263–64.
92. Schlanger, "A Method of Enlarging the Visual Field of the Motion Picture," 507.
93. Christine Grenz, *Trans-Lux: Biography of a Corporation* (Norwalk, CT: Trans-Lux, 1982), 10.
94. George Schutz and F. H. Richardson, "The Trans-Lux System of Operation," *Motion Picture Herald*, Better Theatres sec., May 9, 1931, 12–13; William Mayer, "Trans Lux Rear Stage Projection," *Motion Picture Projectionist*, October 1931, 12–13. For discussions of Trans-Lux theaters within scholarly work, also see Gomery, *Shared Pleasures*, 145–49,

and Raymond Fielding, *The American Newsreel: A Complete History, 1911–1967*, 2nd ed. (Jefferson, NC: McFarland, 2006), 124–25.

95. On such earlier experiments, see Melnick, *American Showman*, 60–64, and Paul, *When Movies Were Theater*, 86–88.
96. "Report of the Projection Screens Committee," 250.
97. Benjamin Schlanger, quoted in Falge and Riddle, "Lighting of Motion Picture Theater Auditoriums," 212. Also see the quote from Mr. Carlson in Falge and Riddle, "Lighting of Motion Picture Theater Auditoriums," 212, and "Report of the Projection Screens Committee," 250.
98. Haidee Wasson, "The Other Small Screen: Moving Images at New York's World Fair, 1939," *Canadian Journal of Film Studies* 21, no. 1 (Spring 2012): 94.
99. Grenz, *Trans-Lux*, 5, 9; Jack Burton, *The Story of Trans-Lux* (New York: Trans-Lux Daylight Picture Screen Corporation, 1929), 13. On the wide use of the Trans-Lux stock ticker by 1931, see Schutz and Richardson, "The Trans-Lux System of Operation," 12.
100. "Trans-Lux," *Time*, March 23, 1931, 46.
101. Burton, *The Story of Trans-Lux*, 34–45.
102. "Report of the Projection Screens Committee," 248.
103. Burton, *The Story of Trans-Lux*, 26–27.
104. Gomery, *Shared Pleasures*, 146.
105. Grenz, *Trans-Lux*, 8; Gomery, *Shared Pleasures*, 144; E. I. Sponable to W. C. Michel, June 24, 1932, box 10 (General Files Fr–Gr), "Grandeur 1932" folder, Earl I. Sponable Papers, Rare Book and Manuscript Library, Columbia University Libraries, New York.
106. Schutz and Richardson, "The Trans-Lux System of Operation," 12–13; "Trans-Lux Theater," 119.
107. Schutz and Richardson, "The Trans-Lux System of Operation," 12; Terry Ramsaye, "Static in Radio City," *Motion Picture Herald*, January 14, 1933, 11.
108. "Report of the Projection Screens Committee," 242–52; Pawley, "Design of Motion Picture Theaters," 38–40.
109. "Report of the Projection Screens Committee," 250.
110. Although the use of perforated sound screens was preferred practice by the time the first Trans-Lux theaters opened in 1931, the Trans-Lux screens designed to transmit light apparently did not transmit sound. Speakers at these theaters were placed below rather than behind the screen. This was considered at the time a less-than-ideal arrangement, though one that was reputedly not uncommon with the RCA Photophone system used by Trans-Lux and made less objectionable by the small size of the auditoriums. See H. F. Hopkins, "Considerations in the Design and Testing of Motion Picture Screens for Sound Picture Work," *JSMPE* 15, no. 3 (September 1930): 320–31; Schutz and Richardson, "The Trans-Lux System of Operation," 13; Mayer, "Trans Lux Rear Stage Projection," 13; and "Report of the Projection Screens Committee," 249.
111. Jennifer Wild, *The Parisian Avant-Garde in the Age of Cinema, 1900–1923* (Oakland: University of California Press, 2015), 44.
112. Grenz, *Trans-Lux*, 37–38.
113. See G. G. Popovici, "Background Projection for Process Cinematography," *JSMPE* 24, no. 2 (February 1935): 105, and G. G. Popovici, "Recent Developments in Background Projection," *JSMPE* 30, no. 5 (May 1938): 539, 540.
114. Pawley, "Design of Motion Picture Theaters," 38.
115. Fielding, *The American Newsreel*, 92.
116. Ray Gingell, "Trans-Lux," *Marquee* 7, no. 3 (1975): 21.

117. Edwin Schallert, "Premiere Jams Broadway," *Los Angeles Times*, February 2, 1931, A7; *Theatre Historical Society of America Annual* 25 (1998): 3. Also see in particular Valentine, *The Show Starts on the Sidewalk*, 53–71.
118. Valentine, *The Show Starts on the Sidewalk*, 47, 78–83.
119. On the Studio Theatre, see "The Unique Studio Theatre," 17–18; "The Studio Theatre: A Machine Age Cinema," *Motion Picture Herald*, Better Theatres sec., August 29, 1931, 14–17; Harold B. Franklin, "Operating the Small 'Automatic' Theatre," *Motion Picture Herald*, Better Theatres sec., November 21, 1931, 12; S. Charles Lee, "The Features of an Automatic Cinema as Found in the Studio Theatre," *Motion Picture Herald*, Better Theatres sec., November 21, 1931, 25, 138; and Valentine, *The Show Starts on the Sidewalk*, 92–94.
120. Valentine, *The Show Starts on the Sidewalk*, 54.
121. "The Los Angeles Theatre," n.d., "California, Los Angeles (Los Angeles Theatre #2)" file, Tom B'hend and Preston Kaufmann Collection, Margaret Herrick Library, Academy of Motion Picture Arts and Sciences, Beverly Hills, CA; "Los Angeles Theatre—History," n.d., http://www.losangelestheatre.com/history.
122. Valentine, *The Show Starts on the Sidewalk*, 53–54; Sam Hall Kaplan, "L.A. Theatre Redefined the 'Movie Palace,'" *Los Angeles Times*, February 21, 1987, D1.
123. Valentine, *The Show Starts on the Sidewalk*, 71.
124. "Los Angeles Theatre—History."
125. William Crouch, "Ultra-modern Facility in Period Design" *Motion Picture Herald*, Better Theatres sec., February 14, 1931, 27–29, 32.
126. On the innovativeness of the restaurant, see "Los Angeles Theater Opens Tomorrow Night," *Los Angeles Times*, January 29, 1931, A8.
127. Crouch, "Ultra-modern Facility in Period Design," 28.
128. "Notes on Equipment of the Los Angeles Theatre," 30.
129. Crouch, "Ultra-modern Facility in Period Design," 28.
130. "Los Angeles Theater Opens Tomorrow Night," A8.
131. Crouch, "Ultra-modern Facility in Period Design," 29, 28.
132. Crouch, "Ultra-modern Facility in Period Design," 29.
133. Crouch, "Ultra-modern Facility in Period Design," 28.
134. "Movie Shown on 2 Screens at Same Time," *Modern Mechanics and Inventions*, April 1931, 58.
135. Blueprints of the Los Angeles Theatre, box 135, S. Charles Lee Papers, Library Special Collections, Charles E. Young Research Library, University of California at Los Angeles.
136. Blueprints of the Los Angeles Theatre, box 137, S. Charles Lee Papers.
137. Blueprints of the Los Angeles Theatre, box 135, S. Charles Lee Papers.
138. Crouch, "Ultra-modern Facility in Period Design," 28. On the cry room in the Tower Theatre, see Marquis Busby, "New First-Run Film House Is Model of Beauty," *Los Angeles Times*, October 14, 1927, A8. I owe knowledge of the Tower Theatre's cry room to Francesco Casetti, "Film as an Environmental Medium," paper presented at the "Ends of Cinema" conference, Center for 21st Century Studies, University of Wisconsin at Milwaukee, May 4, 2018. On Lee's cry rooms in the 1930s more generally, see also Valentine, *The Show Starts on the Sidewalk*, 106.
139. "Movie Shown on 2 Screens at Same Time," 60–61.
140. Kaplan, "L.A. Theatre Redefined the 'Movie Palace'"; Valentine, *The Show Starts on the Sidewalk*, 70.
141. Pawley, "Design of Motion Picture Theaters," 40.
142. Valentine, *The Show Starts on the Sidewalk*, 62.

143. Crouch, "Ultra-modern Facility in Period Design," 27.
144. Pawley, "Design of Motion Picture Theaters," 40.
145. Blueprints, box 135, S. Charles Lee Papers. It is unclear at what point the plan was made to locate the periscope in the promenade rather than the foyer.
146. Blueprints, box 137, S. Charles Lee Papers.
147. Crouch, "Ultra-modern Facility in Period Design," 28.
148. Peter M. Holsken, A.I.A., "Planning the Theatre," *Motion Picture Herald*, Better Theatres sec., May 9, 1931, 75.
149. Schlanger, "Cinemas," 115.
150. "Beverly Theater," *Architectural Forum*, January 1936, 47.
151. Friedberg, *The Virtual Window*, 151.
152. Valentine, *The Show Starts on the Sidewalk*, 62.
153. S. Charles Lee, quoted in Crouch, "Ultra-modern Facility in Period Design," 29, 32.
154. "Los Angeles Theatre—History."
155. "Movie Shown on 2 Screens at Same Time," 59.
156. Michael Cowan, "Taking It to the Street: Screening the Advertising Film in the Weimar Republic," *Screen* 54, no. 4 (Winter 2013): 467, 476, 479, emphasis in original.
157. Mary Ann Doane, *The Desire to Desire: The Woman's Film of the 1940s* (Bloomington: Indiana University Press, 1987), 33. For a discussion of female shoppers that emphasizes both the transparent and the reflective nature of department-store windows, placing more emphasis on women's agency in looking, also see Lauren Rabinovitz, *For the Love of Pleasure: Women, Movies, and Culture in Turn-of-the-Century Chicago* (New Brunswick, NJ: Rutgers University Press,1998), 79.
158. On dish nights and their address to women, see Kathryn H. Fuller-Seeley, "Dish Night at the Movies: Exhibitor Promotions and Female Audiences During the Great Depression," in *Looking Past the Screen: Case Studies in American Film History and Method*, ed. Jon Lewis and Eric Smoodin (Durham, NC: Duke University Press, 2007), 246–75.
159. Susan Smith, "Lee's Picture Palaces: A Part of the Show," *Calendar*, August 26, 1979, 26, "Lee, S. Charles" folder, box T9, Tom B'hend and Preston Kaufmann Collection.
160. In August 1930, for instance, the *Syracuse Herald* not only envisioned theaters of the future showing films by television but also contended that "the local Keith Theater . . . already is electrically equipped for television" (Chester B. Bahn, "Television Soon Will Flash Talkies Through Ether," *Syracuse Herald*, Magazine sec., August 3, 1930). Thanks to Lynn Spigel for drawing my attention to this article.
161. "Television in Theatres with Small Transmitters, Receivers," *Motion Picture Herald*, Better Theatres sec., November 14, 1931, 20.
162. "New Device Permits Simultaneous Image," *Motion Picture Herald*, Better Theatres sec., May 27, 1933, 12.
163. Frederick Kiesler, quoted in "Kiesler's Pursuit of an Idea," *Progressive Architecture* 42 (July 1961): 109. Also see Barbara Lesák, "Visionary of the European Theater," in *Frederick Kiesler*, ed. Lisa Phillips (New York: Whitney Museum of Art, 1989), 40.
164. Frederick Kiesler, quoted in "Kiesler's Pursuit of an Idea," 109. Also see Friedberg, *The Virtual Window*, 137.
165. Katharina Loew, "Magic Mirrors: The Schüfftan Process," in *Special Effects: New Histories/Theories/Contexts*, ed. Dan North, Bob Rehak, and Michael S. Duffy (London: British Film Institute, 2015), 66–67.
166. Harriette Underhill, "'Metropolis' a Cinema 'R. U. R.' Weird and Fantastic but Interesting," *New York Times*, March 7, 1927, 12.

167. Steve Levin, "RKO Roxy, New York City: Long Gone and Almost Totally Forgotten," *Marquee* 34, no. 3 (2002): 20–23; Melnick, *American Showman*, 343–79.
168. Edwin Schallert, "News and Reviews of Stage, Films, and Music," *Los Angeles Times*, March 1, 1933, A7; Mordaunt Hall, "A Fantastic Film in Which a Monstrous Ape Uses Automobiles for Missiles and Climbs a Skyscraper," *New York Times*, March 3, 1933, C12.
169. See "The World's Largest Screen," *Motion Picture Projectionist*, November 1932, 4; "Radio City Premiere Is a Notable Event," *New York Times*, December 28, 1932, 1, 14; "New Movie Palace Gives City a Thrill," *New York Times*, December 30, 1932, 14; Terry Ramsaye, "Gorgeous Theatrics Mark Roxy's Radio City Debut," *Motion Picture Herald*, December 31, 1932, 10–11; George Schutz, "Theatres Are Designed and Equipped on Elaborate Scale," *Motion Picture Herald*, December 31, 1932, 11, 26; "Big Screen in Music Hall," 26; and "The New Pictures," *Time*, January 9, 1933, 47.
170. Philippa Gerry Whiting, "Rockefeller Center Début," *American Magazine of Art* 26, no. 2 (February 1933): 77.
171. Paul, *When Movies Were Theater*, 360 n. 93; see also Ramsaye, "Static in Radio City," 11.
172. Melnick, *American Showman*, 343–45.
173. Schlanger, quoted in Wolf, "Analysis of Theater and Screen Illumination," 544.
174. Schlanger, "Two Late Theatre Forms," 9, 28.
175. Thompson, *The Soundscape of Modernity*, 233, 295, 309–15.
176. Eugene Clute, "Radio City Theatres," *Motion Picture Herald*, Better Theatres sec., January 14, 1933, 10.
177. O. B. Hanson, "The Story of Rockefeller Center IX: The Plan and Construction of the National Broadcasting Company Studios," *Architectural Forum*, August 1932, 157.
178. Clute, "Radio City Theatres," 10.
179. "RKO-Roxy Center Theatre," *Marquee* 8, no. 2 (1976): 15.
180. "Glass Mural for Theatre," source unknown, October 17, 1932, "New York, Manhattan (Center Theatre)" clippings folder (f.2644), Tom B'hend and Preston Kaufmann Collection; Clute, "Radio City Theatres," 19.
181. Clute, "Radio City Theatres," 21.
182. Richard B. Jewell, *RKO Radio Pictures: A Titan Is Born* (Berkeley: University of California Press, 2012), 63; Thomas Doherty, "This Is Where We Came In: The Audible Screen and the Voluble Audience of Early Sound Cinema," in *American Movie Audiences: From the Turn of the Century to the Early Sound Era*, ed. Melvyn Stokes and Richard Maltby (London: British Film Institute, 1999), 147.
183. Paul Young, *The Cinema Dreams Its Rivals: Media Fantasy Films from Radio to the Internet* (Minneapolis: University of Minnesota Press, 2006), 73–135.
184. Ben Schlanger and Jacob Gilston filed their patent for the screen-synchrofield system on May 17, 1937, and demonstrated it to the SMPE that fall. On September 14, 1937, Fred Waller and Ralph Walker filed their patent for a "motion picture theater" that would "employ a plurality of projectors." I believe that Waller began working on his system that summer. Most recollections of the Waller system simply identify the date as 1937. Waller's contribution to the volume *New Screen Techniques* in 1953 says it occurred in "the summer of 1938," and I think it is reasonable to presume that he remembered the season correctly and just got the year wrong in that instance. See Benjamin Schlanger and Jacob Gilston, Screen and Synchronized Light Field, U.S. Patent 2,117,857, filed May 17, 1937, and issued May 17, 1938; Fred Waller and Ralph Walker of the Vitarama Corporation, Motion Picture Theater, U.S. Patent 2,280,206, filed September 14, 1937,

and issued April 21, 1942; Ralph Walker, "The Birth of an Idea," in *New Screen Techniques*, ed. Martin Quigley Jr. (New York: Quigley, 1953), 114; Fred Waller, "Cinerama Goes to War," in *New Screen Techniques*, ed. Quigley, 120; Fred Waller, "The Archeology of Cinerama," *Film History* 5 (1993): 289.

185. Waller, "The Archeology of Cinerama," 297 n. 1; John Belton, *Widescreen Cinema* (Cambridge, MA: Harvard University Press, 1992), 99; Ralph G. Martin, "Mr. Cinerama," *True*, August 1953, 88.
186. Waller, "Cinerama Goes to War," 119, 120.
187. Walker, "The Birth of an Idea," 113.
188. Walker, "The Birth of an Idea," 114.
189. The horizontal curve initially extended 180 degrees, but after experimentation it was trimmed by 7.5 degrees on each side (Waller, "The Archeology of Cinerama," 290–91; Waller, "Cinerama Goes to War," 120).
190. Waller, "The Archeology of Cinerama," 291.
191. Waller, "Cinerama Goes to War," 121.
192. Waller and Walker, U.S. Patent 2,280,206.
193. Walker, "The Birth of an Idea," 116; Waller, "The Archeology of Cinerama," 291–92; Belton, *Widescreen Cinema*, 86.
194. Waller, "The Archeology of Cinerama," 291; Waller, "Cinerama Goes to War," 121.
195. Fred Waller, "The Waller Flexible Gunnery Trainer," *JSMPE* 47, no. 1 (July 1946): 73–74. Also see Waller, "Cinerama Goes to War," 121, and Giles Taylor, "A Military Use for Widescreen Cinema: Training the Body Through Immersive Media," *Velvet Light Trap* 72 (Fall 2013): 17–32.
196. Waller, "The Waller Flexible Gunnery Trainer," 76–78, 80–81; Waller, "Cinerama Goes to War," 124.
197. Waller, "The Archeology of Cinerama," 297 n. 1.
198. A. Gillett, H. Chretien, and J. Tedesco, "Panoramic Screen Projection Equipment Used at the Palace of Light at the International Exposition (Paris 1937)," *JSMPE* 32, no. 5 (May 1939): 530–34.
199. I am indebted here to Francesco Casetti's discussion of the environmental function of screens in "Screening: A Counter-Genealogy of the Silver Screen," paper presented at the Society for Cinema and Media Studies Conference, Chicago, March 24, 2017.
200. See, for instance, Orrin E. Dunlap Jr., "Targets of Sight," no source, December 6, 1936, box 21, "Clippings 1935–1936–1937" folder, Allen Balcom Du Mont Collection, Archives Center, National Museum of American History, Smithsonian Institution, Washington, DC.
201. László Moholy-Nagy, *Painting, Photography, Film*, trans. Janet Seligman (London: Lund Humphries, 1969; orig. pub. in German in 1925), 41; Elcott, "Rooms of Our Time," 35.
202. K. Lönberg-Holm, "New Theatre Architecture in Europe," *Architectural Record*, May 1930, 494; Stefan Jonsson, *Crowds and Democracy: The Idea and Image of the Masses from Revolution to Fascism* (New York: Columbia University Press, 2013), 241.
203. Frederick Kiesler, quoted in Douglas Fox, "The Film Guild Cinema: An Experiment in Theater Design," *Exhibitors Herald-World*, Better Theatres sec., March 16, 1929, 16.
204. Frederick J. Kiesler, "Building a Cinema Theatre," in *Selected Writings*, ed. Siegfried Gohr and Gunda Luyken (Ostfildern bei Stuttgart, Germany: Gerd Hatje, 1996), 17. Also see Laura McGuire, "A Movie House in Space and Time: Frederick Kiesler's Film Arts Guild Cinema, New York, 1929," *Studies in the Decorative Arts* 14, no. 2 (Spring–Summer 2007): 45–78.

205. See Giuliana Bruno, *Atlas of Emotion: Journeys in Art, Architecture, and Film* (New York: Verso, 2002), 45–47; Friedberg, *The Virtual Window*, 117–23, 169–70; and Elcott, "Rooms of Our Time."
206. Fred Turner, *The Democratic Surround: Multimedia and American Liberalism from World War II to the Psychedelic Sixties* (Chicago: University of Chicago Press, 2013), 9, 77–113.

## 4. Extratheatrical Screens in the Long 1930s: Film and Television at Home and in Transit

1. Richard Koszarski, *Hollywood on the Hudson: Film and Television in New York from Griffith to Sarnoff* (New Brunswick, NJ: Rutgers University Press, 2008), 451. The premiere was transmitted from the Capitol Theatre and aired on the NBC station W2XBS. See "Televiews of Pictures," *New York Times*, December 17, 1939, 132, and "Radio 'Eye' Scans Broadway," *New York Times*, December 24, 1939, 94.
2. As of March 1940, an RCA representative estimated that there were 2,500 "television-equipped homes" within a seventy-five-mile radius of NBC's transmitter on the Empire State Building ("Price Cut Sharply on Television Sets," *New York Times*, March 13, 1940, 38).
3. Erik Barnouw, *Tube of Plenty: The Evolution of American Television*, 2nd rev. ed. (New York: Oxford University Press, 1990), 92–93; Gary R. Edgerton, *The Columbia History of American Television* (New York: Columbia University Press, 2007), 15.
4. Haidee Wasson, "Selling Machines: Film and Its Technologies at the New York World's Fair," in *Films That Sell: Moving Pictures and Advertising*, ed. Bo Florin, Nico de Klerk, and Patrick Vonderau (London: British Film Institute, 2016), 56. Also see, especially, Haidee Wasson, *Museum Movies: The Museum of Modern Art and the Birth of Art Cinema* (Berkeley: University of California Press, 2005); Haidee Wasson, "Electric Homes! Automatic Movies! Efficient Entertainment! 16mm and Cinema's Domestication in the 1920s," *Cinema Journal* 48, no. 4 (Summer 2009): 1–21; Charles R. Acland and Haidee Wasson, ed., *Useful Cinema* (Durham, NC: Duke University Press, 2011); and Devin Orgeron, Marsha Orgeron, and Dan Streible, eds., *Learning with the Lights Off: Educational Film in the United States* (New York: Oxford University Press, 2012).
5. Susan J. Douglas, *Listening In: Radio and the American Imagination* (1999; reprint, Minneapolis: University of Minnesota Press, 2004), 52.
6. On cinema's intermedial relationships with radio and television, see Tino Balio, ed., *Hollywood in the Age of Television* (Boston: Unwin Hyman, 1990); Michele Hilmes, *Hollywood and Broadcasting: From Radio to Cable* (Urbana: University of Illinois Press, 1990); Donald Crafton, *The Talkies: American Cinema's Transition to Sound, 1926–1931* (Berkeley: University of California Press, 1997); William Uricchio, "Television, Film, and the Struggle for Media Identity," *Film History* 10, no. 2 (1998): 118–27; Richard Koszarski, "Coming Next Week: Images of Television in Pre-war Motion Pictures," *Film History* 10, no. 2 (1998): 128–40; Paul Young, *The Cinema Dreams Its Rivals: Media Fantasy Films from Radio to the Internet* (Minneapolis: University of Minnesota Press, 2006); and Doron Galili, "Seeing by Electricity: The Emergence of Television and the Modern Mediascape, 1878–1939," Ph.D. diss., University of Chicago, 2011.
7. Patricia R. Zimmermann, *Reel Families: A Social History of Amateur Film* (Bloomington: Indiana University Press, 1995), 5–11; Charles Tepperman, *Amateur Cinema: The Rise of North American Moviemaking, 1923–1960* (Oakland: University of California Press, 2015), 2–5.

8. Joseph H. Udelson, *The Great Television Race: A History of the American Television Industry, 1925–1941* (Tuscaloosa: University of Alabama Press, 1982), 54–55, 65. For this periodization, see Richard Koszarski, introduction to Richard Koszarski and Doron Galili, "Television in the Cinema Before 1939: An International Annotated Database," *Journal of e-Media Studies* 5, no. 1 (2016), http://journals.dartmouth.edu/cgi-bin/WebObjects/Journals.woa/xmlpage/4/article/471. On amateurism and radio, see Susan J. Douglas, *Inventing American Broadcasting, 1899–1922* (Baltimore: Johns Hopkins University Press, 1987), esp. 187–215, 292–303.
9. Lynn Spigel, *Make Room for TV: Television and the Family Ideal in Postwar America* (Chicago: University of Chicago Press, 1992); Douglas, *Listening In*; William Boddy, *New Media and Popular Imagination: Launching Radio, Television, and Digital Media in the United States* (New York: Oxford University Press, 2004); Wasson, "Electric Homes!"
10. See Catherine L. Covert, "'We May Hear Too Much': American Sensibility and the Response to Radio, 1919–1924," in *Mass Media Between the Wars: Perceptions of Cultural Tension, 1918–1941*, ed. Catherine L. Covert and John D. Stevens (Syracuse, NY: Syracuse University Press, 1984), 204–5.
11. See Spigel, *Make Room for TV*, 99–135; Douglas, *Listening In*, 9, 11; Anna McCarthy, *Ambient Television: Visual Culture and Public Space* (Durham, NC: Duke University Press, 2001), 10–20; Haidee Wasson, "The Reel of the Month Club: 16mm Projectors, Home Theaters, and Film Libraries in the 1920s," in *Going to the Movies: Hollywood and the Social Experience of Cinema*, ed. Richard Maltby, Melvyn Stokes, and Robert C. Allen (Exeter, U.K.: University of Exeter Press, 2007), 224–26; and Paddy Scannell, *Television and the Meaning of "Live": An Enquiry Into the Human Situation* (Malden, MA: Polity, 2014), 63–64, 101–2. For an illuminating discussion of the intermedial dynamics of film and television in this regard as well as a historicization of the application of the "window on the world" trope to television, see Galili, "Seeing by Electricity," 70–73, 135–36.
12. McCarthy, *Ambient Television*, 15. Also see Rudolf Arnheim, *Radio: An Art of Sound*, trans. Margaret Ludwig and Herbert Read (New York: Da Capo, 1972; orig. pub. in English in 1936), 276–87; Samuel Weber, *Mass Mediauras: Form, Technics, Media* (Stanford, CA: Stanford University Press, 1996), 108–28; and Paddy Scannell, *Radio, Television, and Modern Life: A Phenomenological Approach* (Cambridge, MA: Blackwell, 1996).
13. Anne Friedberg, *Window Shopping: Cinema and the Postmodern* (Berkeley: University of California Press, 1993), 2, 125–32; Margaret Morse, *Virtualities: Television, Media Art, and Cyberculture* (Bloomington: Indiana University Press, 1998), 101; Scannell, *Television and the Meaning of "Live,"* 175.
14. See William Uricchio, "Storage, Simultaneity, and the Media Technologies of Modernity," in *Allegories of Communication: Intermedial Concerns from Cinema to the Digital*, ed. John Fullerton and Jan Olsson (Rome: John Libbey, 2004), 123–38.
15. See Jane Feuer, "The Concept of Live Television: Ontology as Ideology," in *Regarding Television: Critical Approaches—an Anthology*, ed. E. Ann Kaplan (Frederick, MD: University Publications of America, 1983), 13, and Mary Ann Doane, "Information, Crisis, Catastrophe," in *Logics of Television: Essays in Cultural Criticism*, ed. Patricia Mellencamp (Bloomington: Indiana University Press, 1990), 222. For critiques of the focus on liveness in television, see John Thornton Caldwell, *Televisuality: Style, Crisis, and Authority in American Television* (New Brunswick, NJ: Rutgers University Press, 1995), 27–31, and Mimi White, "The Attractions of Television: Reconsidering

Liveness," in *Mediaspace: Place, Scale, and Culture in a Media Age*, ed. Nick Couldry and Anna McCarthy (New York: Routledge, 2004), 75–91.

16. For a different argument that likewise complicates the distinction between storage media and "live" media, see Anne-Katrin Weber, "Recording on Film, Transmitting by Signals: The Intermediate Film System and Television's Hybridity in the Interwar Period," *Grey Room* 56 (Summer 2014): 6–33.
17. Mary Ann Doane, *The Emergence of Cinematic Time: Modernity, Contingency, the Archive* (Cambridge, MA: Harvard University Press, 2002), esp. 206–32.
18. Jeffrey Sconce, *Haunted Media: Electronic Presence from Telegraphy to Television* (Durham, NC: Duke University Press, 2000), 63.
19. See Ben Singer, "Early Home Cinema and the Edison Home Projecting Kinetoscope," *Film History* 2, no. 1 (Winter 1988): 37–69, and Moya Luckett, "Filming the Family: Home Movie Systems and the Domestication of Spectatorship," *Velvet Light Trap* 36 (Fall 1995): 21–32.
20. See Zimmermann, *Reel Families*, 27–31; Wasson, *Museum Movies*, 44–49, 56; C. E. K. Mees, "The Cine Kodak and Kodascope," *Transactions of the Society of Motion Picture Engineers* (hereafter *TSMPE*), no. 16 (May 1923): 246–51 (the issues of *TSMPE* from the early 1920s have no volume numbers, just issue numbers); and C. E. K. Mees, "A New Substandard Film for Amateur Cinematography," *TSMPE*, no. 16 (May 1923): 252–58.
21. A. F. Victor, "The Motion Picture a Practical Feature of the Home," *TSMPE*, no. 16 (May 1923): 264–65; J. H. McNabb, "The Filmo Automatic Cine-Camera and Cine-Projector," *TSMPE*, no. 18 (May 1924): 127–34.
22. Singer, "Early Home Cinema and the Edison Home Projecting Kinetoscope," 44–45; Wasson, *Museum Movies*, 47. Although the stock used in Edison's Home Projecting Kinetoscope measured 22 millimeters, it included three 5.7mm frames across its width (Singer, "Early Home Cinema and the Edison Home Projecting Kinetoscope," 45–46). On 28mm film and the Pathéscope, see also Anke Mebold and Charles Tepperman, "Resurrecting the Lost History of 28mm Film in North America," *Film History* 15, no. 2 (2003): 137–51.
23. Wasson, *Museum Movies*, 46; Zimmermann, *Reel Families*, 28.
24. Wasson, "The Reel of the Month Club," 222–23.
25. See "Report of the Non-Theatrical Equipment Committee," *Journal of the Society of Motion Picture Engineers* (hereafter *JSMPE*) 23, no. 1 (July 1934): 9–13, and "Report of the Non-Theatrical Equipment Committee," *JSMPE* 24, no. 1 (January 1935): 23–28.
26. Haidee Wasson, "Suitcase Cinema," *Cinema Journal* 51, no. 2 (Winter 2012): 151–52.
27. On educational and industrial film, see, in particular, Orgeron, Orgeron, and Streible, eds., *Learning with the Lights Off*, and Vinzenz Hediger and Patrick Vonderau, eds., *Films That Work: Industrial Film and the Productivity of Media* (Amsterdam: Amsterdam University Press, 2009). For a useful account of extratheatrical cinema from the period, also see the series of articles that Arthur Edwin Krows wrote for *Educational Screen*, beginning with "Motion Pictures—Not for Theatres," *Educational Screen*, September 1938, 211–15. Thanks to Charles Tepperman for drawing my attention to this series.
28. See Herbert C. McKay, "Future Developments in the 16 Mm. Field," *TSMPE* 12, no. 36 (September 1928): 1118, and W. B. Cook, "The 16-Mm. Sound-Film Outlook," *JSMPE* 24, no. 2 (February 1935): 178.
29. J. B. Carrigan and Russell C. Holslag, "An Estimate of the Present Status and Future Development of the Home Talkies," *JSMPE* 16, no. 1 (January 1931): 75–76. Carrigan

and Holslag were the editor and technical editor, respectively, of *Movie Makers* magazine.

30. C. Francis Jenkins, “The Discrola,” *TSMPE*, no. 16 (May 1923): 234–38.
31. Carrigan and Holslag, “An Estimate of the Present Status and Future Development of the Home Talkies,” 73.
32. Wasson, “Electric Homes!” 16; advertisement for Eastman Kodak, *Movie Makers*, December 1929, 806–7; advertisement for Library Kodascope and Cabinet, *Movie Makers*, January 1930, 47.
33. Carrigan and Holslag, “An Estimate of the Present Status and Future Development of the Home Talkies,” 71; Pathé advertisement, *Movie Makers*, December 1930, 736–37. On Pathé’s 9.5mm system, see Alexandra Schneider, “Time Travel with Pathé Baby: The Small-Gauge Film Collection as Historical Archive,” *Film History* 19 (2007): 353–60.
34. Pathé advertisement, *Movie Makers*, August 1931, 447.
35. “News of the Industry for Amateurs and Dealers,” *Movie Makers*, August 1928, 536.
36. See, for instance, Bell and Howell advertisement, *Amateur Movie Makers*, January 1927, inside cover, and Bell and Howell advertisement, *Amateur Movie Makers*, February 1927, back cover.
37. See, for instance, Dr. Kinema [Hiram Percy Maxim], “Bettering Projection,” *Amateur Movie Makers*, June 1927, 12; the tongue-in-cheek Epes W. Sargent, “Helpful Hints for Home Shows,” *Movie Makers*, June 1928, 381; and advertisement for Eastman Kodak, *Movie Makers*, June 1928, 394.
38. F. H. Richardson, “Screen Surfaces,” *Amateur Movie Makers*, May 1928, 324; F. H. Richardson, “Screen Surface Characteristics,” *Movie Makers*, June 1928, 400.
39. Advertisement for Bell and Howell, *Amateur Movie Makers*, April 1928, back cover.
40. Advertisement for Bell and Howell, *Amateur Movie Makers*, September 1927, 25.
41. “News of the Industry,” *Movie Makers*, September 1928, 611. Also see Wasson, “Electric Homes!” 18.
42. Advertisement for Kodacarte, *Movie Makers*, November 1928, 725.
43. Advertisement for Coutard Projection Screen Panel, *Movie Makers*, June 1929, 405.
44. Advertisement for Bell and Howell, *Movie Makers*, May 1929, 314. See also advertisement for Da-Lite, *Movie Makers*, November 1940, 507.
45. See, for instance, “News of the Industry for Amateurs and Dealers,” August 1928, 536; advertisement for Minusa Cine Screen Company, *Movie Makers*, November 1928, 745; and Walter Downs and Aubrey Burnett, “Getting a Longer ‘Throw,’ ” *Movie Makers*, November 1937, 554.
46. For the designation of screens as “automatic,” see advertisement for Minusa Cine Screen Company, *Movie Makers*, December 1928, 829, and advertisement for Brite-Lite Glass Beaded Movie Screen, *Movie Makers*, November 1929, 742.
47. Advertisement for Ray-Flex screen, *Movie Makers*, September 1928, 608.
48. Advertisement for A. C. Hayden Company, *Amateur Movie Makers*, February 1927, inside cover; advertisement for Ray-Flex screen, *Movie Makers*, September 1928, 608. Also see advertisement for Home Film Libraries, *Movie Makers*, August 1928, 536.
49. See Richardson, “Screen Surface Characteristics,” 400, 420; D. F. Lyman, “Relation Between Illumination and Screen Size for Non-theatrical Projection,” *JSMPE* 25, no. 3 (September 1935): 227–38; “News of the Industry,” *Movie Makers*, January 1938, 44; and “The Optical Properties of Commercially Available Screens for 16-Mm. Projection,” *JSMPE* 37, no. 1 (July 1941): 47–56.
50. Bell and Howell advertisement, *Amateur Movie Makers*, April 1927, 30.

51. "News of the Industry for Amateurs and Dealers," August 1928, 536; advertisement for Truvision Projection Screen Corporation, *Movie Makers*, October 1929, 634; Richardson, "Screen Surface Characteristics," 400, 420.
52. Advertisement for Willoughbys, *Movie Makers*, December 1930, 743.
53. "Report of the Non-Theatrical Equipment Committee" (January 1935), 26.
54. Russel C. Holslag, "News of the Industry," *Movie Makers*, December 1933, 518. On the Preview Moviola, see I. Serrurier, "Combined Viewing and Projection Machine with or Without Sound," *JSMPE* 29, no. 6 (December 1937): 674; and David Bordwell and Kristin Thompson, "Technological Change and Classical Film Style," in Tino Balio, *Grand Design: Hollywood as a Modern Business Enterprise, 1930–1939* (Berkeley: University of California Press, 1993), 120.
55. Advertisement for Bell and Howell, *Amateur Movie Makers*, September 1927, 25, emphasis in original.
56. Lyman, "Relation Between Illumination and Screen Size for Non-theatrical Projection," 234.
57. "Report of the Non-Theatrical Equipment Committee" (January 1935), 26.
58. "Report of the Non-Theatrical Equipment Committee" (January 1935), 24.
59. Haidee Wasson, "The Other Small Screen: Moving Images at New York's World Fair, 1939," *Canadian Journal of Film Studies* 21, no. 1 (Spring 2012): 96.
60. Advertisement for Eastman Kodak, *Movie Makers*, December 1929, 806–7; advertisement for Library Kodascope and Cabinet, *Movie Makers*, January 1930, 47.
61. See Wasson, "The Other Small Screen," 94–98; "Progress in the Motion Picture Industry," *JSMPE* 15, no. 6 (December 1930): 782–83; "New Apparatus," *JSMPE* 20, no. 1 (January 1933): 87; and advertisement for Automotion Pictures, Inc., *Movie Makers*, October 1935, 416.
62. H. G. Tasker and A. W. Carpenter, "Motion Pictures with Sound on Standard 16 Mm. Film," *JSMPE* 19, no. 3 (September 1932): 246, 247.
63. See, for instance, C. L. Edson, "Homes for Home Movies," *Movie Makers*, May 1931, 264.
64. See R. Fawn Mitchell, "Perfecting Projection," *Movie Makers*, October 1934, 420, and Ralph G. Wildes, "Pointing Up Projection," *Movie Makers*, February 1935, 67.
65. Barbara Wilinsky, *Sure Seaters: The Emergence of Art House Cinema* (Minneapolis: University of Minnesota Press, 2001), 46–55.
66. Tepperman, *Amateur Cinema*, 27. On such shared billing, see Wilinsky, *Sure Seaters*, 49. On the domestic and public exhibition of amateur cinema in the 1930s, also see Charles Tepperman, "'A Recognized Screen': The New York Annual Movie Parties from Parlor to Public," *Film History* 30, no. 1 (Spring 2018): 58–85.
67. James H. Blauvelt, "Planning Movie Rooms," *Movie Makers*, March 1937, 118–19.
68. Downs and Burnett, "Getting a Longer 'Throw,'" 553.
69. See Jackson M. Hackett, "Choosing a Screen," *Movie Makers*, December 1938, 613; advertisement for Da-Lite screens, *Movie Makers*, September 1939, 456; E. Warren Wood, "Living Room Projection," *Movie Makers*, November 1939, 567; and advertisement for Da-Lite screens, *Movie Makers*, February 1940, 56.
70. Lynn Spigel, "Object Lessons for the Media Home: From Storagewall to Invisible Design," *Public Culture* 24, no. 3 (2012): 573.
71. Wasson, "Electric Homes!" 20.
72. Carrigan and Holslag, "An Estimate of the Present Status and Future Development of the Home Talkies," 75.

73. Carrigan and Holslag, "An Estimate of the Present Status and Future Development of the Home Talkies," 75.
74. Pathé advertisement, *Movie Makers*, August 1931, 447.
75. Carrigan and Holslag, "An Estimate of the Present Status and Future Development of the Home Talkies," 75, 79.
76. See Russell C. Holslag, "Filling the Big Screen," *Movie Makers*, March 1934, 116, and "The Optical Properties of Commercially Available Screens for 16-Mm. Projection," 47. Also see, for instance, R. A. Miller and H. Pfannenstiehl, "A Portable 16 Mm. Sound Picture System," *JSMPE* 19, no. 3 (September 1932): 249.
77. Tasker and Carpenter, "Motion Pictures with Sound on Standard 16 Mm. Film," 245–47.
78. Dr. Kinema, "Bettering Projection," 12. Also see, for instance, Karl A. Barleben Jr., "Pointers on Projection," *Movie Makers*, November 1929, 742. On Maxim's pseudonym, see Tepperman, *Amateur Cinema*, 38.
79. Mees, "The Cine Kodak and Kodascope," 251.
80. See, for instance, advertisement for Eastman Kodak, *Amateur Movie Makers*, January 1927, 33; advertisement for Willoughbys, *Amateur Movie Makers*, January 1928, 9; "News of the Industry for Amateurs and Dealers," *Amateur Movie Makers*, January 1928, 43; advertisement for Truvision, *Movie Makers*, November 1928, 727; advertisement for Bell and Howell, *Movie Makers*, May 1929, 314–15; and advertisement for Da-Lite, *Movie Makers*, February 1931, 101.
81. "Report of the Non-Theatrical Equipment Committee," *JSMPE* 19, no. 2 (August 1932): 200.
82. Advertisement for Eastman Kodak, *Amateur Movie Makers*, April 1927, 37.
83. Advertisement for Arrow Screen Company, *Movie Makers*, May 1929, 319.
84. Zimmermann, *Reel Families*, 31.
85. Advertisement for Bell and Howell, *Amateur Movie Makers*, September 1927, 25.
86. Russell C. Holslag, "News of the Industry," *Movie Makers*, January 1931, 32. On Newcomer's relationship with Chrétien, see John Belton, *Widescreen Cinema* (Cambridge, MA: Harvard University Press, 1992), 128. On Cine-Panor, also see Tepperman, *Amateur Cinema*, 126.
87. Holslag, "News of the Industry," January 1931, 32.
88. Fred Schmid, "Wide-Field Pictures on Narrow Gauge," *American Cinematographer*, May 1931, 36, 37.
89. See, for instance, advertisements for Cine-Panor, *Movie Makers*, January 1931, 39, and June 1931, 341.
90. Advertisement for Cine-Panor, *Movie Makers*, August 1931, 445. Also see, for instance, advertisements for Cine-Panor, *Movie Makers*, September 1931, 505, and November 1931, 621.
91. F. H. Richardson, "Reviewing the Fall Meeting of the S. M. P. E.," *Motion Picture Herald*, Better Theatres sec., October 24, 1931, 47.
92. See Belton, *Widescreen Cinema*, 226.
93. Lyman, "Relation Between Illumination and Screen Size for Non-theatrical Projection," 235; "The Optical Properties of Commercially Available Screens for 16-Mm. Projection," 53.
94. Advertisement for Victor Projector, *Movie Makers*, April 1932, 144.
95. Advertisement for Da-Lite De Luxe Challenger screen, *Movie Makers*, July 1932, 311.
96. Holslag, "Filling the Big Screen," 103.

97. See advertisement for Mogull Holiday Crystal Beaded Screens, *Movie Makers*, December 1934, 543, and advertisement for Da-Lite, *Movie Makers*, May 1936, 215.
98. "Report of the Non-Theatrical Equipment Committee" (January 1935), 23.
99. Advertisement for Da-Lite, *Movie Makers*, February 1940, 56.
100. See advertisement for Da-Lite, *Movie Makers*, February 1940, 56, and the list of companies employing Da-Lite screens in advertisement for Da-Lite, *Movie Makers*, May 1940, 207.
101. See, for instance, "Large Versus Small Picture," *Motion Picture Herald*, Better Theatres sec., May 7, 1932, 32.
102. On 20-foot-wide extratheatrical projection, see "Report of the Non-Theatrical Equipment Committee" (January 1935), 23, and advertisement for Da-Lite, *Movie Makers*, February 1940, 56.
103. See Ben Schlanger, "Motion Picture Theaters," *Architectural Record*, February 1937, 16–24.
104. R. P. May, "16-Mm. Sound-Film Dimensions," *JSMPE* 18, no. 4 (April 1932): 501.
105. Holslag, "Filling the Big Screen," 103, emphasis in original.
106. Holslag, "Filling the Big Screen," 103.
107. "Sees Television Ready for Home," *New York Sun*, June 4, 1938, folder 1 (TV Receivers), box 240, series 53 (Television Receivers), George H. Clark Radioana Collection (hereafter GHCRC), Archives Center, National Museum of American History, Smithsonian Institution, Washington, DC.
108. For an example of a screen that duplicated the look of a theatrical proscenium, see advertisement for the Pro-Vel Screen Unit, *Movie Makers*, December 1938, 631.
109. For the comparison to family albums, see Da-Lite advertisement, *Movie Makers*, April 1939, 155. On the placement of home screens alongside hunting trophies, rifles, and souvenirs of travel, see Edson, "Homes for Home Movies," 265, and James W. Moore, "The Cinema at Home," *Movie Makers*, December 1933, 537.
110. See, in particular, Charles R. Acland, "Curtains, Carts, and the Mobile Screen," *Screen* 50, no. 1 (Spring 2009): 148–66; Wasson, "Suitcase Cinema"; Michael Cowan, "Taking It to the Street: Screening the Advertising Film in the Weimar Republic," *Screen* 54, no. 4 (Winter 2013): 463–79; and Haidee Wasson, "Protocols of Portability," *Film History* 25, nos. 1–2 (2013): 236–47.
111. Anthony Slide, *Before Video: A History of Non-theatrical Film* (New York: Greenwood Press, 1992), 53; Stephen Groening, *Cinema Beyond Territory: Inflight Entertainment and Atmospheres of Globalization* (London: British Film Institute, 2014), 67.
112. On Cinéorama, see Belton, *Widescreen Cinema*, 85. On Hale's Tours, see Lauren Rabinovitz, *For the Love of Pleasure: Women, Movies, and Culture in Turn-of-the-Century Chicago* (New Brunswick, NJ: Rutgers University Press,1998), 145–54.
113. "Ether Circuit for U's 16 mm. Talkers," *Variety*, October 2, 1929, 1; Wasson, "Electric Homes!" 10.
114. Advertisement for Brite-Lite Glass Beaded Movie Screen, *Movie Makers*, December 1929, 835.
115. Wasson, "Electric Homes!" 10. Also see, for instance, "The Camera Reports," *Motion Picture Herald*, August 22, 1931, 19.
116. "Projection on the High Seas," *Motion Picture Projectionist*, January 1932, 20, 39.
117. "Projection on the High Seas," 20. Another image of this arrangement can be found in *Motion Picture Herald*, Better Theatres sec., April 1, 1933, 14.
118. See Emma Widdis, *Visions of a New Land: Soviet Film from the Revolution to the Second World War* (New Haven, CT: Yale University Press, 2003), 14–15, 41–45; Brian

Larkin, *Signal and Noise: Media, Infrastructure, and Urban Culture in Nigeria* (Durham, NC: Duke University Press, 2008), 76–78, 84–86; Zoë Druick, "At the Margins of Cinema History: Mobile Cinema in the British Empire," *Public* 40 (2010): 118–25; and Devin Orgeron, Marsha Orgeron, and Dan Streible, "A History of Learning with the Lights Off," in *Learning with the Lights Off*, ed. Orgeron, Orgeron, and Streible, 59.

119. E. I. Sponable, "Movietone Field Projection Outfit," *TSMPE* 12, no. 36 (September 1928): 1177. Also see Stephanie E. Przybylek, with Marie Eckhardt and Jim Richerson, *Breaking the Silence on Film: The History of the Case Research Lab* (Auburn, NY: Cayuga Museum of History and Art, 1999), 100. On Movietone, see Crafton, *The Talkies*, 89–100.
120. "Portable Commercial Trucks," "Project—Outdoor Projection Truck" folder, box 18 (General Files Projects: Fo–St), Earl I. Sponable Papers (hereafter EISP), Rare Book and Manuscript Library, Columbia University Libraries, New York; sketch of rear-projection arrangement signed J. G. Speer, April 10, 1929, "Project—Outdoor Projection Truck" folder, box 18 (General Files Projects: Fo–St), EISP. On the use of such a set-up in Germany, see Cowan, "Taking It to the Street," 467.
121. Sponable, "Movietone Field Projection Outfit," 1180.
122. Inventories for Projection Trucks nos. 2, 3, 4, "Projection Trucks 1930–31" folder, box 15 (General Files P-Projectors), EISP; D. F. Whiting to Truman Talley, June 26, 1930, "Projector Data 1930–31" folder, box 15 (General Files P-Projectors), EISP.
123. "A Mobile Sound Picture Theater," *Motion Picture Projectionist*, December 1930, 29.
124. "Movietone Is Being Used at the Meetings," *Hudson Register* (New York), October 8, 1930, "Newspaper Clippings" folder, box 14 (General Files Miscellaneous, 1930–1931–Os), EISP; H. F. Jermain to Mr. E. I. Sponable, September 24, 1930, "Projection Trucks 1930–1931" folder, box 15 (General Files P-Projectors), EISP.
125. Sam Kaplan, quoted in "A Mobile Sound Picture Theater," 29.
126. "Metro's New World-Wide Good-Will Ballyhoo Trip," *Variety*, October 11, 1932, 31.
127. "A Reproducer on Wheels," *Motion Picture Projectionist*, November 1932, 34.
128. "A Reproducer on Wheels," 34.
129. On the promotion and use of film for sales and advertising in this period, see Gregory A. Waller, "Projecting the Promise of 16mm," in *Useful Cinema*, ed. Acland and Wasson, 125–48, and Wasson, "Selling Machines." For examples of the marketing of screens for business use, see advertisement for Da-Lite, *Business Screen* 1, no. 1 (1938): 7; and "Screens & Accessories: Motion & Stillfilm Projection," *Business Screen* 1, no. 8 (August 1939): Equipment Review sec., xiii.
130. See Wasson, "Selling Machines," 63–65.
131. Advertisement for Illustravox, *Business Screen* 1, no. 8 (August 1939): 5. Also see, for instance, "Progress in the Motion Picture Industry," *JSMPE* 15, no. 6 (December 1930): 782–83.
132. Advertisement for Automotion Pictures, *Movie Makers*, October 1935, 416. Also see "Continuous Motion Picture Equipment," *Business Screen* 2, no. 8 (September 1940): Equipment Review sec., n.p.
133. See Wasson, "Selling Machines," 54, and Cowan, "Taking It to the Street," 467.
134. See Friedberg, *Window Shopping*, 81; Rabinovitz, *For the Love of Pleasure*, 68–102; Erkki Huhtamo, "Toward a History of Peep Practice," in *A Companion to Early Cinema*, ed. André Gaudreault, Nicolas Dulac, and Santiago Hidalgo (Malden, MA: Wiley-Blackwell, 2012), 44; Mebold and Tepperman, "Resurrecting the Lost History of 28mm Film in North America," 141; and Susan Murray, *Bright Signals: A History of Color Television* (Durham, NC: Duke University Press, 2018), 23.
135. "Motion Pictures in the Department Store," *Business Screen* 1, no. 1 (1938): 32, 34, 32.

136. "How to Use Motion Pictures in the Department Store," *Business Screen* 1, no. 1 (1938): 35–36.
137. "A Survey on the Use of Films at Recent Expositions," *Business Screen* 1, no. 2 (1938): 23–24, 23, 27, 26.
138. "The World's Fair Survey of Motion Pictures and Slidefilms at the Fairs," *Business Screen* 2, no. 1 (September 1939): 21, emphasis removed.
139. Wasson, "Selling Machines," 59. Also see Wasson, "The Other Small Screen," 98.
140. Huhtamo, "Toward a History of Peep Practice," 44.
141. Andrea Kelley, "'A Revolution in the Atmosphere': The Dynamics of Site and Screen in 1940s Soundies," *Cinema Journal* 54, no. 2 (Winter 2015): 72–93.
142. Louis Miller Bailey, "Cinemas for Caravanserais," *Movie Makers*, February 1932, 54, 81.
143. "Drive-in Theater, Camden, New Jersey," *Architectural Record*, March 1934, 235; Mary Morley Cohen, "Forgotten Audiences in the Passion Pits: Drive-in Theatres and Changing Spectator Practices in Post-war America," *Film History* 6, no. 4 (Winter 1994): 475. On drive-ins and spectatorial mobility, also see Anne Friedberg, "Urban Mobility and Cinematic Visuality: The Screens of Los Angeles—Endless Cinema or Private Telematics," *Journal of Visual Culture* 1, no. 2 (2002): 183–204.
144. Margaret Farrand Thorp, *America at the Movies* (1939; reprint, London: Faber and Faber, 1946), 46.
145. "Motion Pictures Between Trains," *Motion Picture Herald*, Better Theatres sec., July 29, 1933, 10. Also see "A News Reel Movie in a Railroad Station," *Architectural Record*, March 1934, 235, and Douglas Gomery, *Shared Pleasures: A History of Movie Presentation in the United States* (Madison: University of Wisconsin Press, 1992), 145.
146. "New York Has Theater for Transients," *Architectural Record*, September 1937, 29.
147. "What the Railroads Are Doing," *Business Screen* 1, no. 4 (1938): 23.
148. Friedberg, *Window Shopping*; Anne Friedberg, *The Virtual Window: From Alberti to Microscoft* (Cambridge, MA: MIT Press, 2006), 160–62.
149. "How to Use Motion Pictures in the Department Store," 35, 36.
150. "United Brings the Skyways to Earth," *Business Screen* 1, no. 4 (1938): 23.
151. "United Brings the Skyways to Earth," 23.
152. Tepperman, *Amateur Cinema*, 62, 32. Also see Zimmermann, *Reel Families*, 73–81.
153. Wasson, "The Reel of the Month Club," 223. Also see, for instance, descriptions of the Pathegram and Kodak Cinegraph libraries in advertisement for De Vry, *Movie Makers*, March 1928, 183, and advertisement for Kodak, *Movie Makers*, December 929, 806.
154. Tom Gunning, "'The Whole World Within Reach': Travel Images Without Borders," in *Virtual Voyages: Cinema and Travel*, ed. Jeffrey Rouff (Durham, NC: Duke University Press, 2006), 27.
155. Gunning, "'The Whole World Within Reach,'" 30.
156. "A Department Store Experiment Opens a Potential Film Field," *Business Screen* 1, no. 8 (August 1939): 16; "Films Via Television," *Business Screen* 2, no. 2 (November 1939): 15. On point-of-purchase television in the period around World War II, see McCarthy, *Ambient Television*, 63–88.
157. James W. Moore, "Amateur Clubs," *Movie Makers*, November 1939, 558.
158. See Edgerton, *Columbia History of American Television*, 5–16.
159. See especially Udelson, *The Great Television Race*; Albert Abramson, *The History of Television, 1880 to 1941* (Jefferson, NC: McFarland, 1987); and R. W. Burns, *Television: An International History of the Formative Years* (London: Institution of Electrical Engineers, 1998). For an account of discourses on television in this period, see also Philip W.

Sewell, *Television in the Age of Radio: Modernity, Imagination, and the Making of a Medium* (New Brunswick, NJ: Rutgers University Press, 2014).

160. Barnouw, *Tube of Plenty*, 61. I am considering "television" to entail the transmission of moving images, but it should be noted that there were earlier demonstrations of the transmission of still images. See Edgerton, *Columbia History of American Television*, 28.
161. See Murray, *Bright Signals*, 11–33.
162. See Thomas Calvert McClary, "What About This Television?" *Forbes* galley proof, July 3, 1931, series 142 (History of Television), box 388, folder 3 (History of Television 1928–1932), GHCRC.
163. Udelson, *The Great Television Race*, 132–58.
164. See William Boddy, *Fifties Television: The Industry and Its Critics* (Urbana: University of Illinois Press, 1990), 34.
165. See, for instance, Friedberg, *Window Shopping*, 133–37.
166. For a discussion of the complexities entailed in thinking about screen technologies in such ways, see Charles R. Acland, "The Crack in the Electric Window," *Cinema Journal* 51, no. 2 (Winter 2012): 167–71.
167. For contemporaneous descriptions of this process, see, for instance, A. Dinsdale, "And Now, We See by Radio!" *Radio Broadcast* 10, no. 2 (December 1926): 139–43, and A. P. Peck, "Television Enters the Home," *Scientific American*, March 1928, 246–47.
168. C. Francis Jenkins, "Radio Photographs, Radio Movies, and Radio Vision," *TSMPE*, no. 16 (May 1923): 78–79.
169. C. Francis Jenkins, "Recent Progress in the Transmission of Motion Pictures by Radio," *TSMPE*, no. 17 (October 1923): 83.
170. Dinsdale, "And Now, We See by Radio!" 142.
171. C. Francis Jenkins to F. G. Taylor, October 11, 1929, folder 5 (Jenkins Television 1928–1948), box 453, series C (Clark Unorganized and/or Duplicate Photographs), GHCRC.
172. C. Francis Jenkins, "Pantomime Pictures by Radio for Home Entertainment," *TSMPE* 12, no. 33 (April 1928): 113–14.
173. On Jenkins's drum receiver and his later reversion to discs, see Burns, *Television*, 204.
174. H. E. Ives, "Two-Way Television," *JSMPE* 16, no. 3 (March 1931): 294. For a discussion of the drawbacks of such lenses, also see A. Dinsdale, "Television Needs New Ideas—and Less Ballyhoo," *Scientific American*, November 1930, 367.
175. C. F. Jenkins Company, "Television: The Eye of Radio," 14, folder 3 (History of Television, 1928–1932), box 388, series 142 (History of Television), GHCRC; D. E. Replogle to C. Francis Jenkins, November 30, 1931, folder 1 (TV Receivers), box 240, series 53 (Television Receivers), GHCRC.
176. "Instructions on Installation and Operation of Model #301, 302, and 303 Jenkins Radiovisors," March 9, 1931, folder 1 (TV Receivers), box 240, series 53 (Television Receivers), GHCRC.
177. Jenkins, "Pantomime Pictures by Radio for Home Entertainment," 115, 116.
178. C. Francis Jenkins to D. E. Replogle, November 27, 1931, emphasis removed, folder 1 (TV Receivers), box 240, series 53 (Television Receivers), GHCRC.
179. "How Large-Screen Television Works," no source, n.d., folder 1 (TV Receivers), box 240, series 53 (Television Receivers), GHCRC.
180. V. K. Zworykin, "Iconoscopes and Kinescopes in Television," *JSMPE* 28, no. 5 (May 1937): 475. On the electronic systems, also see A. P. Peck, "A Real 'Electric Eye,'" *Scientific American*, September 1933, 117, and R. R. Beal, "RCA Developments in Television," *JSMPE* 29, no. 2 (August 1937): 121–43.

181. Zworykin, "Iconoscopes and Kinescopes in Television," 495.
182. "Thru the Laboratory Keyhole," *Radio Today*, May 1936, 16; "Thru the Lab Keyhole," *Radio Today*, August 1936, 22; Beal, "RCA Developments in Television," 140.
183. "Television Testing Apparatus Like Radio Set with Mirror," no source, 1935, "Clippings 1935–1936–1937" folder, box 21, Allen Balcom Du Mont Collection, Archives Center, National Museum of American History, Smithsonian Institution, Washington, DC.
184. "Exhibition of British Television Sets Shows a Trend Toward Direct Viewing," *Broadcasting*, November 1, 1936, 30.
185. See I. J. Kaar, "Road Ahead for Television," *JSMPE* 32, no. 1 (January 1939): 32; "Television: Answers to Your Questions," RCA Brochure, 1939, folder 2 (Television 1928–1944), box 401, series 174 (Television Methods and Systems), GHCRC; and "See! Hear!—American Teleceivers for 1939," *Radio-Craft*, June 1939, 723.
186. Weber, *Mass Mediauras*, 117, emphases removed. See also, for instance, Scannell, *Radio, Television, and Modern Life*, 91.
187. See Doron Galili, "Tom Swift's Three Inventions of Television: Media History and the Technological Imaginary," *View* 4, no. 7 (2015): 60. It might also be noted that the "magic mirror" on Jenkins's 1928 model resembles Albert Robida's futuristic depiction of the "telephonoscope" in 1883. See Jenkins, "Pantomime Pictures by Radio for Home Entertainment," 115, and A. Robida, *Le vingtième siècle* (Paris: Georges Decaux, 1883), 57. On Robida, see Uricchio, "Television, Film, and the Struggle for Media Identity," 118–19.
188. C. E. Huffman, "The Design of a Complete Television System," n.d., folder 2 (Television, 1928–1944), box 401, series 174 (Television Methods and Systems), GHCRC.
189. Dinsdale, "And Now, We See by Radio!" 139. On television's association with the telescope, see also William Uricchio, "There's More to the Camera's Obscura Than Meets the Eye," in *Stop Motion, Fragmentation of Time: Exploring the Roots of Modern Visual Culture*, ed. François Albera, Marta Braun, and André Gaudreault (Lausanne, Switzerland: Éditions Payot Lausanne, 2002), 104, 106–7.
190. Orrin E. Dunlap Jr., *The Outlook for Television* (New York: Harper and Brothers, 1932), 50.
191. Spigel, *Make Room for TV*, 115.
192. Galili, "Tom Swift's Three Inventions of Television," 63.
193. David Sarnoff, statement on television, May 7, 1935, folder 1 (History of Television 1919–1949), box 388, series 142 (History of Television), GHCRC. Also see, for instance, statement by David Sarnoff, December 1931, folder 3 (History of Television 1928–1932), box 388, series 142 (History of Television), GHCRC.
194. Weber, *Mass Mediauras*, 122–23, emphases removed.
195. Albert W. Protzman, "Television Studio Technic," *JSMPE* 33, no. 1 (July 1939): 34–35. On the use of rear projection in postwar television studios, see Lynn Spigel, *TV by Design: Modern Art and the Rise of Network Television* (Chicago: University of Chicago Press, 2008), 115–17. On layering in televisual representation generally, see also Morse, *Virtualities*, 114–15.
196. On cosmoramas, see Huhtamo, "Toward a History of Peep Practice," 40. On shop windows, see Friedberg, *Window Shopping*, 65–68.
197. Indeed, the notion of "listening in" to radio also evoked a form of transgression by suggesting eavesdropping. See Covert, "'We May Hear Too Much,'" 203. On the analogy to radio, see Alfred N. Goldsmith, "Television Terms," ca. 1933, folder 1 (History of Television 1919–1949), box 388, series 142 (History of Television), GHCRC. For the term *looking in*, also see, for instance, A. P. Peck, "Television Advances," *Scientific American*, June 1929, 526.

198. John Ellis, *Visible Fictions: Cinema, Television, Video*, rev. ed. (New York: Routledge, 1992), 137, 167.
199. Spigel, *Make Room for TV*, 104, 100–115.
200. See, for instance, Jenkins, "Pantomime Pictures by Radio for Home Entertainment," 113, and Peck, "Television Advances," 527.
201. Murray, *Bright Signals*, 20–22.
202. "Television: Answers to Your Questions," folder 2 (Television 1928–1944), box 401, series 174 (Television Methods and Systems), GHCRC; "The Story of Television," 1953, folder 7 (Story of Television, 1953), box 44, Du Mont Collection; "Television Receiver," *Pittsburgh Electric League*, n.d., "Clippings 1939" folder, box 21, Du Mont Collection.
203. A. N. Goldsmith, "Television and Motion Pictures," *JSMPE* 25, no. 1 (July 1935): 40–41.
204. Kaar, "Road Ahead for Television," 32. For the call to increase television screen size, also see "Television from the Standpoint of the Motion Picture Industry," *JSMPE* 29, no. 2 (August 1937): 147.
205. Goldsmith, "Television and Motion Pictures," 39. For another discussion balancing such factors, also see, for instance, E. W. Engstrom, "A Study of Television Image Characteristics," *JSMPE* 22, no. 5 (May 1934): 290–313.
206. Goldsmith, "Television and Motion Pictures," 39.
207. "Thru the Laboratory Keyhole," May 1936, 17.
208. "Television from the Standpoint of the Motion Picture Producing Industry," 146; Beal, "RCA Developments in Television," 130; David Sarnoff, quoted in "Television Demonstration at the Fall Convention," *JSMPE* 29, no. 6 (December 1937): 597.
209. Kaar, "Road Ahead for Television," 33–34. On high-definition television, see Barry Fox, "Video Thrills the Hollywood Star," *New Scientist* 100, no. 1380 (October 20, 1983): 208.
210. Dinsdale, "Television Needs New Ideas—and Less Ballyhoo," 366.
211. "Far-off Speakers Seen as Well as Heard Here in Test of Television," *New York Times*, April 8, 1927, 1.
212. C. E. Huffman, "Visit to G.E. Television Demonstration and Laboratories," May 24, 1930, folder 1 (History of Television 1919–1949), box 388, series 142 (History of Television), GHCRC.
213. Huffman, "Visit to G.E. Television Demonstration."
214. McClary, "What About This Television?" Also see, for instance, Jenkins, "Radio Photographs, Radio Movies, and Radio Vision," 86, and C. Francis Jenkins, "Radio Movies and the Theater," *TSMPE* 11, no. 29 (July 1927): 46.
215. J. Weinberger to Dr. Goldsmith, January 23, 1928, folder 2 (History of Television 1922–1929), box 388, series 142 (History of Television), GHCRC.
216. "Television Today," *Radio Today*, September 1935, 23; E. L. Bragdon, "Television Sets Demonstrated," *New York Sun*, August 3, 1935, folder 1 (History of Television 1919–1949), box 388, series 142 (History of Television), GHCRC.
217. See "Television Stages First Real 'Show,'" *New York Times*, July 8, 1936, 21; Will Baltin, "Television Is in the Air—Local Experimenter Picks Up N. Y. Signal," *New Brunswick Sunday Times*, July 4, 1937; and Will Baltin, "Fashion Show Sent Via Television Entertaining and Instructive to Local Men," *New Brunswick Sunday Times*, November 21, 1937.
218. "I Saw Television at Its Best," *New Brunswick Sunday Times*, August 15, 1937.
219. Orrin E. Dunlap, "Televiews Reviewed," *New York Times*, February 21, 1937, 160. Also see, for instance, Martin Codel, "Philco Discloses Its Television Progress," *Broadcasting*, August 15, 1936, 11. On the continuing advocacy for close-ups in the 1940s, see Boddy, *Fifties Television*, 83.

220. Norman Siegel, "New Television to Be Strictly for Homes, Not Movies," *World-Telegram*, February 3, 1938; Norman Siegel, "London Watches Wimbledon Tennis and Televised Plays in Cocktail Bar; 2,000 Sets Sold There in a Year," *World-Telegram*, n.d., folder 1 (History of Television 1919–1949), box 388, series 142 (History of Television), GHCRC. Although the latter item is undated, it is the fifth in a series that *World-Telegram* appears to have been publishing daily since January 31, 1938. For the first installment, see Norman Siegel, "Television Near, but It's Still a Problem Child," *World-Telegram*, January 31, 1938.
221. See also Koszarski, *Hollywood on the Hudson*, 453–54.
222. On the pursuit of theater television, see Boddy, *Fifties Television*, 23–24, and Timothy R. White, "Hollywood's Attempt at Appropriating Television: The Case of Paramount Pictures," in *Hollywood in the Age of Television*, ed. Balio, 149–55. For a detailed account of experiments with large-screen television between 1930 and 1935, see Burns, *Television*, 308–29.
223. RCA's demonstrations, for instance, employed front-projection arrangements with reflective screens. See the depictions in Orrin E. Dunlap, "New Lens Projector Flashes Television on a Screen," *New York Times*, May 16, 1937, 176, and illustration of RCA projection television, folder 3 (Television), box 425, series A (Thomas Coke Knight RCA Photographs), GHCRC (figs. 4.17 and 4.18, this volume).
224. "Far-Off Speakers Seen as Well as Heard Here in a Test of Television," 1.
225. Burns, *Television*, 309; C. E. Huffman, Mr. Barclay Visit to G.E. Television Demonstration and Labs, May 24, 1930, folder 1 (History of Television 1919–1949), box 388, series 142 (History of Television), GHCRC; television timeline, folder 1 (TV Receivers), box 240, series 53 (Television Receivers), GHCRC. See also "Theatrical Television Hardly a Factor in the Radio Market," *Radio Topics* 8, no. 23 (June 2, 1930), folder 3 (History of Television 1928–1932), box 388, series 142 (History of Television), GHCRC.
226. Burns, *Television*, 311.
227. Dunlap, *The Outlook for Television*, 13.
228. Television timeline, folder 1 (TV Receivers), box 240, series 53 (Television Receivers), GHCRC.
229. "Images Thrown on London Screen Are Hailed as 'Marvelous' and Baird Is Cheered," *New York Times*, June 2, 1932, 2; A. T. Stoyanowsky, "A New Process of Television Out of Doors," *JSMPE* 20, no. 5 (May 1933): 438.
230. John Logie Baird, quoted in Burns, *Television*, 320.
231. Image caption, *Radio Today*, November 1935, 7; "Six-Foot Television Screen," *Electronics*, October 1935, folder 2 (Television, 1928–1944), box 401, series 174 (Television Methods and Systems), GHCRC. On the relationship between this screen and Baird's similarly gridded screen of 1930, see Burns, *Television*, 325–26.
232. Uricchio, "Television, Film, and the Struggle for Media Identity," 121–24. See also Wilhelm E. Schrage, "German Television," *Radio News*, July 1935, 9, 60; "Du Mont Sees British Sport by Television," *Newark Ledger*, August 2, 1937; and Burns, *Television*, 326.
233. "Six-Foot Television Screen."
234. Martin J. Porter, "British Prepare to Broadcast: Several Stations Operating in Germany and Russia," *New York Evening Journal*, February 2, 1935.
235. Siegel, "London Watches Wimbledon Tennis."
236. "Across the Atlantic," *Radio Today*, September 1936, 17. See also "Throws Television Picture on Screen," *New York Sun*, September 5, 1936, "Clippings 1935–1936–1937" folder, box 21, Du Mont Collection.

237. See the quote from Mr. Engstrom in A. B. DuMont, "Design Problems in Television Systems and Receivers," *JSMPE* 33, no. 1 (July 1939): 73.
238. "Screen Size Television Popular in London," *New York Sun*, March 13, 1937, folder 2 (Television, 1928–1944), box 401, series 174 (Television Methods and Systems), GHCRC.
239. "Huge Images Made by Television 'Gun,'" no source, May 13, 1937, "Clippings 1935–1936–1937" folder, box 21, Du Mont Collection; "Television from the Standpoint of the Motion Picture Producing Industry," 146–47.
240. "Huge Images Made by Television 'Gun.'" Also see Dunlap, "New Lens Projector Flashes Television on a Screen," 176.
241. "Television Demonstration at the Fall Convention," 600.
242. "New Tube Makes for Large Image," *New York Sun*, November 18, 1939, folder 1 (TV Receivers), box 240, series 53 (Television Receivers), GHCRC.
243. R. W. Stewart, "Demonstrations of Advances in Television Show Higher Entertainment Level," *New York Times*, February 2, 1941, X12; illustration of RCA projection television and accompanying caption, folder 3 (Television), box 425, series A (Thomas Coke Knight RCA Photographs), GHCRC.
244. Morse, *Virtualities*, 99–124; Lynn Spigel, *Welcome to the Dreamhouse: Popular Media and Postwar Suburbs* (Durham, NC: Duke University Press, 2001), 60–103. This work draws on Raymond Williams's related point that broadcasting emerged from the wider context of what he calls "mobile privatization" (*Television: Technology and Cultural Form* [1974; reprint, New York: Routledge, 2003], 19).
245. Arnheim, *Radio*, 279, 280. On Arnheim's writing about television in the 1930s, see Doron Galili, "Television from Afar: Arnheim's Understanding of Media," in *Arnheim for Film and Media Studies*, ed. Scott Higgins (New York: Routledge, 2011), 195–211.
246. Christopher H. Sterling and John Michael Kittross, *Stay Tuned: A History of American Broadcasting*, 3rd ed. (Mahwah, NJ: Lawrence Erlbaum Associates, 2002), 112.
247. "The Chronology of Television, Part II—1931 to 1935," *Television and Short-Wave World*, August 1935, 453, reprinted in Stephen Herbert, *A History of Early TV*, 3 vols. (New York: Routledge, 2004), 2:28.
248. H. R. Lubcke, "The Theatrical Possibilities of Television," *JSMPE* 25, no. 1 (July 1935): 49.
249. David Sarnoff, as told to Donald Furthman Wickets, "You'll Soon See Across the Sea," *Liberty*, February 13, 1937, 18.
250. See "The Opening of Britain's First Television Service," *Television and Short-Wave World*, December 1936, 677–79, reprinted in Herbert, *A History of Early TV*, 2:79–81.
251. Allen DuMont, quoted in "Europe Points the Way in Television," *Short Wave and Television*, December 1937, 448; Allen DuMont, quoted in "London Views the Show," *New York Times*, August 8, 1937, 144.
252. "Screen Size Television Popular in London."
253. Siegel, "London Watches Wimbledon Tennis." On film jukeboxes, see Kelley, "'A Revolution in the Atmosphere,'" 75. On televisions, see McCarthy, *Ambient Television*, 29–62.
254. David Dietz, "Television Due in 1938," *World-Telegram*, November 6, 1936.
255. Orrin E. Dunlap Jr., "Tele-salon of the Air," *Printers Ink Monthly*, November 1937, folder 1 (History of Television 1919–1949), box 388, series 142 (History of Television), GHCRC.
256. "Television in Store Carries Hat Styles," *New York Times*, April 27, 1939, 28; Bloomingdale's advertisement, *New York Times*, April 27, 1939, reprinted in Herbert, *A History of Early TV*, 3:147. See also Koszarski, *Hollywood on the Hudson*, 463.

257. "Television in Store Carries Hat Styles," 28.
258. Ira Hirschmann, quoted in "A Department Store Experiment Opens a Potential Film Field," 16.
259. I. A. Hirschmann, "Television—a New Dimension in Department Store Advertising," *Television*, Spring 1944, 10.
260. "Television Merchandising Due," *New York Times*, July 1, 1939, 20; "Films Via Television," 15.
261. Bud Gamble, "The Television Tour of 88 Department Stores," *Televiser*, Winter 1945, 48.
262. McCarthy, *Ambient Television*, 64–65.
263. Boddy, *Fifties Television*, 20.
264. McCarthy, *Ambient Television*, 65.
265. See Raymond Fielding, *The American Newsreel: A Complete History, 1911–1967*, 2nd ed. (Jefferson, NC: McFarland, 2006), 106.
266. See William Uricchio, "Television as History: Representations of German Television Broadcasting, 1935–1944," in *Framing the Past: The Historiography of German Cinema and Television*, ed. Bruce A. Murray and Christopher J. Wickham (Carbondale: Southern Illinois University Press, 1992), 171.
267. Schrage, "German Television," 60. See also Weber, "Recording on Film, Transmitting by Signals," 10.
268. "The First Real Television 'O.B.,'" *Television and Short-Wave World*, June 1937, 335, 339, reprinted in Herbert, *A History of Early TV*, 2:100–101.
269. NBC Year End Survey, December 20, 1937, folder 1 (History of Television 1919–1949), box 388, series 142 (History of Television), GHCRC; "How RCA Electronic Television Brought Sight to Radio," folder 4 (TV Stations 1926–1944), box 385, series 139 (Television Transmitting Stations), GHCRC.
270. "New Portable Unit to Aid Television," *New York Sun*, December 2, 1939, folder 2 (Television, 1928–1944), box 401, series 174 (Television Methods and Systems), GHCRC.
271. Engstrom, "A Study of Television Image Characteristics," 310.
272. Harry R. Lubcke, "Cinematography for Television," *Business Screen* 2, no. 3 (January 1940): 20.
273. Spigel, *TV by Design*, 115.
274. Engstrom, "A Study of Television Image Characteristics," 303–5, 298–99.
275. Schrage, "German Television," 9; Wilhelm E. Schrage, "A Modern Picture of Television," part II, *Radio-Craft*, May 1935, 684.
276. The article includes a photograph of Felix the Cat, mistakenly captioned as Mickey Mouse. "Television 'Vet' Retired for Age After Last Whirl," *Family Circle*, December 1935, folder 1 (History of Television 1919–1949), box 388, series 142 (History of Television), GHCRC. On NBC's use of Felix the Cat, also see Koszarski, *Hollywood on the Hudson*, 433.
277. Sergei Eisenstein, *Eisenstein on Disney*, ed. Jay Leyda, trans. Alan Upchurch (Calcutta: Seagull Books, 1986), 21, 10, 36–39. On the relation among Mickey Mouse, plasmaticness, and blackface minstrelsy, see Nicholas Sammond, *Birth of an Industry: Blackface Minstrelsy and the Rise of Animation* (Durham, NC: Duke University Press, 2015), 61, 200–201.
278. Eisenstein, *Eisenstein on Disney*, esp. 41–46.
279. "Thru the Laboratory Keyhole," May 1936, 16.
280. "Television Records for 1938," no source, n.d., "Clippings 1939" folder, box 21, Du Mont Collection.

281. "Television Records for 1938."
282. See Baltin, "Fashion Show Sent Via Television," and "Television Demonstration at the Fall Convention," 599–600.
283. Bragdon, "Television Sets Demonstrated."
284. Codel, "Philco Discloses Its Television Progress," 10.
285. Burns, *Television*, 552.
286. "RCA's Television Put on Exhibition," *Broadcasting*, November 1, 1936, 30.
287. Sarnoff, as told to Wickets, "You'll Soon See Across the Sea," 18.
288. "Television Demonstration at the Fall Convention," 596.
289. John Kobler, "Worlds Fair Brought to Radio City by Television," *New York Mirror*, April 21, 1939.
290. Hugo Gernsback, "Presenting—the Multiple-Image Television Receiver," *Radio-Craft*, August 1935, 74, 102. The cover image for this issue of *Radio-Craft*, which illustrates the receiver, is reprinted in Steve Kosareff, *Window to the Future: The Golden Age of Television Marketing and Advertising* (San Francisco: Chronicle Books, 2005), 19. On the Duoscope, see Spigel, *Make Room for TV*, 71–72.
291. Jenkins, "Radio Photographs, Radio Movies, and Radio Vision," 80–81.
292. Koszarski, *Hollywood on the Hudson*, 413.
293. Lubcke, "The Theatrical Possibilities of Television," 47, emphasis in original.
294. Sarnoff, as told to Wickets, "You'll Soon See Across the Sea," 18.
295. Jenkins, "Radio Photographs, Radio Movies, and Radio Vision," 85. Also see Jenkins, "Recent Progress in the Transmission of Motion Pictures by Radio," 84–85.
296. C. Francis Jenkins, "Radio Movies," *TSMPE*, no. 21 (August 1925): 9.
297. Herbert B. Ives, "Telephone Picture Transmission," *TSMPE*, no. 23 (January 1926): 83.
298. Ives, "Two-Way Television," 298–99.
299. Beal, "RCA Developments in Television," 129.

## Coda: Multiplicity, Immersion, and the New Screens

1. John Durham Peters, *The Marvelous Clouds: Toward a Philosophy of Elemental Media* (Chicago: University of Chicago Press, 2015), 15. Peters, it should be noted, also conceives of fire as an "enabling environment for ash and smoke, ink and metal, chemicals and ceramics" (117).
2. Douglas Fox, "The Film Guild Cinema: An Experiment in Theater Design," *Exhibitors Herald-World*, Better Theatres sec., March 16, 1929, 15.
3. See Charles Musser, *The Emergence of Cinema: The American Screen to 1907* (Berkeley: University of California Press, 1990), 181, 183, 443–44.
4. Trans-Lux brochure, "Trans-Lux Projection" file, Theatre Historical Society of America Archives, Pittsburgh.
5. See William Paul, *When Movies Were Theater: Architecture, Exhibition, and the Evolution of American Film* (New York: Columbia University Press, 2016), 89.
6. Sergei Eisenstein, *Eisenstein on Disney*, ed. Jay Leyda, trans. Alan Upchurch (Calcutta: Seagull Books, 1986), 24.
7. Peters, *The Marvelous Clouds*, 115.
8. Eisenstein, *Eisenstein on Disney*, 47, emphasis removed.
9. Stephen J. Pyne, *Fire: A Brief History* (Seattle: University of Washington Press, 2001), 128.
10. Peters, *The Marvelous Clouds*, 126.

11. Myron W. Krueger, *Artificial Reality*, 2nd ed. (Reading, MA: Addison-Wesley, 1991), xiii.
12. For an overview of first-wave virtual-reality technology, see Frank Biocca, "Virtual Reality: A Tutorial," *Journal of Communication* 42, no. 4 (Fall 1992): 23–72.
13. See Howard Rheingold, *Virtual Reality* (New York: Simon and Schuster, 1991), 104–9.
14. Rheingold, *Virtual Reality*, 20–21, 37–43; Frederick P. Brooks Jr., Ming Ouh-Young, James J. Batter, and P. Jerome Kilpatrick, "Project GROPE—Haptic Displays for Scientific Visualization," *Computer Graphics* 24, no. 4 (August 1990): 177–85; Ivan Sutherland, "The Ultimate Display" (1965), in *Multimedia: From Wagner to Virtual Reality*, ed. Randall Packer and Ken Jordan (New York: Norton, 2001), 252–56.
15. Rheingold, *Virtual Reality*, 205–8.
16. Rheingold, *Virtual Reality*, 128, 131–54; Scott Fisher, "Virtual Interface Environments," in *Multimedia*, ed. Packer and Jordan, 257–66.
17. See Jonathan Steuer, "Defining Virtual Reality: Dimensions Determining Telepresence," *Journal of Communication* 42, no. 4 (Fall 1992): 73–93.
18. Steuer, "Defining Virtual Reality," 81–86.
19. Krueger, *Artificial Reality*, 12–64; Daniel Sandin, Thomas DeFanti, and Carolina Cruz-Neira, "A Room with a View" (1991), in *Multimedia*, ed. Packer and Jordan, 286–92.
20. See Fisher, "Virtual Interface Environments," 260–61; Krueger, *Artificial Reality*, 6–8; and Fred Turner, *The Democratic Surround: Multimedia and American Liberalism from World War II to the Psychedelic Sixties* (Chicago: University of Chicago Press, 2013).
21. See Miriam Ross, "The 3-D Aesthetic: *Avatar* and Hyperhaptic Visuality," *Screen* 53, no. 4 (Winter 2012): 381–97, and Ariel Rogers, *Cinematic Appeals: The Experience of New Movie Technologies* (New York: Columbia University Press, 2013), 210–22.
22. See John Belton, *Widescreen Cinema* (Cambridge, MA: Harvard University Press, 1992); Erkki Huhtamo, "Encapsulated Bodies in Motion: Simulators and the Quest for Total Immersion," in *Critical Issues in Electronic Media*, ed. Simon Penny (Albany: State University of New York Press, 1995), 159–86; Alison Griffiths, *Shivers down Your Spine: Cinema, Museums, and the Immersive View* (New York: Columbia University Press, 2008), 79–113; Sara Ross, "Invitation to the Voyage: The Flight Sequence in Contemporary 3D Cinema," *Film History* 24 (2012): 210–20; and Giles Taylor, "A Military Use for Widescreen Cinema: Training the Body Through Immersive Media," *Velvet Light Trap* 72 (Fall 2013): 17–32.
23. Kristen Whissel, *Spectacular Digital Effects: CGI and Contemporary Cinema* (Durham, NC: Duke University Press, 2014), 21–58.
24. See Scott C. Richmond, *Cinema's Bodily Illusions: Flying, Floating, and Hallucinating* (Minneapolis: University of Minnesota Press, 2016), 121–43, and Kristen Whissel, "Parallax Effects: Epistemology, Affect, and Digital 3D Cinema," *Journal of Visual Culture* 15, no. 2 (2016): 233–49.
25. Whissel, *Spectacular Digital Effects*, 21.
26. See Erkki Huhtamo, "Toward a History of Peep Practice," in *A Companion to Early Cinema*, ed. André Gaudreault, Nicolas Dulac, and Santiago Hidalgo (Malden, MA: Wiley-Blackwell, 2012), 32–51.
27. Richmond, *Cinema's Bodily Illusions*, 134–35.
28. Stephen Heath, "Narrative Space," in *Narrative, Apparatus, Ideology: A Film Theory Reader*, ed. Philip Rosen (New York: Columbia University Press, 1986), 393.

29. On the significance of such mapping for the development and operation of virtual-reality technologies, see Biocca, "Virtual Reality," 27, 49–56, and Steuer, "Defining Virtual Reality," 86–87.
30. See the discussion of Oculus's tracking system in its outline of best practices for developers at https://developer.oculus.com/design/latest/concepts/bp_app_tracking.
31. See Stanislav Stanković, *Virtual Reality and Virtual Environments in 10 Lectures* (Williston, VT: Morgan and Claypool, 2016), 92–97.
32. See, for instance, Huhtamo, "Encapsulated Bodies in Motion," 176–77, and Mark B. N. Hansen, *New Philosophy for New Media* (Cambridge, MA: MIT Press, 2004), 161–96. For a more recent discussion, see Sita Popat, "Missing in Action: Embodied Experience and Virtual Reality," *Theatre Journal* 68, no. 3 (September 2016): 357–78.
33. For an argument about the imbrication of body and environment in virtual reality focusing on the body's relation to virtual environments, see Mark B. N. Hansen, "Embodying Virtual Reality: Touch and Self-Movement in the Work of Char Davies," *Critical Matrix* 12, nos. 1–2 (2001): 112–47.
34. See Richmond, *Cinema's Bodily Illusions*, esp. 6–9, 134–35.
35. Oliver Grau, *Virtual Art: From Illusion to Immersion* (Cambridge, MA: MIT Press, 2003), 141. Also see, for instance, Belton, *Widescreen Cinema*, 98, and Philip Lelyveld, "Virtual Reality Primer with an Emphasis on Camera-Captured VR," *SMPTE Motion Imaging Journal* 124, no. 6 (Sept. 2015): 78.
36. Janine Marchessault, "Multi-screens and Future Cinema: The Labyrinth Project at Expo 67," in *Fluid Screens, Expanded Cinema*, ed. Janine Marchessault and Susan Lord (Toronto: University of Toronto Press, 2007), 29–51.
37. Lisa Parks, *Rethinking Media Coverage: Vertical Mediation and the War on Terror* (New York: Routledge, 2018). Also see Hito Steyerl, "In Free Fall: A Thought Experiment on Vertical Perspective," *e-flux journal* 24 (April 2011), https://www.e-flux.com/journal/24/67860/in-free-fall-a-thought-experiment-on-vertical-perspective/.
38. Francesco Casetti, *The Lumière Galaxy: Seven Key Words for the Cinema to Come* (New York: Columbia University Press, 2015), 156, 170.
39. For a discussion of this comparison, see Patrick Jagoda, *Network Aesthetics* (Chicago: University of Chicago Press, 2016), 20–21.
40. For a gloss on these ideas, see Alexander R. Galloway, "Networks," in *Critical Terms for Media Studies*, ed. W. J. T. Mitchell and Mark B. N. Hansen (Chicago: University of Chicago Press, 2010), 280–96.
41. See Michal Levin, *Designing Multi-device Experiences: An Ecosystem Approach to User Experiences Across Devices* (Sebastopol, CA: O'Reilly Media, 2014).
42. See Jennifer Holt and Kevin Sanson, eds., *Connected Viewing: Selling, Streaming, and Sharing Media in the Digital Era* (New York: Routledge, 2014).
43. See Wendy Hui Kyong Chun and Sarah Friedland, "Habits of Leaking: Of Sluts and Network Cards," *differences* 26, no. 2 (2015): 1–28, and Nicole Starosielski, *The Undersea Network* (Durham, NC: Duke University Press, 2015).
44. Jagoda, *Network Aesthetics*, 3.
45. Thomas Elsaesser has similarly argued that attention to imaging practices (especially 3D) in realms such as the military frames contemporary imaging technologies as "technologies of probing and penetration" ("The 'Return' of 3-D: On Some of the Logics and Genealogies of the Image in the Twenty-First Century," *Critical Inquiry* 39, no. 2 [Winter 2013]: 242).
46. Lisa Cartwright, *Screening the Body: Tracing Medicine's Visual Culture* (Minneapolis: University of Minnesota Press, 1995), 111; Yuri Tsivian, "Media Fantasies and

Penetrating Vision: Some Links Between X-Rays, the Microscope, and Film," in *Laboratory of Dreams: The Russian Avant-Garde and Cultural Experiment*, ed. John E. Bowlt and Olga Matich (Stanford, CA: Stanford University Press, 1996), 82.

47. See, in particular, Linda Dalrymple Henderson, "X Rays and the Quest for Invisible Reality in the Art of Kupka, Duchamp, and the Cubists," *Art Journal* 47 (Winter 1988): 323–40, and Tom Gunning, "From the Kaleidoscope to the X-Ray: Urban Spectatorship, Poe, Benjamin, and *Traffic in Souls* (1913)," *Wide Angle* 19, no. 4 (October 1997): 25–61.

# BIBLIOGRAPHY

## Archival Sources

Academy Film Archive. Academy of Motion Picture Arts and Sciences, Beverly Hills, CA.
Archives Center. National Museum of American History, Smithsonian Institution, Washington, DC.
——. Clark, George H. Radioana Collection.
——. Du Mont, Allen Balcom. Collection.
Billy Rose Theatre Division. New York Public Library for the Performing Arts. Lincoln Center. Clippings Files.
Chicago History Museum.
——. Essanay Film Studio. Collection of Visual Materials.
——. Spoor, George, and Essanay Manufacturing Company. Manuscript Collection.
Film and Television Archive. University of California at Los Angeles.
Film Study Center. Museum of Modern Art, New York.
Harry Ransom Center. University of Texas at Austin.
——. Selznick, David O. Collection.
Library Special Collections. Charles E. Young Research Library, University of California at Los Angeles.
——. Lee, S. Charles. Papers.
Margaret Herrick Library. Academy of Motion Picture Arts and Sciences, Beverly Hills, CA.
——. B'hend, Tom, and Preston Kaufmann. Collection.
——. Chamberlin, Audrey. Scrapbooks.
——. Dunn, Linwood G. Papers.
——. Gillespie, A. Arnold. Manuscript.
——. Haver, Ronald. Collection.
——. Hitchcock, Alfred. Collection.
——. MGM Art Department Records.

——. Paramount Pictures Press Sheets.
——. Paramount Pictures Production Records.
——. Turner/MGM Scripts.
——. Vertical File Collection.
Margaret Herrick Library Digital Collections. Academy of Motion Picture Arts and Sciences. http://digitalcollections.oscars.org/.
Paley Center for Media. New York.
Rare Book and Manuscript Library. Columbia University Libraries, New York.
——. Sponable, Earl I. Papers.
Seaver Center for Western History Research. Los Angeles County Museum of Natural History.
Theatre Historical Society of America Archives. Pittsburgh, PA.
Warner Bros. Archives. School of Cinematic Arts, University of Southern California, Los Angeles.

## Periodicals and Trade Journals

*Amateur Movie Makers*
*American Cinematographer*
*American Magazine of Art*
*Architectural Forum*
*Architectural Record*
*Boston Daily Globe*
*Boxoffice*
*Broadcasting*
*Brooklyn Daily Eagle*
*Business Screen*
*Calendar*
*Educational Screen*
*Electronics*
*Exhibitors Herald-World*
*Family Circle*
*Film Daily*
*International Photographer*
*Journal of the Society of Motion Picture Engineers*
*Journal of the Society of Motion Picture and Television Engineers*
*Liberty*
*Los Angeles Times*
*Marquee*
*Modern Mechanics and Inventions*
*Motion Picture Herald*
*Motion Picture News*
*Motion Picture Projectionist*
*Movie Makers*
*Newark Ledger*
*New Scientist*
*New York Herald Tribune*
*New York Mirror*
*New York Sun*
*New York Times*
*Popular Mechanics*
*Printers Ink Monthly*
*Progressive Architecture*
*Radio Broadcast*
*Radio-Craft*
*Radio News*
*Radio Today*
*Radio Topics*
*Scientific American*
*Short Wave and Television*
*SMPTE Motion Imaging Journal*
*Syracuse Herald*
*Technical Bulletin*
*Televiser*
*Television*
*Television and Short-Wave World*
*Theatre Historical Society of America Annual*
*Time*
*Transactions of the Society of Motion Picture Engineers*
*True*
*Variety*
*Washington Tribune*
*World-Telegram*

## Secondary Sources


Abramson, Albert. *The History of Television, 1880 to 1941*. Jefferson, NC: McFarland, 1987.

Acland, Charles R. "The Crack in the Electric Window." *Cinema Journal* 51, no. 2 (Winter 2012): 167–71.

——. "Curtains, Carts, and the Mobile Screen." *Screen* 50, no. 1 (Spring 2009): 148–66.

Acland, Charles, and Haidee Wasson, eds. *Useful Cinema*. Durham, NC: Duke University Press, 2011.

Agamben, Giorgio. *What Is an Apparatus? And Other Essays*. Translated by David Kishik and Stefan Pedatella. Stanford, CA: Stanford University Press, 2009.

Albera, François, and Maria Tortajada, eds. *Cine-Dispositives: Essays in Epistemology Across Media*. Amsterdam: Amsterdam University Press, 2015.

——. "The Dispositive Does Not Exist!" In *Cine-Dispositives: Essays in Epistemology Across Media*, edited by François Albera and Maria Tortajada, 21–44. Amsterdam: Amsterdam University Press, 2015.

Albrecht, Donald. *Designing Dreams: Modern Architecture in the Movies*. New York: Harper and Row, 1986.

Aleiss, Angela. *Making the White Man's Indian: Native Americans and Hollywood Movies*. Westport, CT: Praeger, 2005.

Altman, Rick. "The Evolution of Sound Technology." In *Film Sound: Theory and Practice*, edited by Elisabeth Weis and John Belton, 44–53. New York: Columbia University Press, 1985.

——. "Sound Space." In *Sound Theory Sound Practice*, edited by Rick Altman, 46–64. New York: Routledge, 1992.

Ameri, Amir H. "The Architecture of the Illusive Distance." *Screen* 54, no. 4 (Winter 2013): 439–62.

Appadurai, Arjun. *Modernity at Large: Cultural Dimensions of Globalization*. Minneapolis: University of Minnesota Press, 1996.

Arnheim, Rudolf. *Radio: An Art of Sound*. Translated by Margaret Ludwig and Herbert Read. New York: Da Capo, 1972. Originally published in English in 1936.

Auerbach, Jonathan, and Lisa Gitelman. "Microfilm, Containment, and the Cold War." *American Literary History* 19, no. 3 (Fall 2007): 745–68.

Balio, Tino. *Grand Design: Hollywood as a Modern Business Enterprise, 1930–1939*. Berkeley: University of California Press, 1993.

——, ed. *Hollywood in the Age of Television*. Boston: Unwin Hyman, 1990.

Barnouw, Erik. *Tube of Plenty: The Evolution of American Television*. 2nd rev. ed. New York: Oxford University Press, 1990.

Baudry, Jean-Louis. "The Apparatus: Metapsychological Approaches to the Impression of Reality in the Cinema." In *Narrative, Apparatus, Ideology: A Film Theory Reader*, edited by Philip Rosen, 299–318. New York: Columbia University Press, 1986.

——. "Ideological Effects of the Basic Cinematographic Apparatus." In *Narrative, Apparatus, Ideology: A Film Theory Reader*, edited by Philip Rosen, 286–98. New York: Columbia University Press, 1986.

Bazin, André. *What Is Cinema?* Vol. 1. Edited and translated by Hugh Gray. Berkeley: University of California Press, 1967.

Belton, John. "Fox and 50mm Film." In *Widescreen Worldwide*, edited by John Belton, Sheldon Hall, and Steve Neale, 9–24. New Barnet, Herts, UK: John Libbey, 2010.

——. *Widescreen Cinema*. Cambridge, MA: Harvard University Press, 1992.

Benjamin, Walter. *Selected Writings*. 4 vols. Edited by Michael W. Jennings and others. Translated by Rodney Livingstone and others. Cambridge, MA: Harvard University Press, 1996–2003.

Bennett, Jane. *Vibrant Matter: A Political Ecology of Things*. Durham, NC: Duke University Press, 2010.

Benson-Allott, Caetlin. *Killer Tapes and Shattered Screens: Video Spectatorship from VHS to File Sharing*. Berkeley: University of California Press, 2013.

Berenstein, Rhona J. *Attack of the Leading Ladies: Gender, Sexuality, and Spectatorship in Classic Horror Cinema*. New York: Columbia University Press, 1996.

Biocca, Frank. "Virtual Reality: A Tutorial." *Journal of Communication* 42, no. 4 (Fall 1992): 23–72.

Blau, Eve. "Transparency and the Irreconcilable Contradictions of Modernity." *Praxis* 9 (Fall 2007): 50–59.

Boddy, William. *Fifties Television: The Industry and Its Critics*. Urbana: University of Illinois Press, 1990.

——. *New Media and Popular Imagination: Launching Radio, Television, and Digital Media in the United States*. New York: Oxford University Press, 2004.

Bordwell, David, Janet Staiger, and Kristin Thompson. *The Classical Hollywood Cinema: Film Style and Mode of Production to 1960*. New York: Columbia University Press, 1985.

Bordwell, David, and Kristin Thompson. "Technological Change and Classical Film Style." In Tino Balio, *Grand Design: Hollywood as a Modern Business Enterprise, 1930–1939*, 109–41. Berkeley: University of California Press, 1993.

Bowser, Eileen. *The Transformation of Cinema, 1907–1915*. Berkeley: University of California Press, 1990.

Bronfen, Elisabeth. "Screening and Disclosing Fantasy: Rear Projection in Hitchcock." *Screen* 56, no. 1 (Spring 2015): 1–24.

Brooks, Frederick P., Jr., Ming Ouh-Young, James J. Batter, and P. Jerome Kilpatrick. "Project GROPE—Haptic Displays for Scientific Visualization." *Computer Graphics* 24, no. 4 (August 1990): 177–85.

Brown, Judith. *Glamour in Six Dimensions: Modernism and the Radiance of Form*. Ithaca, NY: Cornell University Press, 2009.

Bruno, Giuliana. *Atlas of Emotion: Journeys in Art, Architecture, and Film*. New York: Verso, 2002.

——. *Surface: Matters of Aesthetics, Materiality, and Media*. Chicago: University of Chicago Press, 2014.

Burns, R. W. *Television: An International History of the Formative Years*. London: Institution of Electrical Engineers, 1998.

Burton, Jack. *The Story of Trans-Lux*. New York: Trans-Lux Daylight Picture Screen Corporation, 1929.

Bush, Vannevar. "As We May Think." 1945. In *The New Media Reader*, edited by Noah Wardrip-Fruin and Nick Montfort, 37–47. Cambridge, MA: MIT Press, 2003.

Caldwell, John Thornton. *Televisuality: Style, Crisis, and Authority in American Television*. New Brunswick, NJ: Rutgers University Press, 1995.

Carroll, Noël. *Interpreting the Moving Image*. New York: Cambridge University Press, 1998.

Cartwright, Lisa. *Screening the Body: Tracing Medicine's Visual Culture*. Minneapolis: University of Minnesota Press, 1995.

Casetti, Francesco. "Film as an Environmental Medium." Paper presented at the "Ends of Cinema" conference, Center for 21st Century Studies, University of Wisconsin at Milwaukee, May 4, 2018.

——. *The Lumière Galaxy: Seven Key Words for the Cinema to Come*. New York: Columbia University Press, 2015.

——. "Notes on a Genealogy of the Excessive Screen." In *Screens*, booklet for Mellon Sawyer Seminar on Genealogies of the Excessive Screen, February–December 2017, Yale University. New Haven, CT: Yale University, 2017. http://dev.screens.yale.edu/sites/default/files/files/Screens_Booklet.pdf.

——. "Screening: A Counter-Genealogy of the Silver Screen." Paper presented at the Society for Cinema and Media Studies Conference, Chicago, March 24, 2017.

Chion, Michel. *Film: A Sound Art*. Translated by Claudia Gorbman. New York: Columbia University Press, 2009.

Chun, Wendy Hui Kyong, and Sarah Friedland. "Habits of Leaking: Of Sluts and Network Cards." *differences* 26, no. 2 (2015): 1–28.

Cohen, Mary Morley. "Forgotten Audiences in the Passion Pits: Drive-in Theatres and Changing Spectator Practices in Post-war America." *Film History* 6, no. 4 (Winter 1994): 470–86.

Colomina, Beatriz. "Enclosed by Images: The Eames' Multimedia Architecture." *Grey Room* 2 (Winter 2001): 6–29.

——. *Privacy and Publicity: Modern Architecture as Mass Media*. Cambridge, MA: MIT Press, 1994.

Comolli, Jean-Louis. "Machines of the Visible." In *The Cinematic Apparatus*, edited by Teresa De Lauretis and Stephen Heath, 121–42. London: Macmillan, 1980.

Covert, Catherine L. "'We May Hear Too Much': American Sensibility and the Response to Radio, 1919–1924." In *Mass Media Between the Wars: Perceptions of Cultural Tension, 1918–1941*, edited by Catherine L. Covert and John D. Stevens, 199–220. Syracuse, NY: Syracuse University Press, 1984.

Cowan, Michael. "Taking It to the Street: Screening the Advertising Film in the Weimar Republic." *Screen* 54, no. 4 (Winter 2013): 463–79.

Crafton, Donald. *Shadow of a Mouse: Performance, Belief, and World-Making in Animation*. Berkeley: University of California Press, 2013.

——. *The Talkies: American Cinema's Transition to Sound, 1926–1931*. Berkeley: University of California Press, 1997.

Cripps, Thomas. *Slow Fade to Black: The Negro in American Film, 1900–1942*. 1977. Reprint. New York: Oxford University Press, 1993.

Danks, Adrian. "Being in Two Places at the Same Time: The Forgotten Geography of Rear-Projection." In *B Is for Bad Cinema: Aesthetics, Politics, and Cultural Value*, edited by Claire Perkins and Constantine Verevis, 65–84. Albany: State University of New York Press, 2014.

Decherney, Peter. "Cult of Attention: An Introduction to Seymour Stern and Harry Alan Potamkin (Contra Kracauer) on the Ideal Movie Theater." *Spectator* 18, no. 2 (Spring–Summer 1998): 18–25.

De Lauretis, Teresa, and Stephen Heath, eds. *The Cinematic Apparatus*. London: Macmillan, 1980.

Deleuze, Gilles. "What Is a *Dispositif*?" In *Two Regimes of Madness: Texts and Interviews 1975–1995*, edited by David Lapoujade, translated by Ames Hodges and Mike Taormina, 338–48. New York: Semiotext(e), 2006.

Doane, Mary Ann. "The Close-Up: Scale and Detail in the Cinema." *differences* 14, no. 3 (Fall 2003): 89–111.

——. *The Desire to Desire: The Woman's Film of the 1940s*. Bloomington: Indiana University Press, 1987.

——. *The Emergence of Cinematic Time: Modernity, Contingency, the Archive*. Cambridge, MA: Harvard University Press, 2002.
——. "Ideology and the Practice of Sound Editing and Mixing." In *Film Sound: Theory and Practice*, edited by Elizabeth Weis and John Belton, 54–63. New York: Columbia University Press, 1985.
——. "Information, Crisis, Catastrophe." In *Logics of Television: Essays in Cultural Criticism*, edited by Patricia Mellencamp, 222–39. Bloomington: Indiana University Press, 1990.
Doherty, Thomas. *Hollywood and Hitler, 1933–1939*. New York: Columbia University Press, 2013.
——. "This Is Where We Came In: The Audible Screen and the Voluble Audience of Early Sound Cinema." In *American Movie Audiences: From the Turn of the Century to the Early Sound Era*, edited by Melvyn Stokes and Richard Maltby, 143–63. London: British Film Institute, 1999.
Douglas, Susan J. *Inventing American Broadcasting, 1899–1922*. Baltimore: Johns Hopkins University Press, 1987.
——. *Listening In: Radio and the American Imagination*. 1999. Reprint. Minneapolis: University of Minnesota Press, 2004.
Druick, Zoë. "At the Margins of Cinema History: Mobile Cinema in the British Empire." *Public* 40 (2010): 118–25.
Dunlap, Orrin E., Jr. *The Outlook for Television*. New York: Harper and Brothers, 1932.
Dyer, Richard. *White*. New York: Routledge, 1997.
Edgerton, Gary R. *The Columbia History of American Television*. New York: Columbia University Press, 2007.
Edwards, Paul N. *The Closed World: Computers and the Politics of Discourse in Cold War America*. Cambridge, MA: MIT Press, 1996.
Eisenstein, Sergei. "The Dynamic Square." In *Film Essays and a Lecture*, edited by Jay Leyda, 48–65. New York: Praeger, 1970.
——. *Eisenstein on Disney*. Edited by Jay Leyda. Translated by Alan Upchurch. Calcutta: Seagull Books, 1986.
Elcott, Noam M. *Artificial Darkness: An Obscure History of Modern Art and Media*. Chicago: University of Chicago Press, 2016.
——. "Rooms of Our Time: László Moholy-Nagy and the Stillbirth of Multi-media Museums." In *Screen/Space: The Projected Image in Contemporary Art*, edited by Tamara Trodd, 25–52. Manchester: Manchester University Press, 2011.
Ellis, John. *Visible Fictions: Cinema, Television, Video*. Rev. ed. New York: Routledge, 1992.
Elsaesser, Thomas. *Film History as Media Archaeology: Tracking Digital Cinema*. Amsterdam: Amsterdam University Press, 2016.
——. "The New Film History as Media Archaeology." *Cinémas* 14, nos. 2–3 (2004): 75–117.
——. "The 'Return' of 3-D: On Some of the Logics and Genealogies of the Image in the Twenty-First Century." *Critical Inquiry* 39, no. 2 (Winter 2013): 217–46.
Erb, Cynthia. *Tracking King Kong: A Hollywood Icon in World Culture*. 2nd ed. Detroit: Wayne State University Press, 2009.
Everett, Anna. *Returning the Gaze: A Genealogy of Black Film Criticism, 1909–1949*. Durham, NC: Duke University Press, 2001.
Feuer, Jane. "The Concept of Live Television: Ontology as Ideology." In *Regarding Television: Critical Approaches—an Anthology*, edited by E. Ann Kaplan, 12–22. Frederick, MD: University Publications of America, 1983.

Fielding, Raymond. *The American Newsreel: A Complete History, 1911–1967*. 2nd ed. Jefferson, NC: McFarland, 2006.
——. *Special Effects Cinematography*. 4th ed. Oxford: Focal Press, 1985.
Fisher, Scott. "Virtual Interface Environments." In *Multimedia: From Wagner to Virtual Reality*, edited by Randall Packer and Ken Jordan, 257–66. New York: Norton, 2001.
Forman, Henry James. *Our Movie Made Children*. New York: Macmillan, 1933.
Foucault, Michel. *Power/Knowledge: Selected Interviews and Other Writings, 1972–1977*. Edited by Colin Gordon. Translated by Colin Gordon, Leo Marshall, John Mepham, and Kate Soper. New York: Pantheon Books, 1980.
Friedberg, Anne. "Urban Mobility and Cinematic Visuality: The Screens of Los Angeles—Endless Cinema or Private Telematics." *Journal of Visual Culture* 1, no. 2 (2002): 183–204.
——. *The Virtual Window: From Alberti to Microscoft*. Cambridge, MA: MIT Press, 2006.
——. *Window Shopping: Cinema and the Postmodern*. Berkeley: University of California Press, 1993.
Fuller-Seeley, Kathryn H. "Dish Night at the Movies: Exhibitor Promotions and Female Audiences During the Great Depression." In *Looking Past the Screen: Case Studies in American Film History and Method*, edited by Jon Lewis and Eric Smoodin, 246–75. Durham, NC: Duke University Press, 2007.
Galili, Doron. "Seeing by Electricity: The Emergence of Television and the Modern Mediascape, 1878–1939." PhD diss., University of Chicago, 2011.
——. "Television from Afar: Arnheim's Understanding of Media." In *Arnheim for Film and Media Studies*, edited by Scott Higgins, 195–211. New York: Routledge, 2011.
——. "Tom Swift's Three Inventions of Television: Media History and the Technological Imaginary." *View* 4, no. 7 (2015): 54–67.
Galloway, Alexander R. *The Interface Effect*. Malden, MA: Polity, 2012.
——. "Networks." In *Critical Terms for Media Studies*, edited by W. J. T. Mitchell and Mark B. N. Hansen, 280–96. Chicago: University of Chicago Press, 2010.
Giedion, Sigfried. *Space, Time, and Architecture: The Growth of a New Tradition*. 5th rev. ed. Cambridge, MA: Harvard University Press, 2008.
Gillespie, A. Arnold. *The Wizard of MGM: Memoirs of A. Arnold Gillespie, Art Director/Head of Special Effects from 1924–1965*. Edited by Philip J. Riley and Robert A. Welch. Duncan, OK: BearManor Media, 2011.
Goldner, Orville, and George E. Turner. *The Making of "King Kong": The Story Behind a Film Classic*. South Brunswick, NJ: Barnes, 1975.
Gomery, Douglas. *The Coming of Sound: A History*. New York: Routledge, 2005.
——. "The Coming of Sound: Technological Change in the American Film Industry." In *Film Sound: Theory and Practice*, edited by Elisabeth Weis and John Belton, 5–24. New York: Columbia University Press, 1985.
——. *Shared Pleasures: A History of Movie Presentation in the United States*. Madison: University of Wisconsin Press, 1992.
Gorbman, Claudia. *Unheard Melodies: Narrative Film Music*. Bloomington: Indiana University Press, 1987.
Grau, Oliver. *Virtual Art: From Illusion to Immersion*. Cambridge, MA: MIT Press, 2003.
Grenz, Christine. *Trans-Lux: Biography of a Corporation*. Norwalk, CT: Trans-Lux Corporation, 1982.
Griffiths, Alison. *Shivers down Your Spine: Cinema, Museums, and the Immersive View*. New York: Columbia University Press, 2008.
Groening, Stephen. *Cinema Beyond Territory: Inflight Entertainment and Atmospheres of Globalization*. London: British Film Institute, 2014.

Gunning, Tom. "An Aesthetic of Astonishment: Early Film and the (In)Credulous Spectator." *Art and Text* 34 (Spring 1989): 31–45.
——. *D. W. Griffith and the Origins of American Narrative Film: The Early Years at Biograph*. Urbana: University of Illinois Press, 1991.
——. "Fritz Lang Calling: The Telephone and the Circuits of Modernity." In *Allegories of Communication: Intermedial Concerns from Cinema to the Digital*, edited by John Fullerton and Jan Olsson, 19–37. Rome: John Libbey, 2004.
——. "From the Kaleidoscope to the X-Ray: Urban Spectatorship, Poe, Benjamin, and *Traffic in Souls* (1913)." *Wide Angle* 19, no. 4 (October 1997): 25–61.
——. "Phantasmagoria and the Manufacturing of Wonder: Towards a Cultural Optics of the Cinematic Apparatus." In *The Cinema: A New Technology for the 20th Century*, edited by André Gaudreault, Catherine Russell, and Pierre Véronneau, 31–44. Lausanne, Switzerland: Éditions Payot Lausanne, 2004.
——. "'The Whole World Within Reach': Travel Images Without Borders." In *Virtual Voyages: Cinema and Travel*, edited by Jeffrey Rouff, 25–41. Durham, NC: Duke University Press, 2006.
Hall, Ben M. *The Best Remaining Seats: The Story of the Golden Age of the Movie Palace*. New York: Bramhall House, 1961.
Hansen, Mark B. N. "Embodying Virtual Reality: Touch and Self-Movement in the Work of Char Davies." *Critical Matrix* 12, nos. 1–2 (2001): 112–47.
——. *New Philosophy for New Media*. Cambridge, MA: MIT Press, 2004.
Hansen, Miriam Bratu. *Babel and Babylon: Spectatorship in American Silent Film*. Cambridge, MA: Harvard University Press, 1991.
——. *Cinema and Experience: Siegfried Kracauer, Walter Benjamin, and Theodor W. Adorno*. Berkeley: University of California Press, 2012.
——. "Early Cinema, Late Cinema: Permutations of the Public Sphere." *Screen* 34, no. 3 (Autumn 1993): 197–210.
——. "The Mass Production of the Senses: Classical Cinema as Vernacular Modernism." *Modernism/modernity* 6, no. 2 (April 1999): 59–77.
——. "Vernacular Modernism: Tracking Cinema on a Global Scale." In *World Cinemas, Transnational Perspectives*, edited by Nataša Ďurovičová and Kathleen Newman, 287–314. New York: Routledge, 2010.
Hark, Ina Rae, ed. *American Cinema of the 1930s: Themes and Variations*. New Brunswick, NJ: Rutgers University Press, 2007.
Haver, Ronald. *David O. Selznick's Hollywood*. New York: Knopf, 1980.
Heath, Stephen. "Narrative Space." In *Narrative, Apparatus, Ideology: A Film Theory Reader*, edited by Philip Rosen, 379–420. New York: Columbia University Press, 1986.
Hediger, Vinzenz, and Patrick Vonderau, eds. *Films That Work: Industrial Film and the Productivity of Media*. Amsterdam: Amsterdam University Press, 2009.
Henderson, Linda Dalrymple. "X Rays and the Quest for Invisible Reality in the Art of Kupka, Duchamp, and the Cubists." *Art Journal* 47 (Winter 1988): 323–40.
Herbert, Stephen. *A History of Early TV*. 3 vols. New York: Routledge, 2004.
Higgins, Scott. *Harnessing the Technicolor Rainbow: Color Design in the 1930s*. Austin: University of Texas Press, 2007.
Hilmes, Michele. *Hollywood and Broadcasting: From Radio to Cable*. Urbana: University of Illinois Press, 1990.
Holt, Jennifer, and Kevin Sanson, eds. *Connected Viewing: Selling, Streaming, and Sharing Media in the Digital Era*. New York: Routledge, 2014.

Huhtamo, Erkki. "Elements of Screenology: Toward an Archaeology of the Screen." *Iconics* 7 (2004): 31–82.

——. "Encapsulated Bodies in Motion: Simulators and the Quest for Total Immersion." In *Critical Issues in Electronic Media*, edited by Simon Penny, 159–86. Albany: State University of New York Press, 1995.

——. "Toward a History of Peep Practice." In *A Companion to Early Cinema*, edited by André Gaudreault, Nicholas Dulac, and Santiago Hidalgo, 32–51. Malden, MA: Wiley-Blackwell, 2012.

Jacobs, Lea. *The Wages of Sin: Censorship and the Fallen Woman Film, 1928–1942*. Berkeley: University of California Press, 1995.

Jagoda, Patrick. *Network Aesthetics*. Chicago: University of Chicago Press, 2016.

Jewell, Richard B. *The Golden Age of Cinema: Hollywood, 1929–1945*. Malden, MA: Blackwell, 2007.

——. *RKO Radio Pictures: A Titan Is Born*. Berkeley: University of California Press, 2012.

Johnson, Katie N. "Brutus Jones's Remains: The Case of Jules Bledsoe." *Eugene O'Neill Review* 36, no. 1 (2015): 1–28.

Johnston, Andrew Robert. "The Color of Prometheus: Thomas Wilfred's Lumia and the Projection of Transcendence." In *Color and the Moving Image: History, Theory, Aesthetics, Archive*, edited by Simon Brown, Sarah Street, and Liz Watkins, 67–78. New York: Routledge, 2013.

Jonsson, Stefan. *Crowds and Democracy: The Idea and Image of the Masses from Revolution to Fascism*. New York: Columbia University Press, 2013.

Jowett, Garth S., Ian C. Jarvie, and Kathryn H. Fuller. *Children and the Movies: Media Influence and the Payne Fund Controversy*. New York: Cambridge University Press, 1996.

Jurca, Catherine. *Hollywood 1938: Motion Pictures' Greatest Year*. Berkeley: University of California Press, 2012.

Keating, Patrick. *Hollywood Lighting: From the Silent Era to Film Noir*. New York: Columbia University Press, 2010.

Keil, Charlie, and Kristen Whissel, eds. *Editing and Special/Visual Effects*. New Brunswick, NJ: Rutgers University Press, 2016.

Kelley, Andrea. "'A Revolution in the Atmosphere': The Dynamics of Site and Screen in 1940s Soundies." *Cinema Journal* 54, no. 2 (Winter 2015): 72–93.

Kern, Stephen. *The Culture of Time and Space, 1880–1918*. Cambridge, MA: Harvard University Press, 2003.

Kessler, Frank. "The Cinema of Attractions as *Dispositif*." In *The Cinema of Attractions Reloaded*, edited by Wanda Strauven, 57–69. Amsterdam: Amsterdam University Press, 2006.

——. "Notes on *Dispositif*." May 2010. http://www.frankkessler.nl/wp-content/uploads/2010/05/Dispositif-Notes.pdf.

Kiesler, Frederick J. *Selected Writings*. Edited by Siegfried Gohr and Gunda Luyken. Ostfildern bei Stuttgart, Germany: Gerd Hatje, 1996.

Klinger, Barbara. *Beyond the Multiplex: Cinema, New Technologies, and the Home*. Berkeley: University of California Press, 2006.

Knight, Arthur. *Disintegrating the Musical: Black Performance and American Musical Film*. Durham, NC: Duke University Press, 2002.

Kosareff, Steve. *Window to the Future: The Golden Age of Television Marketing and Advertising*. San Francisco: Chronicle Books, 2005.

Koszarski, Richard. "Coming Next Week: Images of Television in Pre-war Motion Pictures." *Film History* 10, no. 2 (1998): 128–40.

——. *An Evening's Entertainment: The Age of the Silent Feature Picture, 1915–1928.* Berkeley: University of California Press, 1990.

——. *Hollywood on the Hudson: Film and Television in New York from Griffith to Sarnoff.* New Brunswick, NJ: Rutgers University Press, 2008.

Koszarski, Richard, and Doron Galili. "Television in the Cinema Before 1939: An International Annotated Database." *Journal of e-Media Studies* 5, no. 1 (2016). https://journals.dartmouth.edu/cgi-bin/WebObjects/Journals.woa/1/xmlpage/4/article/471.

Kracauer, Siegfried. *The Mass Ornament: Weimar Essays.* Edited and translated by Thomas Y. Levin. Cambridge, MA: Harvard University Press, 1995.

Krueger, Myron W. *Artificial Reality.* 2nd ed. Reading, MA: Addison-Wesley, 1991.

Larkin, Brian. *Signal and Noise: Media, Infrastructure, and Urban Culture in Nigeria.* Durham, NC: Duke University Press, 2008.

Lastra, James. *Sound Technology and the American Cinema: Perception, Representation, Modernity.* New York: Columbia University Press, 2000.

Latour, Bruno. *Reassembling the Social: An Introduction to Actor-Network-Theory.* New York: Oxford University Press, 2005.

Le Corbusier. "Twentieth Century Building and Twentieth Century Living." 1930. In *Raumplan Versus Plan Libre*, edited by Max Risselada, 145–49. Delft, Netherlands: Delft University Press, 1988.

Lesák, Barbara. "Visionary of the European Theater." In *Frederick Kiesler*, edited by Lisa Phillips, 37–45. New York: Whitney Museum of Art, 1989.

Levin, Michal. *Designing Multi-device Experiences: An Ecosystem Approach to User Experiences Across Devices.* Sebastopol, CA: O'Reilly Media, 2014.

Loew, Katharina. "Magic Mirrors: The Schüfftan Process." In *Special Effects: New Histories/Theories/Contexts*, edited by Dan North, Bob Rehak, and Michael S. Duffy, 62–77. London: British Film Institute, 2015.

Lott, Eric. *Love and Theft: Blackface Minstrelsy and the American Working Class.* 1993. Reprint. New York: Oxford University Press, 2013.

Luckett, Moya. "Filming the Family: Home Movie Systems and the Domestication of Spectatorship." *Velvet Light Trap* 36 (Fall 1995): 21–32.

Maltby, Richard. "The Production Code and the Hayes Office." In Tino Balio, *Grand Design: Hollywood as a Modern Business Enterprise, 1930–1939*, 37–72. Berkeley: University of California Press, 1993.

Manovich, Lev. *The Language of New Media.* Cambridge, MA: MIT Press, 2001.

Marchessault, Janine. "Multi-screens and Future Cinema: The Labyrinth Project at Expo 67." In *Fluid Screens, Expanded Cinema*, edited by Janine Marchessault and Susan Lord, 29–51. Toronto: University of Toronto Press, 2007.

Massey, Doreen. *For Space.* Thousand Oaks, CA: Sage, 2005.

May, Larry. *The Big Tomorrow: Hollywood and the Politics of the American Way.* Chicago: University of Chicago Press, 2000.

Mayne, Judith. "'King Kong' and the Ideology of Spectacle." *Quarterly Review of Film and Video* 1, no. 4 (1976): 373–87.

——. "Paradoxes of Spectatorship." In *Viewing Positions: Ways of Seeing Film*, edited by Linda Williams, 155–83. New Brunswick, NJ: Rutgers University Press, 1995.

McCarthy, Anna. *Ambient Television: Visual Culture and Public Space.* Durham, NC: Duke University Press, 2001.

McGuire, Laura. "A Movie House in Space and Time: Frederick Kiesler's Film Arts Guild Cinema, New York, 1929." *Studies in the Decorative Arts* 14, no. 2 (Spring–Summer 2007): 45–78.

Mebold, Anke, and Charles Tepperman. "Resurrecting the Lost History of 28mm Film in North America." *Film History* 15, no. 2 (2003): 137–51.

Melnick, Ross. *American Showman: Samuel "Roxy" Rothafel and the Birth of the Entertainment Industry*. New York: Columbia University Press, 2012.

Metz, Christian. *The Imaginary Signifier: Psychoanalysis and the Cinema*. 1977. Translated by Celia Britton, Annwyl Williams, Ben Brewster, and Alfred Guzzetti. Bloomington: Indiana University Press, 1982.

——. *Impersonal Enunciation, or the Place of Film*. Translated by Cormac Deane. New York: Columbia University Press, 2016. Originally published in French in 1991.

——. "*Trucage* and the Film." Translated by Françoise Meltzer. *Critical Inquiry* 3, no. 4 (Summer 1977): 657–75. Originally published in French in 1972.

Moholy-Nagy, László. "Light-Architecture." *Industrial Arts* 1, no. 1 (Spring 1936): 15–17.

——. *Painting, Photography, Film*. Translated by Janet Seligman. London: Lund Humphries, 1969. Originally published in German in 1925.

Morse, Margaret. *Virtualities: Television, Media Art, and Cyberculture*. Bloomington: Indiana University Press, 1998.

Mulvey, Laura. "A Clumsy Sublime." *Film Quarterly* 60, no. 3 (Spring 2007): 3.

——. "Rear-Projection and the Paradoxes of Hollywood Realism." In *Theorizing World Cinema*, edited by Lúcia Nagib, Chris Perriam, and Rajinder Kumar Dudrah, 207–20. London: I. B. Tauris, 2012.

Murray, Susan. *Bright Signals: A History of Color Television*. Durham, NC: Duke University Press, 2018.

Musser, Charles. *The Emergence of Cinema: The American Screen to 1907*. Berkeley: University of California Press, 1990.

North, Dan. *Performing Illusions: Cinema, Special Effects, and the Virtual Actor*. New York: Wallflower Press, 2008.

——. "The Silent Screen, 1895–1927: Special/Visual Effects." In *Editing and Special/Visual Effects*, edited by Charlie Keil and Kristen Whissel, 37–50. New Brunswick, NJ: Rutgers University Press, 2016.

North, Dan, Bob Rehak, and Michael S. Duffy, eds. *Special Effects: New Histories/Theories/Contexts*. London: British Film Institute, 2015.

Nyce, James M., and Paul Kahn, eds. *From Memex to Hypertext: Vannevar Bush and the Mind's Machines*. San Diego, CA: Academic Press, 1991.

——. "A Machine for the Mind: Vannevar Bush's Memex." In *From Memex to Hypertext: Vannevar Bush and the Mind's Machines*, edited by James M. Nyce and Paul Kahn, 39–66. San Diego, CA: Academic Press, 1991.

Orgeron, Devin, Marsha Orgeron, and Dan Streible. "A History of Learning with the Lights Off." In *Learning with the Lights Off: Educational Film in the United States*, edited by Devin Orgeron, Marsha Orgeron, and Dan Streible, 15–66. New York: Oxford University Press, 2012.

——, eds. *Learning with the Lights Off: Educational Film in the United States*. New York: Oxford University Press, 2012.

Païni, Dominique. "The Wandering Gaze: Hitchcock's Use of Transparencies." In *Hitchcock and Art: Fatal Coincidences*, edited by Dominique Païni and Guy Cogeval, 51–78. Montreal: Montreal Museum of Fine Arts, 2000.

Parks, Lisa. *Rethinking Media Coverage: Vertical Mediation and the War on Terror*. New York: Routledge, 2018.

Paul, William. "The Aesthetics of Emergence." *Film History* 5, no. 3 (1993): 321–55.

——. "Screening Space: Architecture, Technology, and the Motion Picture Screen." *Michigan Quarterly Review* 35, no. 1 (Winter 1996): 143–73.

——. *When Movies Were Theater: Architecture, Exhibition, and the Evolution of American Film*. New York: Columbia University Press, 2016.

Pedullà, Gabriele. *In Broad Daylight: Movies and Spectators After the Cinema*. Translated by Patricia Gaborik. New York: Verso, 2012.

Peters, John Durham. *The Marvelous Clouds: Toward a Philosophy of Elemental Media*. Chicago: University of Chicago Press, 2015.

Petty, Miriam J. *Stealing the Show: African American Performers and Audiences in 1930s Hollywood*. Oakland: University of California Press, 2016.

Popat, Sita. "Missing in Action: Embodied Experience and Virtual Reality." *Theatre Journal* 68, no. 3 (September 2016): 357–78.

Prince, Stephen. *Digital Visual Effects in Cinema: The Seduction of Reality*. New Brunswick, NJ: Rutgers University Press, 2012.

Przybylek, Stephanie E., with Marie Eckhardt and Jim Richerson. *Breaking the Silence on Film: The History of the Case Research Lab*. Auburn, NY: Cayuga Museum of History and Art, 1999.

Purse, Lisa. *Digital Imaging in Popular Cinema*. Edinburgh: Edinburgh University Press, 2013.

Pyne, Stephen J. *Fire: A Brief History*. Seattle: University of Washington Press, 2001.

Rabinovitz, Lauren. *For the Love of Pleasure: Women, Movies, and Culture in Turn-of-the-Century Chicago*. New Brunswick, NJ: Rutgers University Press, 1998.

Raheja, Michelle H. *Reservation Reelism: Redfacing, Visual Sovereignty, and Representations of Native Americans in Film*. Lincoln: University of Nebraska Press, 2010.

Rheingold, Howard. *Virtual Reality*. New York: Simon and Schuster, 1991.

Richardson, F. H. *F. H. Richardson's Bluebook of Projection*. 6th ed. New York: Quigley, 1935.

——. *Motion Picture Handbook: A Guide for Managers and Operators of Motion Picture Theaters*. 2nd ed. New York: Moving Picture World, 1912.

——. *Richardson's Handbook of Projection: The Blue Book of Projection*. 5th ed. 3 vols. New York: Chalmers, 1927–1930.

Richmond, Scott C. *Cinema's Bodily Illusions: Flying, Floating, and Hallucinating*. Minneapolis: University of Minnesota Press, 2016.

Rickitt, Richard. *Special Effects: The History and Technique*. New York: Billboard Books, 2007.

Robida, A. *Le vingtième siècle*. Paris: Georges Decaux, 1883.

Rodowick, D. N. *The Virtual Life of Film*. Cambridge, MA: Harvard University Press, 2007.

Rogers, Ariel. *Cinematic Appeals: The Experience of New Movie Technologies*. New York: Columbia University Press, 2013.

Rogin, Michael. *Blackface, White Noise: Jewish Immigrants in the Hollywood Melting Pot*. Berkeley: University of California Press, 1996.

Rony, Fatimah Tobing. *The Third Eye: Race, Cinema, and Ethnographic Spectacle*. Durham, NC: Duke University Press, 1996.

Rosen, Philip, ed. *Narrative, Apparatus, Ideology: A Film Theory Reader*. New York: Columbia University Press, 1986.

Ross, Miriam. "The 3-D Aesthetic: *Avatar* and Hyperhaptic Visuality." *Screen* 53, no. 4 (Winter 2012): 381–97.

Ross, Sara. "Invitation to the Voyage: The Flight Sequence in Contemporary 3D Cinema." *Film History* 24 (2012): 210–20.

Salt, Barry. *Film Style and Technology: History and Analysis*. 2nd ed. London: Starword, 1992.

Sammond, Nicholas. *Birth of an Industry: Blackface Minstrelsy and the Rise of American Animation*. Durham, NC: Duke University Press, 2015.

Sandin, Daniel, Thomas DeFanti, and Carolina Cruz-Neira. "A Room with a View." 1991. In *Multimedia: From Wagner to Virtual Reality*, edited by Randall Packer and Ken Jordan, 286–92. New York: Norton, 2001.

Scannell, Paddy. *Radio, Television, and Modern Life: A Phenomenological Approach*. Cambridge, MA: Blackwell, 1996.

——. *Television and the Meaning of "Live": An Enquiry Into the Human Situation*. Malden, MA: Polity, 2014.

Schatz, Thomas. *The Genius of the System: Hollywood Filmmaking in the Studio Era*. New York: Pantheon, 1988.

Schneider, Alexandra. "Time Travel with Pathé Baby: The Small-Gauge Film Collection as Historical Archive." *Film History* 19 (2007): 353–60.

Sconce, Jeffrey. *Haunted Media: Electronic Presence from Telegraphy to Television*. Durham, NC: Duke University Press, 2000.

Selznick, David O. *Memo from David O. Selznick*. Edited by Rudy Behlmer. New York: Random House, 2000.

Sewell, Philip W. *Television in the Age of Radio: Modernity, Imagination, and the Making of a Medium*. New Brunswick, NJ: Rutgers University Press, 2014.

Shohat, Ella, and Robert Stam. *Unthinking Eurocentrism: Multiculturalism and the Media*. 2nd ed. New York: Routledge, 2014.

Singer, Ben. "Early Home Cinema and the Edison Home Projecting Kinetoscope." *Film History* 2, no. 1 (Winter 1988): 37–69.

Singer, Beverly R. *Wiping the War Paint off the Lens: Native American Film and Video*. Minneapolis: University of Minnesota Press, 2001.

Sklar, Robert. *Movie-Made America: A Cultural History of American Movies*. Rev. ed. New York: Vintage, 1994.

Slide, Anthony. *Before Video: A History of Non-theatrical Film*. New York: Greenwood Press, 1992.

Slowik, Michael. *After the Silents: Hollywood Film Music in the Early Sound Era, 1926–1934*. New York: Columbia University Press, 2014.

Smith, Jacob. *The Thrill Makers: Celebrity, Masculinity, and Stunt Performance*. Berkeley: University of California Press, 2012.

Smoodin, Eric. *Regarding Frank Capra: Audience, Celebrity, and American Film Studies, 1930–1960*. Durham, NC: Duke University Press, 2004.

Snead, James A. *White Screens/Black Images: Hollywood from the Dark Side*. Edited by Colin McCabe and Cornel West. New York: Routledge, 1994.

Sobchack, Vivian. *The Address of the Eye: A Phenomenology of Film Experience*. Princeton, NJ: Princeton University Press, 1992.

——. "*Detour*: Driving in Back Projection, or Forestalled by Film Noir." In *Kiss the Blood off My Hands: On Classic Film Noir*, edited by Robert Miklitsch, 113–29. Urbana: University of Illinois Press, 2014.

Spadoni, Robert. *Uncanny Bodies: The Coming of Sound Film and the Origins of the Horror Genre*. Berkeley: University of California Press, 2007.

Spigel, Lynn. *Make Room for TV: Television and the Family Ideal in Postwar America*. Chicago: University of Chicago Press, 1992.

——. "Object Lessons for the Media Home: From Storagewall to Invisible Design." *Public Culture* 24, no. 3 (2012): 535–76.
——. *TV by Design: Modern Art and the Rise of Network Television*. Chicago: University of Chicago Press, 2008.
——. *Welcome to the Dreamhouse: Popular Media and Postwar Suburbs*. Durham, NC: Duke University Press, 2001.
Sprengler, Christine. *Hitchcock and Contemporary Art*. New York: Palgrave Macmillan, 2014.
Stadel, Luke. "Natural Sound in the Early Talkie Western." *Music, Sound, and the Moving Image* 5, no. 2 (Autumn 2011): 113–36.
Stanković, Stanislav. *Virtual Reality and Virtual Environments in 10 Lectures*. Williston, VT: Morgan and Claypool, 2016.
Starosielski, Nicole. *The Undersea Network*. Durham, NC: Duke University Press, 2015.
Sterling, Christopher H., and John Michael Kittross. *Stay Tuned: A History of American Broadcasting*. 3rd ed. Mahwah, NJ: Lawrence Erlbaum Associates, 2002.
Sterne, Jonathan, and Tara Rodgers. "The Poetics of Signal Processing." *differences* 22, nos. 2–3 (2011): 31–53.
Steuer, Jonathan. "Defining Virtual Reality: Dimensions Determining Telepresence." *Journal of Communication* 42, no. 4 (Fall 1992): 73–93.
Stewart, Jacqueline Najuma. *Migrating to the Movies: Cinema and Black Urban Modernity*. Berkeley: University of California Press, 2005.
Steyerl, Hito. "In Free Fall: A Thought Experiment on Vertical Perspective." *e-flux journal* 24 (April 2011). https://www.e-flux.com/journal/24/67860/in-free-fall-a-thought-experiment-on-vertical-perspective.
Straw, Will. "Proliferating Screens." *Screen* 41, no. 1 (Spring 2000): 115–19.
——. "Pulling Apart the Apparatus." *Recherches Sémiotiques / Semiotic Inquiry* 31 (2011): 59–73.
Sutherland, Ivan. "The Ultimate Display." 1965. In *Multimedia: From Wagner to Virtual Reality*, edited by Randall Packer and Ken Jordan, 252–56. New York: Norton, 2001.
Szczepaniak-Gillece, Jocelyn. *The Optical Vacuum: Spectatorship and Modernized American Theater Architecture*. New York: Oxford University Press, 2018.
——. "Revisiting the Apparatus: The Theatre Chair and Cinematic Spectatorship." *Screen* 57, no. 3 (Autumn 2016): 253–76.
Taylor, Clyde. "The Re-birth of the Aesthetic in Cinema." In *The Birth of Whiteness: Race and the Emergence of U.S. Cinema*, edited by Daniel Bernardi, 15–37. New Brunswick, NJ: Rutgers University Press, 1996.
Taylor, Giles. "A Military Use for Widescreen Cinema: Training the Body Through Immersive Media." *Velvet Light Trap* 72 (Fall 2013): 17–32.
Tepperman, Charles. *Amateur Cinema: The Rise of North American Moviemaking, 1923–1960*. Oakland: University of California Press, 2015.
——. "'A Recognized Screen': The New York Annual Movie Parties from Parlor to Public." *Film History* 30, no. 1 (Spring 2018): 58–85.
Thompson, Emily. *The Soundscape of Modernity: Architectural Acoustics and the Culture of Listening in America, 1900–1933*. Cambridge, MA: MIT Press, 2002.
Thorp, Margaret Farrand. *America at the Movies*. 1939. Reprint. London: Faber and Faber, 1946.
Tolson, Melvin B. *Caviar and Cabbage: Selected Columns by Melvin B. Tolson from the "Washington Tribune," 1937–1944*. Edited by Robert M. Farnsworth. Columbia: University of Missouri Press, 1982.

Tsivian, Yuri. "Media Fantasies and Penetrating Vision: Some Links Between X-Rays, the Microscope, and Film." In *Laboratory of Dreams: The Russian Avant-Garde and Cultural Experiment*, edited by John E. Bowlt and Olga Matich, 81–99. Stanford, CA: Stanford University Press, 1996.

Turner, Fred. *The Democratic Surround: Multimedia and American Liberalism from World War II to the Psychedelic Sixties*. Chicago: University of Chicago Press, 2013.

Turner, George E., ed. *The ASC Treasury of Visual Effects*. Hollywood, CA: American Society of Cinematographers, 1983.

——. "The Evolution of Special Visual Effects." in *The ASC Treasury of Visual Effects*, edited by George E. Turner, 15–82. Hollywood, CA: American Society of Cinematographers, 1983.

Turnock, Julie. *Plastic Reality: Special Effects, Technology, and the Emergence of 1970s Blockbuster Aesthetics*. New York: Columbia University Press, 2015.

——. "The Screen on the Set: The Problem of Classical-Studio Rear Projection." *Cinema Journal* 51, no. 2 (Winter 2012): 157–62.

Udelson, Joseph H. *The Great Television Race: A History of the American Television Industry, 1925–1941*. Tuscaloosa: University of Alabama Press, 1982.

Uricchio, William. "Historicizing Media in Transition." In *Rethinking Media Change: The Aesthetics of Transition*, edited by David Thorburn and Henry Jenkins, 23–38. Cambridge, MA: MIT Press, 2003.

——. "Storage, Simultaneity, and the Media Technologies of Modernity." In *Allegories of Communication: Intermedial Concerns from Cinema to the Digital*, edited by John Fullerton and Jan Olsson, 123–38. Rome: John Libbey, 2004.

——. "Television as History: Representations of German Television Broadcasting, 1935–1944." In *Framing the Past: The Historiography of German Cinema and Television*, edited by Bruce A. Murray and Christopher J. Wickham, 167–96. Carbondale: Southern Illinois University Press, 1992.

——. "Television, Film, and the Struggle for Media Identity." *Film History* 10, no. 2 (1998): 118–27.

——. "There's More to the Camera's Obscura Than Meets the Eye." In *Stop Motion, Fragmentation of Time: Exploring the Roots of Modern Visual Culture*, edited by François Albera, Marta Braun, and André Gaudreault, 103–17. Lausanne, Switzerland: Éditions Payot Lausanne, 2002.

Valentine, Maggie. *The Show Starts on the Sidewalk: An Architectural History of the Movie Theatre, Starring S. Charles Lee*. New Haven, CT: Yale University Press, 1996.

Verhoeff, Nanna. *Mobile Screens: The Visual Regime of Navigation*. Amsterdam: Amsterdam University Press, 2012.

Vertrees, Alan David. *Selznick's Vision: "Gone with the Wind" and Hollywood Filmmaking*. Austin: University of Texas Press, 1997.

Vidler, Anthony. *The Architectural Uncanny: Essays in the Modern Unhomely*. Cambridge, MA: MIT Press, 1992.

Walker, Ralph. "The Birth of an Idea." In *New Screen Techniques*, edited by Martin Quigley Jr., 112–17. New York: Quigley, 1953.

Waller, Fred. "The Archeology of Cinerama." *Film History* 5 (1993): 289–97.

——. "Cinerama Goes to War." In *New Screen Techniques*, edited by Martin Quigley Jr., 119–26. New York: Quigley, 1953.

Waller, Gregory A. *Main Street Amusements: Movies and Commercial Entertainment in a Southern City, 1896–1930*. Washington, DC: Smithsonian Institution Press, 1995.

——. "Projecting the Promise of 16mm." In *Useful Cinema*, edited by Charles Acland and Haidee Wasson, 125–48. Durham, NC: Duke University Press, 2011.
Wasson, Haidee. "Electric Homes! Automatic Movies! Efficient Entertainment! 16mm and Cinema's Domestication in the 1920s." *Cinema Journal* 48, no. 4 (Summer 2009): 1–21.
——. Introduction to "In Focus: Screen Technologies," edited by Haidee Wasson. Special issue of *Cinema Journal* 51, no. 2 (Winter 2012): 143–46.
——. *Museum Movies: The Museum of Modern Art and the Birth of Art Cinema*. Berkeley: University of California Press, 2005.
——. "The Networked Screen: Moving Images, Materiality, and the Aesthetics of Size." In *Fluid Screens, Expanded Cinema*, edited by Janine Marchessault and Susan Lord, 74–95. Toronto: University of Toronto Press, 2007.
——. "The Other Small Screen: Moving Images at New York's World Fair, 1939." *Canadian Journal of Film Studies* 21, no. 1 (Spring 2012): 81–103.
——. "Protocols of Portability." *Film History* 25, nos. 1–2 (2013): 236–47.
——. "The Reel of the Month Club: 16mm Projectors, Home Theaters, and Film Libraries in the 1920s." In *Going to the Movies: Hollywood and the Social Experience of Cinema*, edited by Richard Maltby, Melvyn Stokes, and Robert C. Allen, 217–34. Exeter, UK: University of Exeter Press, 2007.
——. "Selling Machines: Film and Its Technologies at the New York World's Fair." In *Films That Sell: Moving Pictures and Advertising*, edited by Bo Florin, Nico de Klerk, and Patrick Vonderau, 54–70. London: British Film Institute, 2016.
——. "Suitcase Cinema." *Cinema Journal* 51, no. 2 (Winter 2012): 148–52.
Weber, Anne-Katrin. "Recording on Film, Transmitting by Signals: The Intermediate Film System and Television's Hybridity in the Interwar Period." *Grey Room* 56 (Summer 2014): 6–33.
Weber, Samuel. *Mass Mediauras: Form, Technics, Media*. Stanford, CA: Stanford University Press, 1996.
Whissel, Kristen. "Parallax Effects: Epistemology, Affect, and Digital 3D Cinema." *Journal of Visual Culture* 15, no. 2 (2016): 233–49.
——. *Spectacular Digital Effects: CGI and Contemporary Cinema*. Durham, NC: Duke University Press, 2014.
White, Mimi. "The Attractions of Television: Reconsidering Liveness." In *Mediaspace: Place, Scale, and Culture in a Media Age*, edited by Nick Couldry and Anna McCarthy, 75–91. New York: Routledge, 2004.
White, Timothy R. "Hollywood's Attempt at Appropriating Television: The Case of Paramount Pictures." In *Hollywood in the Age of Television*, edited by Tino Balio, 149–55. Boston: Unwin Hyman, 1990.
Widdis, Emma. *Visions of a New Land: Soviet Film from the Revolution to the Second World War*. New Haven, CT: Yale University Press, 2003.
Wild, Jennifer. *The Parisian Avant-Garde in the Age of Cinema, 1900–1923*. Oakland: University of California Press, 2015.
Wilinsky, Barbara. *Sure Seaters: The Emergence of Art House Cinema*. Minneapolis: University of Minnesota Press, 2001.
Williams, Linda. Introduction to *Viewing Positions: Ways of Seeing Film*, edited by Linda Williams, 1–20. New Brunswick, NJ: Rutgers University Press, 1995.
Williams, Raymond. *Television: Technology and Cultural Form*. 1974. Reprint. New York: Routledge, 2003.
Wilson, Steve. *The Making of "Gone with the Wind."* Austin: University of Texas Press, 2014.

Wolfe, Charles. "The Poetics and Politics of Nonfiction: Documentary Film." In Tino Balio, *Grand Design: Hollywood as a Modern Business Enterprise, 1930–1939*, 351–86. Berkeley: University of California Press, 1993.

Young, Paul. *The Cinema Dreams Its Rivals: Media Fantasy Films from Radio to the Internet.* Minneapolis: University of Minnesota Press, 2006.

Zimmermann, Patricia R. *Reel Families: A Social History of Amateur Film.* Bloomington: Indiana University Press, 1995.

Zone, Ray. *Stereoscopic Cinema and the Origins of 3-D Film, 1838–1952.* Lexington: University Press of Kentucky, 2007.

# INDEX

## FILM AND CULTURE

Edited by John Belton

A series of Columbia University Press

*What Made Pistachio Nuts? Early Sound Comedy and the Vaudeville Aesthetic*
Henry Jenkins
*Showstoppers: Busby Berkeley and the Tradition of Spectacle*
Martin Rubin
*Projections of War: Hollywood, American Culture, and World War II*
Thomas Doherty
*Laughing Screaming: Modern Hollywood Horror and Comedy*
William Paul
*Laughing Hysterically: American Screen Comedy of the 1950s*
Ed Sikov
*Primitive Passions: Visuality, Sexuality, Ethnography, and Contemporary Chinese Cinema*
Rey Chow
*The Cinema of Max Ophuls: Magisterial Vision and the Figure of Woman*
Susan M. White
*Black Women as Cultural Readers*
Jacqueline Bobo
*Picturing Japaneseness: Monumental Style, National Identity, Japanese Film*
Darrell William Davis
*Attack of the Leading Ladies: Gender, Sexuality, and Spectatorship in Classic Horror Cinema*
Rhona J. Berenstein
*This Mad Masquerade: Stardom and Masculinity in the Jazz Age*
Gaylyn Studlar
*Sexual Politics and Narrative Film: Hollywood and Beyond*
Robin Wood
*The Sounds of Commerce: Marketing Popular Film Music*
Jeff Smith
*Orson Welles, Shakespeare, and Popular Culture*
Michael Anderegg
*Pre-Code Hollywood: Sex, Immorality, and Insurrection in American Cinema, 1930–1934*
Thomas Doherty
*Sound Technology and the American Cinema: Perception, Representation, Modernity*
James Lastra
*Melodrama and Modernity: Early Sensational Cinema and Its Contexts*
Ben Singer
*Wondrous Difference: Cinema, Anthropology, and Turn-of-the-Century Visual Culture*
Alison Griffiths
*Hearst Over Hollywood: Power, Passion, and Propaganda in the Movies*
Louis Pizzitola
*Masculine Interests: Homoerotics in Hollywood Film*
Robert Lang
*Special Effects: Still in Search of Wonder*
Michele Pierson
*Designing Women: Cinema, Art Deco, and the Female Form*
Lucy Fischer

*Cold War, Cool Medium: Television, McCarthyism, and American Culture*
Thomas Doherty
*Katharine Hepburn: Star as Feminist*
Andrew Britton
*Silent Film Sound*
Rick Altman
*Home in Hollywood: The Imaginary Geography of Cinema*
Elisabeth Bronfen
*Hollywood and the Culture Elite: How the Movies Became American*
Peter Decherney
*Taiwan Film Directors: A Treasure Island*
Emilie Yueh-yu Yeh and Darrell William Davis
*Shocking Representation: Historical Trauma, National Cinema, and the Modern Horror Film*
Adam Lowenstein
*China on Screen: Cinema and Nation*
Chris Berry and Mary Farquhar
*The New European Cinema: Redrawing the Map*
Rosalind Galt
*George Gallup in Hollywood*
Susan Ohmer
*Electric Sounds: Technological Change and the Rise of Corporate Mass Media*
Steve J. Wurtzler
*The Impossible David Lynch*
Todd McGowan
*Sentimental Fabulations, Contemporary Chinese Films: Attachment in the Age of Global Visibility*
Rey Chow
*Hitchcock's Romantic Irony*
Richard Allen
*Intelligence Work: The Politics of American Documentary*
Jonathan Kahana
*Eye of the Century: Film, Experience, Modernity*
Francesco Casetti
*Shivers Down Your Spine: Cinema, Museums, and the Immersive View*
Alison Griffiths
*Weimar Cinema: An Essential Guide to Classic Films of the Era*
Edited by Noah Isenberg
*African Film and Literature: Adapting Violence to the Screen*
Lindiwe Dovey
*Film, A Sound Art*
Michel Chion
*Film Studies: An Introduction*
Ed Sikov
*Hollywood Lighting from the Silent Era to Film Noir*
Patrick Keating
*Levinas and the Cinema of Redemption: Time, Ethics, and the Feminine*
Sam B. Girgus
*Counter-Archive: Film, the Everyday, and Albert Kahn's Archives de la Planète*
Paula Amad
*Indie: An American Film Culture*
Michael Z. Newman
*Pretty: Film and the Decorative Image*
Rosalind Galt

*Film and Stereotype: A Challenge for Cinema and Theory*
Jörg Schweinitz
*Chinese Women's Cinema: Transnational Contexts*
Edited by Lingzhen Wang
*Hideous Progeny: Disability, Eugenics, and Classic Horror Cinema*
Angela M. Smith
*Hollywood's Copyright Wars: From Edison to the Internet*
Peter Decherney
*Electric Dreamland: Amusement Parks, Movies, and American Modernity*
Lauren Rabinovitz
*Where Film Meets Philosophy: Godard, Resnais, and Experiments in Cinematic Thinking*
Hunter Vaughan
*The Utopia of Film: Cinema and Its Futures in Godard, Kluge, and Tahimik*
Christopher Pavsek
*Hollywood and Hitler, 1933–1939*
Thomas Doherty
*Cinematic Appeals: The Experience of New Movie Technologies*
Ariel Rogers
*Continental Strangers: German Exile Cinema, 1933–1951*
Gerd Gemünden
*Deathwatch: American Film, Technology, and the End of Life*
C. Scott Combs
*After the Silents: Hollywood Film Music in the Early Sound Era, 1926–1934*
Michael Slowik
*"It's the Pictures That Got Small": Charles Brackett on Billy Wilder and Hollywood's Golden Age*
Edited by Anthony Slide
*Plastic Reality: Special Effects, Technology, and the Emergence of 1970s Blockbuster Aesthetics*
Julie A. Turnock
*Maya Deren: Incomplete Control*
Sarah Keller
*Dreaming of Cinema: Spectatorship, Surrealism, and the Age of Digital Media*
Adam Lowenstein
*Motion(less) Pictures: The Cinema of Stasis*
Justin Remes
*The Lumière Galaxy: Seven Key Words for the Cinema to Come*
Francesco Casetti
*The End of Cinema? A Medium in Crisis in the Digital Age*
André Gaudreault and Philippe Marion
*Studios Before the System: Architecture, Technology, and the Emergence of Cinematic Space*
Brian R. Jacobson
*Impersonal Enunciation, or the Place of Film*
Christian Metz
*When Movies Were Theater: Architecture, Exhibition, and the Evolution of American Film*
William Paul
*Carceral Fantasies: Cinema and Prison in Early Twentieth-Century America*
Alison Griffiths
*Unspeakable Histories: Film and the Experience of Catastrophe*
William Guynn
*Reform Cinema in Iran: Film and Political Change in the Islamic Republic*
Blake Atwood
*Exception Taken: How France Has Defied Hollywood's New World Order*
Jonathan Buchsbaum

*After Uniqueness: A History of Film and Video Art in Circulation*
Erika Balsom
*Words on Screen*
Michel Chion
*Essays on the Essay Film*
Edited by Nora M. Alter and Timothy Corrigan
*The Essay Film After Fact and Fiction*
Nora Alter
*Specters of Slapstick and Silent Film Comediennes*
Maggie Hennefeld
*Melodrama Unbound: Across History, Media, and National Cultures*
Edited by Christine Gledhill and Linda Williams
*Show Trial: Hollywood, HUAC, and the Birth of the Blacklist*
Thomas Doherty
*Cinema/Politics/Philosophy*
Nico Baumbach
*The Dynamic Frame: Camera Movement in Classical Hollywood*
Patrick Keating
*Hollywood's Dirtiest Secret: The Hidden Environmental Costs of the Movies*
Hunter Vaughan
*Chromatic Modernity: Color, Cinema, and Media of the 1920s*
Sarah Street and Joshua Yumibe
*Rewriting Indie Cinema: Improvisation, Psychodrama, and the Screenplay*
J. J. Murphy